Activated Carbon

Activated Carbon

Solutions for Improving Water Quality

Zaid K. Chowdhury
R. Scott Summers
Garret P. Westerhoff
Brian J. Leto
Kirk O. Nowack
Christopher J. Corwin

Laurel B. Passantino, Technical Editor

American Water Works Association

Activated Carbon: Solutions for Improving Water Quality

Disclaimer:
The authors, contributors, editors, and publisher do not assume responsibility for the validity of the content or any consequences of its use. In no event will AWWA be liable for direct, indirect, special, incidental, or consequential damages arising out of the use of information presented in this book. In particular, AWWA will not be responsible for any costs, including, but not limited to, those incurred as a result of lost revenue. In no event shall AWWA's liability exceed the amount paid for the purchase of this book.

AWWA Sr. Manager of Editorial Development and Production: Gay Porter De Nileon
AWWA Sr. Technical Editor/Project Manager: Martha Ripley Gray
Cover Art: Cheryl Armstrong, AWWA Sr. Production Editor
Production: Cheryl Armstrong; Sheryl Tongue, Stonehill Graphics

Library of Congress Cataloging-in-Publication Data
Chowdhury, Zaid K.
Activated carbon : solutions for improving water quality / Zaid K. Chowdhury, R. Scott Summers, Garret P. Westerhoff, Brian J. Leto, Kirk O. Nowack, Christopher J. Corwin ; Laurel B. Passantino, technical editor.
pages cm
Includes bibliographical references and index.
ISBN 978-1-58321-907-2 (hardback) -- ISBN 1-58321-907-2 (hardback) -- ISBN 978-1-61300-202-5 (electronic) -- ISBN 1-61300-202-5 (electronic) 1. Water--Purification--Adsorption. 2. Water--Purification--Organic compounds removal. 3. Carbon, Activated. 4. Water quality. I. Title.
TD449.5.C46 2012
628.1'66--dc23
2012039033

Printed in the United States of America

ISBN-13, print: 978-1-58321-907-2
ISBN-13, electronic: 978-1-61300-202-5
ISBN-10, print: 1-58321-907-2
ISBN-10, electronic: 1-61300-202-5

American Water Works Association

6666 West Quincy Avenue
Denver, CO 80235-3098
303.794.7711
www.awwa.org

Contents

List of Figures

List of Tables

Authors and Editor

Zaid K. Chowdhury, PhD, PE, BCEE

As the Director of Applied Research for the Water Division of ARCADIS, Zaid K. Chowdhury serves as the top water treatment expert for the firm and is responsible for resolving challenging technical issues in the area of water treatment. His long tenure with the firm and active involvement in the professional and research circles has earned him a considerable reputation in the industry as an expert in various water treatment technologies including GAC. His direct involvement in the selection of GAC technologies for water treatment plants and in the evaluation and design of GAC facilities at many major water plants, including those for the cities of Scottsdale and Phoenix in Arizona, have helped him develop a keen understanding of GAC applications for adsorption and biological treatment media.

In addition, Chowdhury has been involved in numerous water treatment–related studies and research projects as well as in projects that augmented the knowledge base on treatment technologies relevant to various water quality regulations, including the Disinfectants and Disinfection By-Products Rule, the Long-Term 2 Enhanced Surface Water Treatment Rule, and the Arsenic Rule.

A licensed Professional Engineer and a Board Certified Environmental Engineer, Chowdhury holds a BS degree in civil engineering from the Bangladesh University of Engineering and Technology and MS and PhD degrees in civil engineering from the University of Arizona.

R. Scott Summers, PhD

Since 1998, Dr. Summers has been a professor of environmental engineering at the University of Colorado. He spent two years as a research associate at the German Water Works Research Institute (DVGW) at the University of Karlsruhe, 10 years as a professor at the University of Cincinnati, and was a Fulbright Professor at the University of Crete (Greece). He has been the Principal Investigator or co-PI of more than 30 research projects, 20 of which have focused on activated carbon. He has served as the research advisor for 25 PhD and 50 MS students and has coauthored more than 250 publications and 150 presentations.

Dr. Summers is the coauthor or editor of several major activated carbon publications: "Adsorption of Organic Compounds by Activated Carbon" (Knappe D.R.U., Snoeyink V.L.) in AWWA's *Water Quality and Treatment*, 6th edition, 2010; USEPA's *ICR Manual for Bench- and Pilot-Scale Treatment Studies: Granular Activated Carbon Precursor Removal Studies* (Hooper S.M., Hong S.), 1996; *Activated Carbon For Water Treatment*, (Sontheimer H., Crittenden J.C.), DVGW, Karlsruhe, Germany, 1988; and USEPA's *Adsorption Techniques in Drinking Water Treatment* (Roberts P.V., Regli S.),

EPA 570/9-84-005, 1984. He served as the technical director of Summers & Hooper Inc. (1995–2000) overseeing 20 bench- and pilot-scale GAC treatment studies. He also has served as a technical consultant on more than 30 GAC or PAC treatment studies and to USEPA's regulatory development and review process (1992–present).

He holds BS and MS degrees from the University of Cincinnati and a PhD from Stanford University in Civil and Environmental Engineering.

Garret P. Westerhoff, PE, BCEE, Member NAE

Former chair and chief executive officer of Malcolm Pirnie and longtime director of Malcolm Pirnie's drinking water programs, Garret P. Westerhoff is a leading national expert with more than 40 years' experience on water treatment systems, water resources planning, and design and management of water systems. He has pioneered the applications of innovative technologies such as high-rate and direct filtration and led design of the nation's first major installation of granular activated carbon (GAC) treatment to remove organics from drinking water for Cincinnati, Ohio.

Westerhoff is a member of the National Academy of Engineering and has held leadership positions in national and international professional societies. He has authored more than 100 articles on water quality, water treatment, alternative methods of project delivery, and water and wastewater utility management. He was the lead author of two books published by The American Water Works Association (AWWA): *The Changing Water Utility: Creative Approaches to Effectiveness and Efficiency* (1998) and *The Evolving Water Utility: Pathways to Higher Performance* (2003).

A licensed Professional Engineer in 14 states and a Professional Planner in New Jersey, Westerhoff holds BS and MS degrees in engineering from the New Jersey Institute of Technology, is an Honorary Member of American Water Works Association, and is a former member of National Research Council's Water Science and Technology Board.

Brian J. Leto, PE

Brian Leto is a senior process design engineer in Malcolm Pirnie/ARCADIS. Leto has contributed to design and construction of activated carbon applications including gravity applications of post-filter GAC contactors and filter adsorbers, pressure applications of GAC adsorption, and powdered activated carbon (PAC) storage and feed systems. He has conducted pilot studies of variations of activated carbon filtration including biologically active filters and adsorption and operated liquid and vapor phase activated carbon systems; trained operations staff regarding the operation of activated carbon processes; and optimized activated carbon operations after construction. In addition to his activated carbon experience, Leto has contributed to planning, design, and construction of multiple drinking water treatment and wastewater treatment and reuse projects during his 14-year career with Malcolm Pirnie/ARCADIS, including conventional and reverse osmosis water treatment plants and Class A+ wastewater reuse facilities.

Leto is a licensed Professional Engineer, a board-certified Environmental Engineer, a certified Construction Documents Technologist, and a member of the American Water Works Association. He holds a BS in environmental engineering from Rensselaer Polytechnic Institute and a MS in civil engineering from the University of Texas at Austin.

Kirk O. Nowack, PhD

Kirk Nowack is a senior process engineer with Malcolm Pirnie, the Water Division of ARCADIS. He currently leads a national team within Malcolm Pirnie that provides technical guidance and support for projects involving activated carbon. Nowack has extensive experience in all facets of planning, designing, and optimizing activated carbon treatment systems. He has also conducted research in the areas of activated carbon production and adsorption performance, and his findings have been published in *Carbon*, *Environmental Science and Technology*, and the *Journal American Water Works Association*. He led the development of a thermal technique that greatly enhances the adsorption capacity of commercially available activated carbons, and this technique has since been patented. Nowack previously worked as a water quality manager for the Pennsylvania-American Water Company, where he also served as a licensed water treatment plant operator. He is a member of the American Water Works Association and holds a PhD in environmental engineering from the Pennsylvania State University.

Christopher J. Corwin, PhD, PE

Chris Corwin earned his PhD from the University of Colorado at Boulder, where he investigated the removal of trace organic compounds from drinking waters with granular activated carbon. His research on activated carbon has been published in *Environmental Science & Technology*, *Water Research*, and the *Journal American Water Works Association*. During his graduate studies, Corwin participated in more than 25 bench-scale activated carbon evaluations for utilities across the country. These included the evaluation of PAC and GAC for the control of organic matter, disinfection by-product formation, taste-and-odor compounds, specific organic contaminants, and micropollutants. In the eight years prior to pursuing his advanced degrees, he acquired a diverse portfolio of experience in civil and environmental design.

Corwin is a licensed Professional Engineer in the state of Colorado and is a member of the American Water Works Association. He holds a BS in civil engineering from the University of Kentucky and MS and PhD degrees in civil engineering from the University of Colorado.

Laurel B. Passantino, PE (Technical Editor)

While working on this book, Laurel Passantino was a senior project engineer in the Drinking Water Process and Planning Engineering Group of Malcolm Pirnie/ARCADIS. She has authored or been a contributing author to over 30 publications and presentations in the area of water quality and treatment. Passantino has been involved in all aspects of drinking water treatment projects, including treatment process evaluations, water quality master planning, infrastructure master planning, regulatory compliance, and pilot studies. In particular, she has experience in coagulation and filtration, high-rate clarification, disinfection by-product formation and control, granular media filtration, granular activated carbon adsorption, arsenic treatment, and ultraviolet disinfection. Her activated carbon experience includes master planning for several communities in the Phoenix area as well as preliminary design for the second largest post-filter granular activated carbon facility in the United States, which is located in Scottsdale, Ariz.

continued next page

Passantino is a licensed Professional Engineer in the states of Arizona and Pennsylvania and is a member of the American Water Works Association. She holds a BS in civil engineering from Duke University and a MS in civil and environmental engineering from the University of New Hampshire.

Acknowledgments

As the book's coauthors, we assume responsibility for the research, writing, coordinating, and reviewing of the entire contents of the book. Throughout the process, we had the assistance of our technical editor, Laurel Passantino, to keep us on track and bring the prose of a diverse group of authors into an integrated text. We would also like to thank Jessica Cunningham, Kelley Newman, Alex Rivas, Andrea Odegard-Begay, Steve Porteous, and Andrea Traviglia for their contributions to the manuscript; and Amit Chattopadhyay, Jim Dettmer, Richard Pohlman, and Peter Tymkiw for providing peer review of early drafts.

We also recognize the many utility leaders who participated in an opinion survey conducted in December 2001 and January 2002 for Malcolm Pirnie Inc. by the BTI Consulting Group (Boston), others who offered advice and suggestions during interviews, and those who contributed material for the case studies that serve as the foundation for much of the book.

Special thanks go to Richard Brady, PE, BCEE, Richard Brady & Associates; and Robert M. Clark, PhD, US Environmental Protection Agency (retired), for reviewing the entire manuscript.

Finally, we thank our advocates and supporters at the American Water Works Association. Special thanks to Gay Porter De Nileon, Martha Ripley Gray, Cheryl Armstrong, Daniel Feldman, and Sheryl Tongue at Stonehill Graphics.

Without the help of these individuals, this book would not have been published, and we are grateful to all of them. Though bringing this project to completion required the help of many, we reserve for ourselves responsibility for any shortcomings the book may have. It is our hope that *Activated Carbon: Solutions for Improving Water Quality* will help its readers meet the challenges they face in bringing high-quality drinking water to their consumers at reasonable costs of service.

Zaid K. Chowdhury
R. Scott Summers
Garret P. Westerhoff
Brian J. Leto
Kirk O. Nowack
Christopher J. Corwin

Preface

Determining the appropriate treatment regime to provide safe drinking water for customers continues to challenge drinking water purveyors. Source water quality is becoming more compromised as global demand increases, high-quality sources are depleted, and improvements in analytical methods reveal increasingly lower concentrations of contaminants in treated water. Regulatory agencies struggle with adequate resources to make sound scientific judgments regarding safe levels of contaminants in drinking water while media reports of detected levels of chemicals and microbes leave customers apprehensive about the safety of what comes out of their faucets every day.

Maintaining microbiological quality continues to be a cornerstone of water treatment as reinforced by the unfortunate incidence of waterborne disease in impoverished and developing nations. Nevertheless, conventional disinfectants used effectively in treatment for more than a century produce by-products that may have long-term chronic health effects, and sources degraded by anthropogenic inputs increase the portfolio of chemical contaminants that must be addressed. Personal care products and pharmaceuticals in drinking water are reported with increasing frequency in the global media, while the effects on humans remain unresolved.

Through all of this uncertainty, proactive measures that can reduce a wide variety of contaminants to low concentrations through multi-objective treatment remain an important element of robust and reliable drinking water production. Activated carbon, one of the oldest treatment technologies, is once again demonstrating its value in these challenging times. Activated carbon is simple to operate as an adsorption medium, serves as a proactive barrier for contamination, and does not produce by-products from its use. It can be reactivated and reused. And it removes compounds that customers can perceive with their senses—taste-and-odor compounds—as well as reduces a suite of potentially harmful chemical contaminants to low concentrations.

Many books have been written about granular activated carbon. Some focus on the theory of performance and removal mechanisms while others focus on design features. This book focuses on solutions. It describes the challenges facing water providers to provide safe water that is acceptable to their customers, utility experiences using activated carbon, activated carbon applications, and design and procurement approaches. The appendices include detailed case studies and a life-cycle assessment demonstrating favorable sustainability considerations for activated carbon when compared with other treatment technologies.

Never before has all of this information been brought together in one location. The what, why, where, and how of this treatment technology are connected in this book and demonstrate why activated carbon has maintained its status as an integral treatment technology in the quest for pure water over millennia.

Enjoy the story!

Introduction

Water purveyors throughout the globe have been, and continue to be, challenged to support existing and growing populations with an adequate and safe water supply. Historically, communities developed where water supplies were available and abundant. For example, the settlements resulting from the westward migration in the United States in the 1800s were often determined according to where water supplies were found. Now, however, there are few new locations where safe water supplies are available and abundant, either in the United States or globally. Instead, the challenge of maintaining and protecting drinking water supplies from further degradation is high on the minds of water purveyors and environmentally minded individuals and groups. Existing supplies continue to be threatened by microbiological and chemical contaminants introduced by increasing populations and associated economic development as well as by natural sources of contamination.

Water scarcity is a constant area of concern in major metropolitan areas in arid regions, and climate change is affecting how regions that previously had sufficient resources view their supplies into the future. These scarcity issues are driving water purveyors to use lower-quality water sources to meet increasing demand. At the same time, improved analytical techniques are able to detect compounds at lower and lower concentrations, either revealing contaminants that previously had not been detected or indicating the presence of contaminants that have been recently introduced into the water supply. Although health effects of many of these micropollutants are not currently known and may not be known for decades or longer, consumers are rightfully concerned about their presence in drinking water, and water purveyors must respond.

In the fundamental charge to protect public health, water purveyors rely on a combination of treatment and watershed protection to meet water quality goals and regulations. In meeting these goals, water purveyors should consider both the quantity and quality of the supply and choose suitable treatment approaches. The approaches are often a combination of physical, chemical, adsorption, and biological processes. The challenge is to determine the best combination of processes that protect public health and meet customer desires and regulatory requirements for water quality while doing so in a financially responsible manner.

The Case for Activated Carbon

Activated carbon is an adsorption medium and its use is considered an advanced technique for meeting many water quality demands. Treatment with activated carbon is not new and has in fact been used for thousands of years to improve the quality of drinking water. It has been used in various forms (powdered and granular) around the globe in a multi-objective manner, removing heterogeneous compounds that produce color and are precursors to contaminants upon disinfection, trace organic and inorganic contaminants, and taste-and-odor compounds. Activated carbon also has the flexibility to be operated in both adsorption and biological modes. In the latter, it provides a large surface area for organisms to populate and biologically degrade contaminants. Utilities may implement activated carbon for several reasons, including regulatory compliance, positioning for future regulations, public health protection and customer confidence, and sustainability considerations.

Compliance With Existing Regulations

For most water systems, the biggest driver for implementing activated carbon treatment is to gain compliance with water quality regulations. The US Congress originally passed the Safe Drinking Water Act (SDWA) in 1974 to protect public health by regulating the nation's public drinking water supplies. The law was subsequently amended in 1986 and 1996. The two categories of drinking water standards in the SDWA are:

1. *Primary Standards:* Legally enforceable standards that limit the levels of specific hazardous contaminants having an adverse effect on human health.
2. *Secondary Standards:* Nonenforceable guidelines for nonhazardous contaminants that may cause cosmetic effects (such as skin or tooth discoloration) or aesthetic effects (such as taste, odor, or color) in drinking water. USEPA recommends secondary standards to water systems but does not require systems to comply unless the state chooses to require compliance.

Several individual regulations fall under the umbrella of the SDWA. The following components are most likely to influence a water purveyor's decision to implement activated carbon.

Disinfectants and Disinfection By-products (D/DBP) Rule

For more than 100 years, the practice of disinfecting drinking water using chlorine and its compounds has protected consumers from waterborne diseases by inactivating pathogens. However, disinfectants react with organic matter in the water supply, and many of the by-products formed are of concern to public health. The primary objective of the D/DBP Rule, which was promulgated in two stages, is to reduce exposure of drinking water consumers to DBPs such as total trihalomethanes (TTHMs), the sum of five haloacetic acids (HAA5), bromate, and chlorite while still providing adequate disinfection. The rule also contains requirements for removing DBP precursors, as demonstrated by total organic carbon (TOC) removal using a treatment technique termed *enhanced coagulation*, which means achieving additional TOC removal by adding increased amounts of coagulant over what is required for turbidity removal.

To adequately protect public health from many waterborne diseases, the practice of disinfection must be continued. Activated carbon helps water purveyors comply with the D/DBP Rule by providing an additional removal mechanism for TOC. Because TOC is a major contributor to TTHM and HAA formation upon chlorination, reducing the TOC will also reduce the formation potential of these DBPs.

Enhanced Surface Water Treatment Rule

The Enhanced Surface Water Treatment Rule (ESWTR) also has two stages, corresponding to the two stages of the D/DBP Rule. The objective of the ESWTR is to confirm that disinfection is not compromised by utilities in their effort to minimize DBP formation. Compliance is demonstrated by providing treatment processes that remove or inactivate microorganisms. The final stage of this rule, the Long-Term 2 Enhanced Surface Water Treatment Rule, requires additional removal or inactivation of *Cryptosporidium*, depending on the source water quality. The USEPA's treatment toolbox for *Cryptosporidium* grants 0.5 log removal when granular activated carbon (GAC) filters or contactors are placed in series with another filtration process (granular media or membrane).

Unregulated Contaminant Monitoring Rule (UCMR) and Contaminant Candidate List (CCL)

The CCL is used by USEPA to identify contaminants that may be regulated in future, and the UCMR is used to gather data on the occurrence of unregulated contaminants in drinking water systems. Although there are no enforceable standards for UCMR contaminants, data are collected and reported to USEPA to assist with future regulatory policymaking. The first CCL was published in March 1998 and contained 60 contaminants under regulatory consideration. Based on the data from the first monitoring cycle of the UCMR, USEPA published the second drinking water CCL (CCL2) in 2005. The list carried forward 51 of the original 60 contaminants, and 9 were removed because sufficient data were collected and indicated that further regulating action was not required. In addition to the CCL2 list, the USEPA published the UCMR2, which required monitoring of 26 contaminants.

The third CCL (CCL3) was published as a draft in February 2008 and was finalized in 2009. After evaluating approximately 7,500 potential contaminants based on occurrence, production, and toxicology, an expert panel under the direction of the National Research Council (NRC), National Drinking Water Advisory Council (NDWAC), and Science Advisory Board (SAB) helped USEPA systematically narrow down the list of potential contaminants in the CCL3 to 104 chemicals and 12 microbiological contaminants. UCMR3 was proposed in February 2011. When it is finalized, this rule will require monitoring of 30 contaminants during the 2013–2015 time frame.

Many of the contaminants on the CCL and monitored in the UCMR can be effectively removed using activated carbon. Consequently, should any of them be regulated either individually or as a class of contaminants in the future, activated carbon will become an important part of the process train for many utilities.

Positioning for Future Regulations

The continued pressure to improve water quality is mounting as source waters are challenged with a variety of micropollutants. These contaminants include those being detected because of improved analytical methods and those being introduced into source waters at higher concentrations, such as personal care products and pharmaceutically active compounds. Although effective for reducing concentrations of these micropollutants, chemical oxidation does not convert them into carbon dioxide and water, and it is often unknown what compounds form in their place. Therefore, true removal processes are being revisited with renewed vigor. Membrane processes are improving, but only higher-pressure options such as nanofiltration and reverse osmosis can address most of these micropollutants. The following emerging issues are likely to affect utility strategies for using activated carbon in their treatment systems.

MTBE and Perchlorate

Both MTBE (methyl-*tert*-butyl-ether) and perchlorate continue to gain public interest and deserve more scrutiny. As two of the original contaminants on the CCL in March 1998, MTBE and perchlorate have undergone UCMR monitoring to determine the viability of regulating the chemicals or removing them from the list.

In UCMR monitoring of more than 3,400 systems, MTBE was detected in only 0.5 percent of the systems, and perchlorate was detected in nearly 4.5 percent of samples. It is still unclear whether the MTBE concern is in localized regions or widespread throughout the country. It is very likely that if MTBE detections are localized, the monitoring would spark state regulation rather than federal. While the USEPA is still in the process of revising its MTBE risk assessment, California has set an enforceable standard of 14 µg/L for MTBE. Because perchlorate is more widespread, it is more likely to warrant federal regulation.

Endocrine Disrupting Compounds (EDCs) and Pharmaceuticals

Pharmaceuticals, personal care products, and some household compounds are starting to appear in drinking water systems around the United States and in Europe. Some of these compounds are known to be endocrine disrupting compounds (EDCs), but their significance in drinking water is still not clear. Future monitoring and testing are needed to determine which of these compounds, if any, pose a threat to human health and at what dose. At that point, monitoring water systems for such compounds and evaluating ways to remove the compounds from the water may be necessary.

Nitrogenous Disinfection By-Products

The potential exists for future regulation of nitrogenous disinfection by-products (N-DBPs). Many N-DBPs can be found in treated drinking water; however, the most common ones include the various species of nitrosamines, particularly *N*-nitrosodimethylamine (i.e., NDMA) and halonitromethanes. Six of the nine possible nitrosamines are currently included in the CCL3 list. Based on the results of UCMR2 monitoring, NDMA is the most commonly occurring nitrosamine in drinking water. Because of the significant occurrence and the associated high carcinogenic potency of NDMA, it is anticipated that USEPA will consider developing a regulation for NDMA in the near future.

Carcinogenic Volatile Organic Compounds (cVOCs)

Under the auspices of six-year review, USEPA is currently reviewing the standards for trichloroethylene (TCE) and perchlorethylene (PCE). With USEPA's strategic direction to regulate contaminants by groups, USEPA is considering the revised TCE/PCE standards in a combined regulation for carcinogenic VOCs. Eight different cVOCs are currently regulated, and USEPA is considering regulating up to eight more in the group of cVOCs. A group regulation for cVOCs is expected to be proposed in 2013. Although the regulatory limits for specific VOCs are not known at the time of preparing this book, it is widely anticipated that the current regulatory limits for TCE and PCE will be lowered from the current limits of 5 µg/L. The existing limits were based on the limits of the analytical techniques available at the time; however, because analytical methods have been refined, lower detection limits are feasible compared to when the first VOC regulations were developed.

Activated carbon is the most widely accepted technology used to adsorb many of the organic compounds of concern. Many water utilities around the world are currently using activated carbon for removal of Natural Organic Matter (NOM), Synthetic Organic Chemicals (SOCs), and taste-and-odor compounds. It also helps with N-DBPs because systems that lower their TTHM and HAA formation potential are less likely to use chloramine, thereby avoiding formation of N-DBPs.

Public Health Protection and Customer Confidence

The state and federal drinking water regulations that provide legally enforceable standards are the foundation for a water utility's public health commitment to its customers and the public. However, meeting the standards does not result in zero risk; rather, the standards are based on peer-reviewed science, including data on how often the regulated contaminant occurs in the environment, how humans are exposed to it, the health effects of exposure, and cost considerations. A water system can elect to provide treatment to a quality higher than that required by a standard. However, most systems find it difficult to obtain the financial resources that may be needed to provide treatment levels above those required to comply with state and federal standards.

Science is continually identifying the presence of additional chemicals in the drinking water supply, often in minute concentrations. While evidence is lacking that many of these pose a significant threat to public health, customers may become concerned at the presence of these compounds in their water supply, especially when reported by various media outlets. Because we are in an era of information overload, multitasking, and sound bites, few people have the time, desire, or even sufficient technical expertise to fully examine and form their own educated opinion on all of the issues and challenges facing them today. Hence, opinions are often based on perceptions formed by instincts and input received from a variety of sources.

Much of the media only focus on water issues during droughts, floods, proposed rate increases, reported failure, inefficiencies, or health emergencies. Seldom is there a positive story of how well a water utility is performing, the quality of life it supports, or the health protection it is providing. With most of the media information reporting the negative, it is understandable that many customers are biased with negative perceptions and concerns about the quality of their water.

To counter these negative perceptions, it is critical for utilities to provide their customers with outstanding customer service and to become trusted partners in the goal to protect public health. During a Gallup Organization's Drinking Water Customer Satisfaction Survey for the USEPA of 1,000 households nationwide in 2002, general drinking water consumer knowledge and public confidence with information sources were assessed. Findings from the survey demonstrated that Americans recognize the importance of receiving information on aspects of their drinking water and value being informed. This accentuates the need for honest, unbiased information reaching the customer. Another way for utilities to demonstrate their commitment to understanding customer concerns is to provide additional treatment barriers for unregulated contaminants such as micropollutants, taste- and odor-causing (T&O) compounds, or aesthetic issues. However, the decision to implement additional treatment must be sensitive to the ability of the community to afford the increased level of treatment.

The bottom line is that while most of our water treatment systems do a good job with the technologies they have in place, a broad spectrum of chemicals in a water supply remain that are not being removed or reduced to the degree they could be by using activated carbon technologies. Although implementation of GAC treatment technologies costs money and will result in increased water rates, implementation needs to be considered in the light of improved public health protection. The current economic conditions may inhibit the ability to incur these costs; however, the value should be considered in strategic long-range planning.

Sustainability Considerations

It is sometimes thought that the use of a GAC treatment technology would result in a significant environmental burden. This needs to be considered in light of other options that could be used to achieve a desired treatment effect. In an effort to reduce the environmental burdens associated with producing drinking water, many water utilities have begun evaluating the sustainability of potential treatment scenarios prior to implementation. As a water utility evaluates the potential use of various processes to reduce disinfection by-products, one approach to assessing sustainability is a life-cycle assessment (LCA). LCA, which is often referred to as *cradle-to-grave*, is a systematic approach that follows the International Organization of Standardization (ISO) 14040 standard to quantify potential environmental burdens of a product or process over its lifetime.

Appendix A contains an example illustrating the use of LCA to evaluate the environmental impacts of typical processes to reduce DBPs. Three treatment technology scenarios were evaluated in the example: (1) GAC filter adsorbers, (2) GAC post-filter contactors, and (3) enhanced coagulation followed by disinfection using chloramines. In the appendix A example, no single scenario had significantly lower results across all LCA categories and sustainability measures analyzed, challenging the thought that GAC results in a significant environmental burden.

Concerns With Activated Carbon

Despite its merits, activated carbon has yet to be accepted as a "baseline" process in water treatment. Nevertheless, the USEPA recognized the significant benefits of activated carbon in its seminal 1986 Amendments to the Safe Drinking Water Act and chose GAC as a best available technology (BAT) for treating a suite of chemical contaminants. Since that time, some water purveyors have chosen to implement GAC as a treatment technique, while others have found different and often less expensive ways to meet federal and state water quality requirements. The use of GAC treatment is expensive, both in initial capital cost as well as in on going operational costs associated with reactivating and replacing the media. The information in this book describes methods and techniques to minimize these costs. The cost of GAC treatment should be considered in light of the benefits accrued by significant removal of a broad spectrum of organic contaminants from a water supply and the ability to furnish customers with a water quality that not only meets current regulations but also reduces risks that may result from currently unregulated contaminants and the risks of unintended consequences that may be associated with other treatment methods. The end result is the ability to furnish customers with a very high-quality water that is likely to improve their quality of life and protect human health.

Book Organization

Many books and book chapters have been written on activated carbon treatment. Most focus on removal mechanisms and capabilities, with some discussion of process approaches and applications. This book puts the fundamentals of activated carbon treatment, adsorption applications, and design of systems in the context of today's and tomorrow's water quality concerns, presenting the reader with a holistic view of the role of activated carbon in the water treatment process.

The intent of the book organization is to serve the needs of various water utility leaders, managers, and professionals; water treatment providers, scientists, students, and engineers; and water utility consultants in three distinct areas related to planning for and designing activated carbon systems:

PART 1 *Activated Carbon Adsorption Technologies* covers fundamentals and is targeted at assisting engineers and students who will use the book to gain a basic understanding and knowledge of activated carbon technologies for drinking water technologies.

PART 2 *Adsorption Applications* is for those who will benefit from approaches to planning the use of activated carbon treatment. In addition, the numerous case studies presented in part 2 demonstrate how and where activated carbon has been successfully implemented to solve specific water quality challenges.

PART 3 *Design and Procurement of Activated Carbon Systems* provides practical approaches to designers and system operators for effective and efficient design and use of activated carbon technologies as well as strategies for procuring and implementing the systems.

To address the concerns of sustainability related to the reactivation process, the authors included an appendix on sustainability, illustrating the full life-cycle assessment of activated carbon compared to other options for reducing TOC and complying with the Stage 2 D/DBP Rule. A similar process could be used to evaluate removal of micropollutants, comparing activated carbon to technologies such as reverse osmosis membranes and advanced oxidation processes using ozone.

Seventeen case studies comprise appendix B at the end of this book. These case studies will be useful to those readers seeking further practical information and experience from others using activated carbon technologies to improve the quality of their drinking water.

The book is not intended to be a textbook, although instructors can use portions of it to give students information on the basic mechanisms of the technology and practical guidance for training them as practicing engineers. It is assumed that different readers of the book will seek out relevant sections of the book as their needs dictate. Consequently, the book is written such that each of the three parts could be useful for a given audience without detailed study of the remainder.

List of Abbreviations and Acronyms

AC activated carbon
AOC assimilable organic carbon
AOP advanced oxidation process
AWWA American Water Works Association
BAC biologically enhanced activated carbon
BAF biologically active filter
BAT best available technology
BDOC biodegradable dissolved organic carbon
BOM biodegradable organic matter
BTEX benzene, toluene, ethyl benzene, and xylene
BV bed volume
BWWB Water Works Board of the City of Birmingham
CAP Central Arizona Project
CCL Contaminant Candidate List
CCL2 second Contaminant Candidate List
CCL3 third Contaminant Candidate List
CFSTR continuous flow stirred tank reactor
CGTF Central Ground Water Treatment Facility
CIP capital improvements program
CLSA closed-loop stripping analysis
CM construction manager
CMBR completely mixed batch reactor
CMWC Consolidated Mutual Water Company
CSO combined sewer overflow
cVOCs carcinogenic volatile organic compounds
D/DBP Disinfectants and Disinfection By-products
DBP disinfection by-product
DDT dichlorodiphenyl trichloroethane
DHS Department of Health and Safety (California)
DOC dissolved organic carbon
DOM dissolved organic matter
EBCT empty bed contact time
EDC endocrine disrupting compound
EfOM effluent organic matter
ESWTR Enhanced Surface Water Treatment Rule
EMT external mass transfer
GAC granular activated carbon
GC-MS gas chromatography and mass spectrometry
GCWW Greater Cincinnati Water Works
HAA5 sum of five haloacetic acids
HLR hydraulic loading rate
HNM halonitromethanes
HOCs hydrophobic organic compounds
HSDM homogenous surface diffusion model
IAST ideal adsorbed solution theory
IMT internal mass transfer
ISO International Organization of Standardization
LCA life-cycle assessment
MCL maximum contaminant level
MF microfiltration
MIB 2-methylisoborneol
MIEX® magnetic ion exchange
MSDBA mulitstage diffused bubble aeration
MTBE methyl-tert-butyl-ether
MTZ mass transfer zone
N-DBPs nitrogenous disinfection by-products
NDMA N-nitrosodimethylamine

NDWAC National Drinking Water Advisory Council
NJDEP New Jersey Department of Environmental Protection
NOM natural organic matter
NRC National Research Council
O&M operations and maintenance
OTC odor threshold concentration
PAC powdered activated carbon
PAH polycyclic aromatic hydrocarbon
PCB polychlorinated biphenyl
PCE perchlorethylene
PD-RSSCT proportional diffusivity RSSCT
PFOA perfluorooctanoic acid
PFR plug flow reactor
PhAC pharmaceutically active compound
PPCP pharmaceuticals and personal care products
PSDM pore and surface diffusion model
pzc point-of-zero-charge
RSSCT rapid small-scale column test
SAB Science Advisory Board
SCADA supervisory control and data acquisition
SDS simulated distribution system
SDWA Safe Drinking Water Act
SOCs synthetic organic chemicals
SUVA specific ultraviolet absorbance
T&O taste- and odor-causing
TCE trichloroethylene
TDS total dissolved solids
TOC total organic carbon
TTHMs total trihalomethanes
TTHMFP TTHM formation potential
UCMR Unregulated Contaminant Monitoring Rule
UF ultrafiltration
USEPA United States Environmental Protection Agency
USGS US Geological Society
UV ultraviolet
UVA ultraviolet absorbance
VOCs volatile organic compound
WQMP Water Quality Master Plan
WTP water treatment plant
WWTP wastewater treatment plant

PART 1

ACTIVATED CARBON ADSORPTION TECHNOLOGIES

For drinking water treatment, activated carbon is primarily used in two ways: fixed bed reactors with granular activated carbon (GAC) and slurry application of the smaller powdered activated carbon (PAC). In general, the manufacturing process is similar for both products. PAC is applied in a manner such that it cannot be economically recovered from the water treatment stream, so it is removed with the process solid residuals. For surface water treatment, GAC can be applied as media in a filter, termed a *filter adsorber*, or applied in a reactor positioned after the filter, a post-filter contactor. For direct groundwater treatment, reactors similar to post-filter contactors are used. After the adsorption capacity of GAC is exhausted, it must be removed from the fixed-bed reactor and reactivated either on-site or off-site.

While the physical and chemical characteristics of activated carbon can give some insight into the potential adsorption behavior, actual adsorption tests with the specific activated carbon and the compound of interest in the actual water yield the most useful information. These tests are described in the next chapters and include: (a) batch equilibrium capacity tests, i.e., "isotherms"; and (b) methods that assess both capacity and kinetics, including PAC jar tests and GAC bench-scale and pilot-scale column tests.

One critical factor that affects the adsorption efficiency is the influence of the background matrix on the adsorption capacity of the target compound, whether it is naturally occurring compounds like the odorants 2-methylisoborneol (MIB) and geosmin, or anthropogenic compounds like trichloroethylene and atrazine. Increases in the concentration of the compounds that compose the background matrix decrease the adsorption capacity for the target compound. Activated carbon can also be used to control the formation of disinfection by-products (DBPs) through the adsorption of dissolved organic matter (DOM). The critical factor here is understanding the effect of the influent DOM concentration on the adsorption efficiency.

The objective of part 1 of this book is to provide an understanding of the fundamentals of adsorption by activated carbon from a practical perspective. The chapters include adsorption fundamentals (chapter 1), the basics of PAC systems (chapter 2), and the basics of GAC systems (chapter 3). The approach taken herein is a semi-quantitative one that intends to educate the reader on the factors that affect activated carbon adsorption. The approach is based on well-established principles, and the reader is referred to several existing monographs on activated carbon adsorption that provide the extensive literature review and the quantitative basis that underline the approach taken herein (e.g., Sontheimer et al. 1988 and Summers et al. 2010).

References

Summers, R.S., D.R.U. Knappe, and V.L. Snoeyink. 2010. Adsorption of organic compounds by acivated carbon. In *Water quality and treatment*, 6th ed. J.K. Edzwald, ed. New York: McGraw-Hill and American Water Works Association.

Sontheimer H., J.C. Crittenden, and R.S. Summers. 1988. *Activated carbon for water treatment*, 2nd ed. Karlsruhe, Germany: DVWG Forschungsstelle.

Chapter 1

Fundamentals of Adsorption

Adsorption is used in water treatment to remove dissolved contaminants by physical phase separation in either a steady-state (PAC) or a nonsteady-state (GAC) process. Adsorbates accumulate on the surface of the adsorbent until the adsorption capacity is consumed. In the PAC process, the PAC is removed from the water treatment stream as a process solids residual. In the GAC process, once the effluent concentration of the adsorbate reaches the treatment objective, the GAC must be replaced in order to continue satisfactory performance. The high internal surface area of activated carbon provides enough adsorption capacity to make it an economical choice of adsorbent. However, regardless of how much surface area is provided for adsorption, different target compounds have various affinities for the surface phase versus the water phase, making adsorption dependent on the properties of the target compound and the background matrix.

This chapter first provides an introduction to the process of manufacturing activated carbon. Then the physical-chemical process fundamentals of adsorption systems are discussed.

Activated Carbon Manufacturing and Properties

Activated carbons are manufactured from a wide range of carbonaceous materials. The most common precursor materials for use in drinking water treatment are bituminous coal, lignite coal, coconut shells, and wood. The choice of raw material affects the internal pore structure, surface area distribution, and surface chemistry. Commercial activated carbon production is a two-step process of carbonization of the precursor materials and activation. In the pyrolytic carbonization process, the temperature of the raw material is raised to the range of 500 to 800 °C in the absence of oxygen. Volatile organic matter of the raw material is thermally released, and the carbon atoms realign to form a more crystalline structure. The carbonized product at this point in the process is heavily influenced by the raw materials used. For activated carbon products used in water treatment, a thermal or physical activation process then follows in which the temperature of the carbonized product is increased to the range of 850 to 1,000 °C in the presence of an oxidizing agent, typically steam or carbon dioxide. Activation increases the pore sizes and creates a continuous pore structure, which increases the micropore volume (pore width < 2 nm) and the internal surface area. The mesopore and macropore volumes (pore width > 2 nm) are critical to the internal transport of the compound to the micropore surface area, which is where most of the adsorption occurs. Activated carbon can be produced in a

Table 1-1. Typical activated carbon physical characteristics

Porosity	Surface area (m^2/g)	Particle density (kg/m^3)	Apparent dry density (kg/m^3)		Average particle diameter (mm)	
			GAC	PAC	GAC	PAC
0.5–0.8	500–1,200	600–850	300–650	200–750	0.6–3.0	0.01–0.03

* 1 $lb/ft^3 = 16\ kg/m^3$

direct activation process in which the raw material is crushed and then activated, or in a reagglomeration process in which the raw material is crushed, reagglomerated, crushed again, and then activated. Table 1-1 lists some typical physical properties of commercial activated carbon used in water treatment.

The apparent density of GAC is the as-received density of unstratified dry material. The bed density is the density after the GAC is placed in the filter or contactor, backwashed, and drained. Bed density is generally about 90 percent of the apparent density due to stratification.

Once the adsorption capacity of GAC is exhausted, it is removed from the filter or contactor and the adsorption capacity is restored in a reactivation process. The reactivation requires conditions very similar to the activation process used in the original production. During reactivation, some adsorbed compounds are thermally desorbed, but most compounds, including the adsorbed DOM, are oxidized on the GAC surface.

The surface chemistry of activated carbon is a product of the raw materials and the activation process. The noncarbon content of activated carbon varies but is typically in the 10 to 20 percent range and is mostly made up of surface oxygen functional groups and minerals (e.g., ash), both of which can negatively affect the adsorption of some compounds.

Adsorption

Adsorption is the accumulation of a substance (the adsorbate) on the surface of a solid (the adsorbent). While most, but not all, organic compounds of interest in drinking water quality are adsorbable by activated carbon, it is the degree of adsorbability that is important, as it dictates the cost of using activated carbon. For example, DOM consists of a mixture of compounds of which about 90 percent are adsorbable to some degree. Large-molecular-weight compounds are excluded from internal adsorption sites, and hydrophilic DOM compounds of any size are not attracted to the adsorbent surface. The most strongly adsorbed DOM components are low- and medium-molecular-weight hydrophobic compounds.

Adsorption of compounds of interest in drinking water by activated carbon occurs mostly by nonspecific physical adsorption as opposed to site-specific chemical adsorption. Physical adsorption is caused by nonspecific secondary binding mechanisms, in which electrons are shared between the adsorbate and the adsorbent rather than electrons being transferred, as occurs in chemical adsorption. Physical adsorption can be reversed in response to a change in the concentration gradient between the bulk liquid phase and the adsorbed solid phase. Because activated carbon has a nonhomogeneous surface with specific function groups, a limited degree of chemical adsorption for some compounds is possible.

Batch Adsorption Kinetics

As illustrated in Figure 1-1, removal of a compound by activated carbon is a three-step process: external mass transfer (EMT), internal mass transfer (IMT), and adsorption. The first step is external mass transfer of the compound from the bulk liquid phase through the hydrodynamic layer that surrounds the activated carbon particle. External mass transfer, or film diffusion, can be important and control the overall kinetics of the adsorbate uptake when the flow through a GAC bed is very low or when the mixing in a PAC reactor, like the flocculation basin, is low. After external mass transfer delivers the adsorbate to the external surface of the activated carbon particle, internal mass transfer conveys the adsorbate to an adsorption site on the surface of the activated carbon where adsorption occurs. The adsorption reaction of the compound at the activated carbon surface is very fast and is not considered to be rate limiting. Thus, the overall uptake is thought of as a two-step mass transfer process of external and internal mass transfer. Because external diffusion and internal diffusion occur in series, the slowest step is rate limiting. Both external and internal mass transfers are diffusion-based phenomena where the driving force is the concentration gradient of the adsorbate: the difference between the bulk concentration, C_b, and the internal concentration at equilibrium, C_e. Internal mass transfer is often the slowest and thus controls the overall uptake of an adsorbate. Internal mass transfer may occur through the pore fluid or along the pore wall in parallel processes. Both pore diffusion and surface diffusion are controlled by both the activated carbon pore structure and the adsorbate properties. Because the processes occur in parallel, the faster step, which is typically pore diffusion under most conditions relevant to water treatment, controls the overall uptake of an adsorbate.

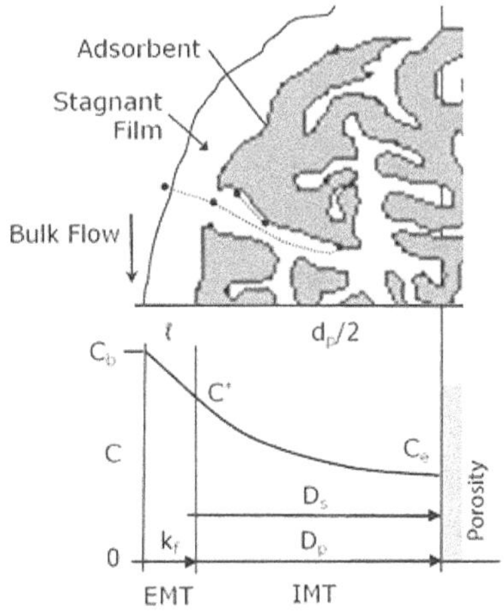

Figure 1-1. Illustration of the mass transfer–adsorption process

The kinetics of the overall adsorbate uptake can be illustrated using a completely mixed batch reactor (CMBR) such as shown in Figure 1-2, where V is the reactor volume, M_{GAC} is the mass of activated carbon added, and C_0 is the initial liquid phase adsorbate concentration. Mixing in the reactor is important as the external mass transfer resistance can be reduced by increasing the mixing enough to reduce the thickness of the stagnant film surrounding the particle.

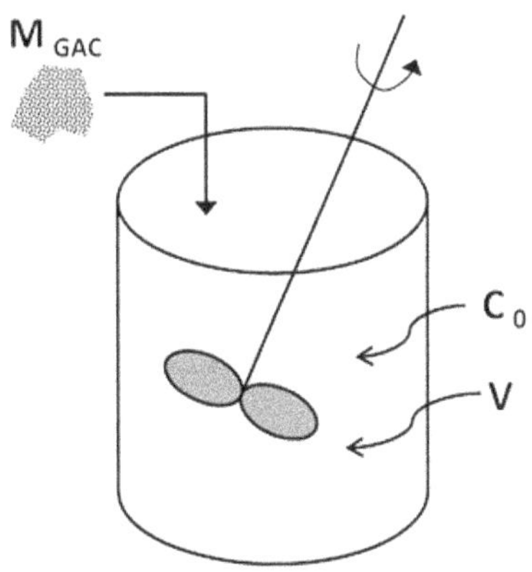

Figure 1-2. Activated carbon adsorption in a completely mixed batch reactor

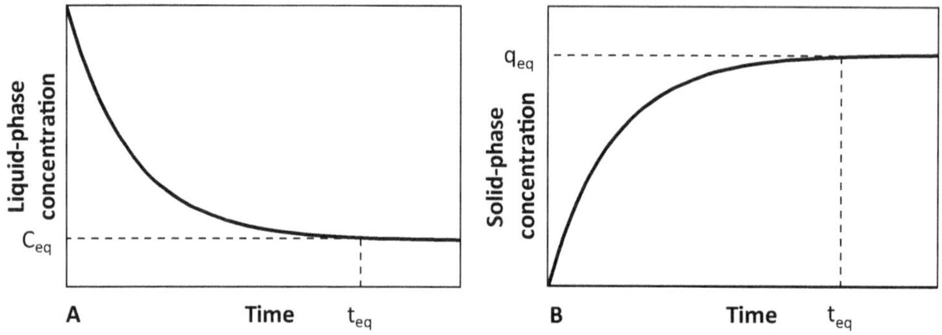

Figure 1-3. Adsorbate uptake kinetics in a CMBR shown as both (a) liquid-phase and (b) solid-phase concentrations

After the addition of the activated carbon, the liquid phase concentration of the adsorbate in the reactor will decrease as a function of time. Eventually, the net uptake of the adsorbate reaches zero. The adsorbate-adsorbent pair is now at equilibrium. The liquid phase concentration at this point is termed the *equilibrium liquid-phase concentration*, C_{eq}, and the time is termed the *time to equilibrium*, t_{eq}. Figure 1-3 illustrates the liquid-phase concentration history in a batch test along with the kinetics of the adsorbate concentration on the adsorbent surface, solid-phase concentration, q. The solid phase concentration at equilibrium, q_{eq}, is attained at the same time as C_{eq}.

As the adsorbent dose increases, the equilibrium liquid-phase concentration decreases as shown in Figure 1-4a. In addition to C_{eq} being affected by the dose increase, the time to equilibrium is also affected. Other factors that affect the kinetics or time to equilibrium, but not the equilibrium liquid-phase concentration, include the size of the adsorbent, expressed here as the diameter (d_{AC}); diffusion coefficient of the adsorbate (D); initial liquid phase concentration; and amount of shear or mixing imparted on the external surface of the adsorbent. The adsorbent size affects the internal mass transfer. As the adsorbent size decreases, the internal diffusion path length the adsorbate must travel to an adsorption site decreases, and the adsorption sites can be reached more quickly. The same mass of smaller adsorbent particles

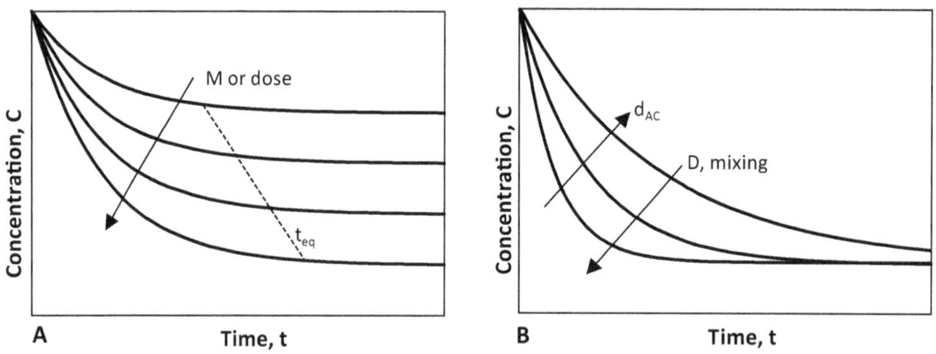

Figure 1-4. Adsorbate uptake kinetics and equilibrium liquid-phase concentration in a CMBR as affected by (a) the adsorbent dose and (b) adsorbent size (d_{AC}), adsorbate diffusivity (D), and system mixing or shear

achieves equilibrium state more rapidly, lower t_{eq}, than that of large adsorbent particles. As the diffusion coefficient and the concentration of the adsorbate increases, the flux into the adsorbent increases and the adsorption sites can be reached more quickly, also decreasing the time to equilibrium. Factors affecting time to equilibrium are summarized in Eq 1-1:

$$t_{eq} = f\,(d_{AC}^{2}, D^{-1}, C_0^{-1}) \qquad \text{(Eq 1-1)}$$

When external mass transfer is an important factor, increasing the mixing in the CMBR or the fluid velocity decreases the boundary layer around the particle, thus increasing the overall uptake kinetics.

Adsorption Equilibrium, or Isotherm

Mass Balance

As was shown in Figure 1-4a, increasing the mass of activated carbon added to a batch reactor will decrease the equilibrium liquid-phase concentration, and the equilibrium solid-phase concentration increases by mass balance as shown in Eq 1-2:

$$q_{eq} = (C_0 - C_{eq}) \times V/m = (C_0 - C_{eq})/\text{dose} \qquad \text{(Eq 1-2)}$$

If an experiment is run that varies the adsorbent dose and/or initial concentration, the equilibrium solid-phase concentration can be calculated, and a relationship with the equilibrium liquid-phase concentration can be established. This adsorption equilibrium relationship, often termed *isotherm*, allows the adsorption capacity of the adsorbent to be calculated at any given liquid-phase concentration.

Example 1-1. Different masses of activated carbon are added to bottles that contain an adsorbate of interest to yield the doses noted in Table 1-2. Two different initial concentrations of the adsorbate were used. To facilitate the mass transfer, small activated carbon sizes were used, and the bottles were shaken. Preliminary experiments determined the time needed to achieve equilibrium (Figure 1-3). After waiting the prescribed equilibrium time, activated carbon was filtered from the solution, and the adsorbent equilibrium liquid-phase concentration was measured. Calculate the equilibrium solid-phase concentration at each dose and plot as a function of the equilibrium liquid-phase concentration.

Solution. Using the mass balance (Eq. 1-2), the equilibrium solid-phase concentration was calculated. The results are shown in Table 1-2 and Figure 1-5.

The shape of the isotherm plot is typical of the adsorption of target compounds by activated carbon from laboratory clean water. As can be seen for single-solute isotherms, the equilibrium relationship is not a function of the initial concentration.

Equilibrium Adsorption Isotherm Models

To be more useful, equilibrium isotherm relationships can be fit with model equations. The two most common are the Langmuir and Freundlich models. The Langmuir model is more fundamental, has conceptual endpoints as it asymptotically approaches a maximum solid-phase concentration at the very high liquid-phase concentration range, and has a linear equilibrium relationship at the very low–concentration ranges. However, experience has shown that the empirical Freundlich model (Eq 1-3), which

Table 1-2. Example isotherm results

Initial concentration, C_0 2.00 mg/L			Initial concentration, C_0 0.500 mg/L		
Dose (mg/L)	C_{eq} (mg/L)	q_{eq} (mg/g AC)	Dose (mg/L)	C_{eq} (mg/L)	q_{eq} (mg/g AC)
5	1.53	92.77	5	0.295	40.74
10	1.18	81.47	10	0.18	31.82
25	0.575	56.87	25	0.056	17.75
50	0.224	35.50	65	0.01	7.50
100	0.066	19.27	150	0.002	3.35
200	0.0175	9.92	335	0.0004	1.50
400	0.0044	4.97	610	0.00012	0.82
800	0.00111	2.50			
1,600	0.00028	1.25			
3,000	0.00008	0.67			

is a power function, works best. The Freundlich model can be derived theoretically from the Langmuir model assuming a distribution of adsorption site energies.

$$q_{eq} = K_F\, C_{eq}^{\ 1/n} \qquad \text{(Eq 1-3)}$$

Where

K_F = Freundlich constant
$1/n$ = Freundlich exponent

To simplify the notation in this chapter, the equilibrium liquid-phase and solid-phase concentrations are expressed as shown in Eq 1-4:

$$q = K_F\, C^{1/n} \qquad \text{(Eq 1-4)}$$

To fit the Freundlich parameters to the data, the Freundlich equation can be linearized as shown in Eq 1-5:

$$\log q = \log K_F + 1/n \log C \qquad \text{(Eq 1-5)}$$

When the isotherm data are plotted on a log scale as shown in Figure 1-5b, the slope of the line is the 1/n value, and K_F is evaluated on the y-axis at a concentration of 1 mg/L. For data expressed in the units shown, the value of 1/n is 0.5 and K_F is 75 $(mg/g)/(mg/L)^{1/n}$. The value of K_F depends on the units of the isotherm data. For the same data as above, but if C is in units of µg/L instead of mg/L, the value of K_F is 2.37 $(mg/g)/(\mu g/L)^{1/n}$.

The value of the Freundlich constant (K_F) can be useful in gauging the difference in the adsorption capacity of a given activated carbon for a range of compounds or

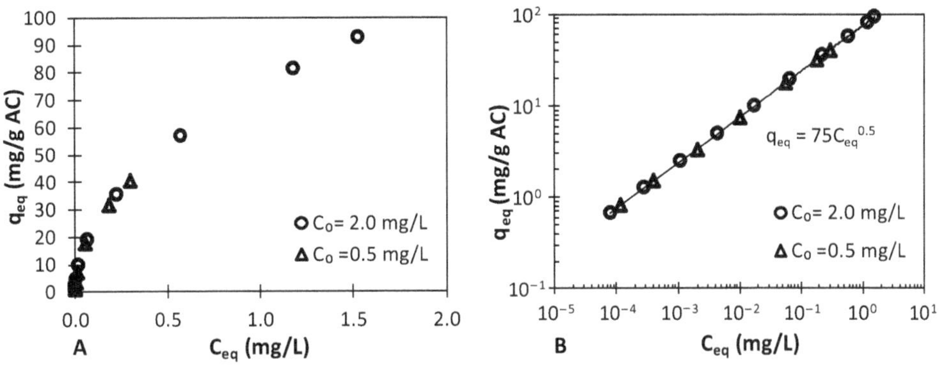

Figure 1-5. Example isotherm (a) arithmetic plot and (b) logarithmic plot

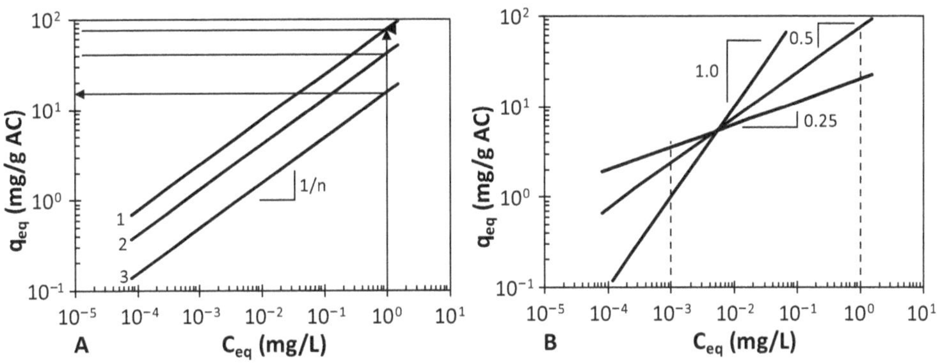

Figure 1-6. Effect of Freundlich parameters (a) K_F and (b) 1/n on isotherm

the adsorption capacity of a given compound for a range of activated carbons, as long as the Freundlich exponent (1/n) value is relatively constant. Figure 1-6a illustrates three isotherms that all have a 1/n value of 0.50. The larger the K_F value, the higher the adsorption capacity for a given adsorbate-adsorbent pair.

However, when the 1/n values of the different isotherms are significantly different, as shown in Figure 1-6b, the K_F value is only useful in the concentration range around 1, in this case 1 mg/L. The isotherm plots converge at lower concentration ranges, and to compare adsorption capacities both the K_F and 1/n values are needed, or comparisons can be made on the basis of a solid-phase concentration at a given liquid-phase concentration (q_C). For example, $q_{0.001}$ can be use to represent the solid-phase concentration at a liquid-phase concentration of 0.001 mg/L (1 μg/L). For this example, the isotherm that yielded the lowest capacity in the mg/L range now has the highest capacity in the μg/L range.

Summers et al. (2010) have compiled a list of single-solute Freundlich parameters for more than 70 compounds of interest; the K_F values range from 100 to 0.1(mg/g)/(μg/L)$^{1/n}$, and the 1/n values range from approximately 0.8 to 0.1. In general, for the K_F values expressed in the units (mg/g)/(μg/L)$^{1/n}$ and in a concentration range around 1 μg/L, compounds with K_F values above 10 (mg/g)/(μg/L)$^{1/n}$ can be considered well removed by activated carbon, and compounds with K_F values below 1 (mg/g)/(μg/L)$^{1/n}$ can be considered poorly removed by activated carbon. These Freundlich parameter values have been generated from isotherms with a single type of activated carbon, and the adsorption capacity for these compounds may be increased or decreased by the use of other activated carbon products.

A major limitation of the Freundlich model is that it should only be used to model the equilibrium data in the concentration range for which it was developed; it does not asymptotically approach a maximum adsorption capacity on the high concentration end, nor does it approach a linear isotherm on the low concentration range.

Competitive Adsorption

Caution must be exercised in not overusing single-solute Freundlich parameter values, as they are affected by the background matrix in the water to be treated. If other adsorbates are present in the water, as they are in nearly all practical cases, then the adsorption capacity for a specific compound decreases, even if the background contains only weaker adsorbing compounds. Figure 1-7 illustrates competitive adsorption behavior for a binary system in which Compound 1 is more strongly adsorbed than Compound 2 (i.e., $K_{F,1} > K_{F,2}$). In Figure 1-7a, the effect of the more weakly adsorbed Compound 2 on the adsorption of Compound 1 is shown. Increasing the initial concentration of Compound 2 causes the adsorption capacity of Compound 1 to decrease relative to the single-solute isotherm at $C_{2,0} = 0$. Figure 1-7b shows that increasing the initial concentration of more strongly adsorbing Compound 1 causes the adsorption capacity of Compound 2 to decrease relative to the single-solute isotherm at $C_{1,0} = 0$. As expected, the presence of the same amount of Compound 1 is more detrimental to the adsorption of Compound 2 (Figure 1-7b) than the other way around (Figure 1-7a). While not illustrated in Figure 1-7, the slope and shape of the isotherm in a mixture can also be affected.

Multisolute systems, like the adsorption of a target compound in the presence of DOM, add additional complexity to adsorption behavior. Competing adsorbates in the same molecular size range as the target adsorbate compete directly for adsorption sites (Li et al. 2003). Larger competing adsorbates may not be able to access micropores because of size exclusion. However, size exclusion may also result in partial or complete blockage of a pore. Partial blockage of a pore may slow kinetics of the target adsorbate, whereas complete blockage of a pore results in surface area behind the

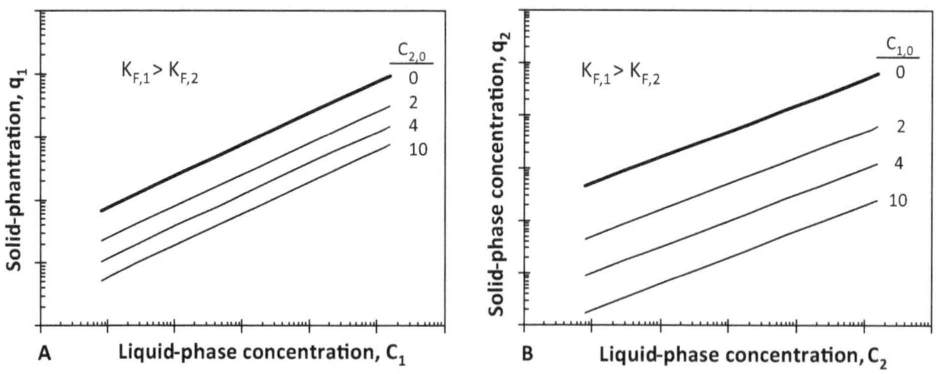

Figure 1-7. Equilibrium isotherm in a binary mixture (a) strongly adsorbing Compound 1 and (b) weakly adsorbing Compound 2

blockage not being available for further adsorption. Because of its tendency to adsorb at multiple sites, adsorbed DOM is nondisplaceable to a large degree (Summers et al. 2010). Pore blockage and nondisplaceable adsorption cannot be described by ideal adsorbed solution theory (IAST). Therefore, reductions in adsorption capacity due to these mechanisms are often termed *fouling*.

DOM Equilibrium Adsorption Behavior

An understanding of DOM adsorption is important, as DOM serves as a precursor for the formation of DBPs. DOM is a heterogeneous mixture of organic compounds and normally dominated by hydrophobic organic acids such as humic and fulvic acids but also contains hydrophilic compounds that are weakly adsorbable or nonadsorbable. To measure DOM collectively, the total organic carbon (TOC) concentration is measured. Figure 1-8 illustrates the adsorption equilibrium relationship of DOM, as measured by TOC, at two initial DOM concentrations. The adsorption behavior of both isotherms displays a nonadsorbable fraction of about 10 percent, and as the equilibrium liquid-phase concentration increases, so does the adsorbability of the DOM in a nonlinear manner. The behavior shown in Figure 1-8 cannot be modeled by the Freundlich or Langmuir isotherms. When the initial DOM concentration decreases to a TOC = 2.5 mg/L, the adsorption capacity increases in all concentration ranges. The complex adsorption behavior of the DOM mixture limits the use of the isotherm in comparing the DOM adsorption behavior from different waters. Because the initial DOM concentration must be the same, it requires the use of complex adsorption models.

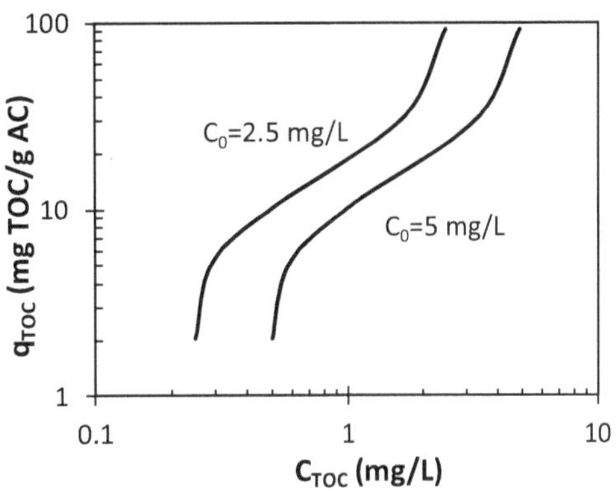

Figure 1-8. Equilibrium adsorption isotherms of DOM at two different initial concentrations

Several other adsorbate properties can also affect the adsorption equilibrium relationship. Large-molecular-weight molecules (i.e., > 1,000 Daltons) are often larger in size than the micropores where much of the surface area is located; thus size exclusion can limit the amount adsorbed. Weak organic bases and acids, like humic and fulvic acids in DOM, can be affected by pH because adsorption capacity is higher in the neutral form than the ionized form.

Ideal Adsorption Reactors

The above adsorption fundamentals were developed for CMBRs that have no volumetric flow, Q, into the reactor. Reactors used in water treatment are flow-through reactors and behave in the envelope of performance of the two idealized flow-through reactors: the plug flow reactor (PFR) and the continuous flow stirred tank reactor (CFSTR), shown in Figure 1-9. The difference in behavior is critical in understanding activated carbon performance. These will be described in detail in the next two chapters, but an overview at this point is instructive.

The mixing required to suspend PAC media also destroys the spatial concentration gradient of the target compound, yielding CFSTR behavior. In the ideal CFSTR

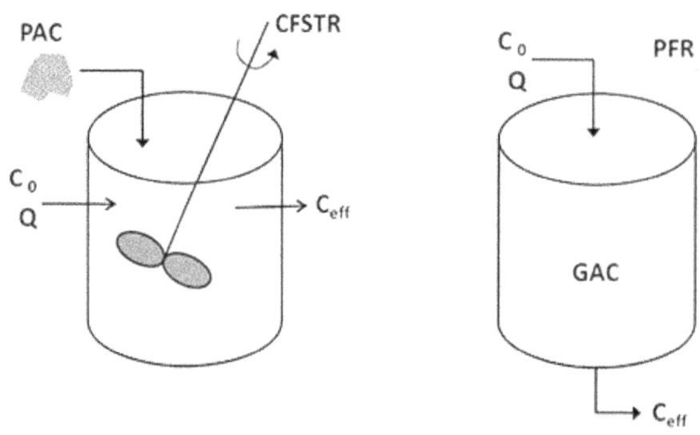

Figure 1-9. Flow-through reactors: CFSTR for PAC and PFR for GAC

(Figure 1-9), the mixing conditions make the concentration of the target compound equal to the effluent concentration, C_{eff}, at all points in the reactor. On the other end of the spectrum, the generally laminar flow through a packed GAC bed maintains the concentration gradient from the reactor influent to the effluent and yields PFR behavior. With long enough run times, the GAC will come to equilibrium with the influent concentration, C_0. The adsorbent in the CFSTR comes to equilibrium with the effluent concentration in Eq 1-4, whereas a PFR will come to equilibrium with the influent concentration.

Example 1-2. Calculate the solid phase concentration, q_{eq}, of atrazine in laboratory clean water at equilibrium in a PFR and a CFSTR. The influent concentration of atrazine is 50 µg/L, K_F is 38.7 $(mg/g)(L/µg)^{1/n}$, and 1/n is 0.291. The treatment objective or desired effluent concentration for atrazine is the maximum contaminant level (MCL) of 3 µg/L.

Solution. The solid-phase concentration is calculated directly from Eq 1-3 or 1-4. The PFR comes to equilibrium with the influent concentration of 50 µg/L:

$$q_{eq\,PFR} = K_F \times C_0^{1/n} = 38.7\,\frac{mg}{g} \times \left(\frac{L}{\mu g}\right)^{1/n} \times \left[50\,\frac{\mu g}{L}\right]^{0.291} = 121\,\frac{mg\ atrazine}{g_{AC}}$$

The CFSTR comes to equilibrium with the effluent concentration, in this case 3 µg/L:

$$q_{eqCFSTR} = K_F \times C_e^{1/n} = 38.7\,\frac{mg}{g} \times \left(\frac{L}{\mu g}\right)^{1/n} \times \left[3\,\frac{\mu g}{L}\right]^{0.291} = 53.3\,\frac{mg\ atrazine}{g_{AC}}$$

Thus the GAC in the PFR adsorbs over twice the amount of atrazine per mass of activated carbon compared to the PAC in a CFSTR. The PFR is a more efficient reactor.

Another way of comparing the efficiency of the PAC in a CFSTR to GAC in a PFR is the use rate. Table 1-3 defines the use rate in two different forms for each reactor: mass of AC per volume of water treated or mass of AC used per day. As such, lower use rates are more favorable. Note that the ideal PFR removes all the contaminant until breakthrough is reached; therefore, the effluent concentration is zero.

These calculated use rates are representative of single-solute systems (no background natural organic matter [NOM]). Typical use rates in drinking water treatment may be more than an order of magnitude higher.

Example 1-3. For the problem given in Example 1-2, calculate use rate in kg/1,000 m^3 and in kg/day if the flow rate is 4.0 × 10^4 m^3/day (10.6 mgd).

Solution. See Table 1-3. The GAC system requires less than half the amount of AC compared to the PAC system. *Note:* in practical use, the adsorption capacity for

Table 1-3. Use rate definitions using isotherm parameters and Example 1-3 calculations

Use Rate	PAC–CFSTR	GAC–PFR
Mass of AC/volume water treated	$(C_0 - C_{eff})/q_{eqCFSTR}$ $= (C_0 - C_{eff})/K_F \times C_{eff}^{1/n}$	$(C_0 - C_{eff})/q_{eq\,PFR}$ $= (C_0 - C_{eff})/K_F \times C_0^{1/n}$
Mass of AC/day	$Q \times (C_0 - C_{eff})/\, q_{eq\,CFSTR}$ $= Q \times (C_0 - C_{eff})/\, K_F \times C_{eff}^{1/n}$	$Q \times (C_0 - C_{eff})/\, q_{eq\,PFR}$ $= Q \times (C_0 - C_{eff})/\, K_F \times C_0^{1/n}$

Example 1-3

Use Rate	PAC–CFSTR	GAC–PFR
Mass of AC/volume water treated	= (50 µg/L – 3 µg/L) / 53.3 µg atrazine/mg AC	= (50 µg/L – 0 µg/L) / 121 µg atrazine/ mg AC
	= 0.88 mg AC/L	= 0.41 mg AC/L
	= 0.88 g AC/m^3	= 0.41 g AC/m^3
	= 0.88 kg AC/1,000 m^3	= 0.41 kg AC/1,000 m^3
	= 0.0074 lb/1,000 gal	= 0.0035 lb/1,000 gal
Mass of AC/day	= Q × (50 µg/L – 3 µg/L) / 53.3 µg atrazine/mg AC	= Q × (50 µg/L – 0 µg/L) / 121 µg atrazine/mg AC
	= 4.0×10^4 m^3/day × 0.88 g AC/m^3	= $4.0 \times 10^4 m^3$/day × 0.41 g AC/m^3
	= 35 kg/day	= 16 kg/day
	= 78 lb/day	= 36 lb/day

the target compound is much lower due to the competitive effects from background organic matter, resulting in higher use rates.

References

Li, Q.L., V.L. Snoeyink, B.J. Marinas, and C. Campos. 2003. Elucidating competitive adsorption mechanisms of atrazine and NOM using model compounds. *Water Res.* 37(4):773–784.

Summers, R.S., D.R.U. Knappe, and V.L. Snoeyink. 2010. Adsorption of organic compounds by activated carbon. In *Water quality and treatment*, 6th ed. J.K. Edzwald, ed. New York: McGraw-Hill and American Water Works Association.

Chapter 2

Powdered Activated Carbon Systems

Powdered activated carbon (PAC) can be an effective and economical treatment approach under the appropriate conditions. PAC is contacted with the contaminated water as a slurry, adsorbs the contaminants, and is removed with particles in the solids management process. PAC is not recovered and reactivated, thus its carbon use rate (lb/day) or dose (lb/MG) can be high. However, the low capital cost associated with the PAC feed system combined with the ability to apply PAC seasonally or for periodic issues can make it an economic alternative based on an annualized cost. The most common PAC use is to control seasonal occurrences of taste-and-odor compounds and pesticides. Many utilities that treat surface waters affected by municipal and industrial wastewater discharges and agricultural runoff also keep PAC on-site for the emergency control of periodic spikes and spills. In a few cases, PAC has also been used for the periodic reduction of cyanotoxins, DOM for DBP control, and micropollutants (Summers et al. 2010).

PAC is applied as a slurry, and it is effective in removing adsorbable compounds while it is suspended in the water. However, once it is removed from suspension and is not in contact with the contaminated water, it is no longer effective. As was shown in chapter 1, the full capacity of the PAC is achieved when the equilibrium concentration, C_{eq}, is reached (Figure 1-3). Several factors affect the time that it takes to reach equilibrium, t_{eq} (Figure 1-4), including the PAC size (d_{AC}) and mixing. This chapter explores the influence of these and other operating factors, as well as the adsorbate and background matrix chemistry, on the effective use of PAC.

Continuous Flow Stirred Tank Reactor Performance

The most common point of PAC application is before or at the rapid mixer (Figure 2-1). The rapid mixer and the subsequent flocculation basins are designed to keep coagulated particles in suspension such that if PAC is added at the rapid mixer, it will remain in suspension through the rapid mixer, the flocculation basin, and part of the sedimentation basin. The rapid mixer can be modeled as continuous flow stirred tank reactor (CFSTR), and most flocculation basins can be modeled as a series of CFSTRs. While CFSTRs are not efficient reactors from a chemical conversion perspective, they achieve the main objective of rapid mixers and flocculation basins: disperse the coagulants in the water so that they come into contact with the raw water particles and subsequently promote floc growth.

Figure 2-2 shows a diagram of a CFSTR. The PAC is added on a continuous basis and on average is in contact with the CFSTR for a residence time (τ) defined as

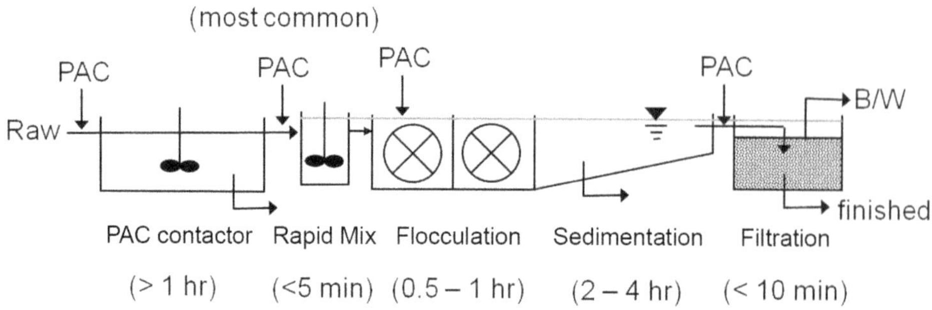

Figure 2-1. PAC application points in a conventional surface water treatment plant

the ratio of the reactor volume (V) to the volumetric flow rate (Q). C_0 is the influent liquid-phase adsorbate concentration, and C is the effluent liquid-phase adsorbate concentration. The dose of the PAC is calculated as PAC mass flow rate (PAC mass added per time divided by Q).

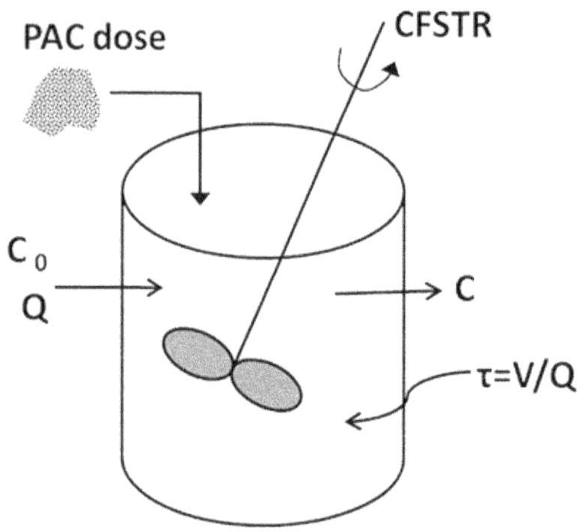

Figure 2-2. Continuous PAC addition to a CFSTR

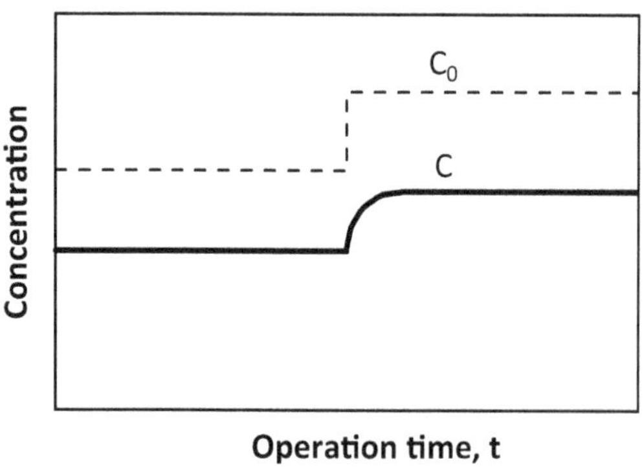

Figure 2-3. Steady-state CFSTR performance followed by an increase in the influent concentration and a transition in the effluent concentration to new steady-state condition

The added PAC will remove the adsorbate to varying degrees, resulting in a lower effluent concentration (C) relative to the influent concentration (C_0). After a short startup time that lasts 3 to 4 times the reactor residence time, the effluent concentration remains at a steady-state level as long as there are no changes in any other operating factor or critical influent water quality parameters. Steady-state performance for a CFSTR is shown in Figure 2-3 under two influent concentration conditions. If changes occur in the important influent and operating conditions, then the CFSTR performance is affected. Figure 2-3 illustrates that an increase in the CFSTR influent concentration results in an increase in the effluent concentration with a small transition period prior to achieving a new steady-state condition. The reactor will remain in this new steady-state condition until changes in critical influent or operating factors occur. It is important not to confuse this steady-state performance with equilibrium. Reactor residence times are generally shorter than the time to reach equilibrium.

Factors Affecting PAC Performance

In addition to the influent concentration, several other factors influence the performance of PAC in a CFSTR and are illustrated in Eq 2-1 and Figure 2-4, where the steady-state effluent concentration is shown as a function of increasing the parameter value.

Reactor effluent concentration, $C = f(\text{dose}^{-X}, \tau^{-Y}, K_F^{-Z}, t_{eq}, C_0, C_{2,0})$ (Eq 2-1)

Increasing the PAC dose will decrease the reactor effluent concentration as more adsorbate can be removed from solution. However, the percent effluent concentration decrease is not proportional to an equivalent percent increase in PAC dose, as shown in Figure 2-5. For the high TOC water (TOC = 10 mg/L), a 40 percent decrease in the influent geosmin concentration can be achieved with a dose of 5 mg/L, but it takes about 17 mg/L of PAC to yield 90 percent removal.

If the residence time in the CFSTR is longer than the time to equilibrium ($\tau > t_{eq}$), the full capacity of the PAC is realized, and increasing the residence time will not improve performance. However, the t_{eq} for most PAC-adsorbate pairs is on the order of hours, while the residence times in most water treatment plants are less than that. Steps can be taken to decrease the t_{eq} as much as practicable. Decreasing the PAC size significantly decreases t_{eq}, as shown in chapter 1. However, the PAC must be removed from solution and the decrease in t_{eq} achieved by decreasing d_{PAC} is offset by the difficulty in removing the smaller PAC by sedimentation.

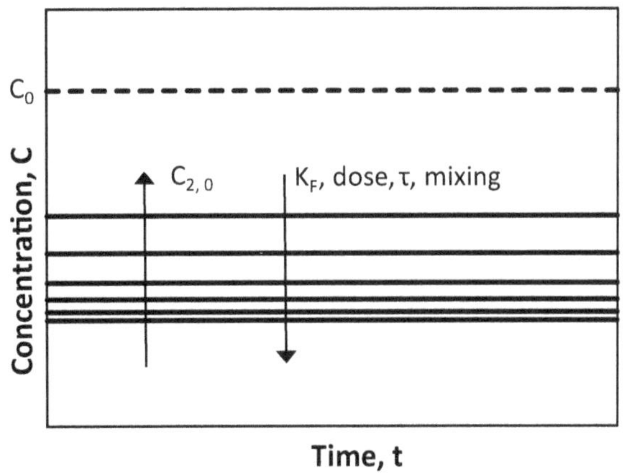

Figure 2-4. Effect of operating, adsorbate, and background factors on steady-state effluent concentration of a PAC–CFSTR system

If external mass transfer is an important factor, increasing the mixing in the CFSTR decreases the boundary layer around the particle, thus increasing the overall uptake kinetics and subsequently decreasing t_{eq}. The effectiveness of increasing mixing is also constrained. Most rapid mixers already have sufficient mixing to minimize external mass transfer, and increasing mixing in well-run flocculators would decrease the promotion of floc particle growth, which is the primary objective of flocculation. As shown previously in Figure 1-4, the uptake kinetic curve is an exponential decay, and to build a reactor solely for PAC contact that yields $\tau = t_{eq}$ would yield an uneconomically large volume.

In addition to operating factors, the adsorbate properties and background matrix also affect the PAC performance. As the overall adsorbability of the PAC-adsorbate pair

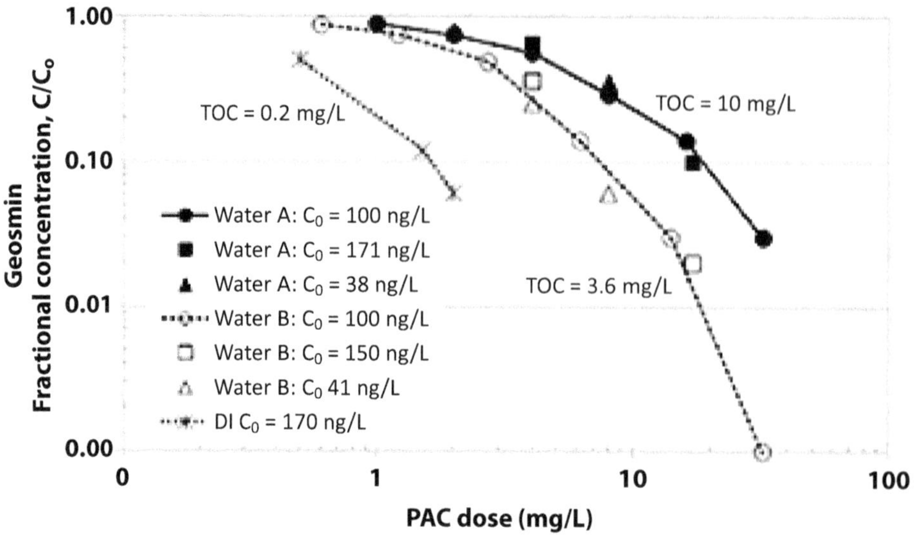

Figure 2-5. Effect of PAC dose, background DOM, and initial geosmin concentration on fractional removal (adapted from Graham et al. 2000)

increases, as measured by an increase in the Freundlich K_F value, the reactor effluent concentration deceases. Thus, PACs with a higher K_F value for a given contaminant will yield lower effluent concentrations. In addition, for a given PAC, the contaminant with the higher K_F value will have a lower effluent concentration. As was shown in Figure 2-3, an increase in the influent adsorbate concentration will yield a higher effluent concentration for contaminants of interest in drinking water treatment.

Multisolute systems, such as one that affects the adsorption of a target compound in the presence of DOM, add additional complexity to PAC treatment. In addition to directly competing with the target compound for adsorption sites, the large-molecular-weight DOM can block pores and can irreversibly "foul" the activated carbon surface, thus reducing the available surface area for target compound adsorption. Thus, an increase in the influent concentration of a competing substance ($C_{2,0}$) will cause the effluent concentration of target compound to increase in a manner similar to an increase in the target compound influent concentration.

An example of an increase in background DOM on the adsorbability of geosmin is shown in Figure 2-5 for the same PAC-adsorbate (geosmin) pair. The best geosmin removal was found in distilled water, and a dose of 1.5 mg/L was able to achieve a 90 percent removal. When a water with a TOC = 3.6 mg/L (Water B) was tested, a dose of 8 mg/L was needed to reach 90 percent removal; when another water with a TOC of 10 mg/L (Water A) was tested, 17 mg/L of PAC was needed to reach 90 percent removal. Another perspective is that as the background TOC concentration increases from 0.2 to 3.6 to 10 mg/L, the removal of geosmin decreases from >90 percent to 40 percent to 25 percent, respectively, at a PAC dose of 2 mg/L.

While an increase in the influent concentration of the target compound will yield an increase in the effluent concentration, it does not affect the fractional removal if the target compound is in the low μg/L range, which is demonstrated in Figure 2-5 for two waters. Most of the data for both waters were gathered at an initial geosmin concentration of 100 ng/L. These data are shown by the circle symbols and connected by the solid and dashed lines. Two additional runs were made using each water: one at a higher initial concentration (170 ng/L) and another at a lower initial concentration (approximately 40 ng/L). The PAC performance on a fractional concentration basis was the same and was independent of initial geosmin concentration. Constant fractional removal was found in both waters and has also been reported for the odorant MIB, the pesticide atrazine, the antibiotic sulfamethoxazole, and the hormone estradiol at concentrations in the low to sub µg/L range (Summers et al. 2010).

PAC Application

Figure 2-1 shows the common points of PAC application in conventional surface water treatment plants. The PAC can be metered into the water in the dry, as-received form or premixed into treated plant water and added as a slurry. In a 1997 survey (Graham et al. 2000), it was found that 23 percent of the utilities added PAC either in a presedimentation basin or in a PAC contactor prior to the coagulation rapid-mix stage. About half of the utilities added PAC immediately before or directly into the rapid mixer, with 10 percent adding it in the flocculator and 7 percent adding it in the sedimentation basin. About 25 percent of the utilities had the ability to add PAC at multiple points. PAC is removed from suspension with the settled solids in the presedimentation basin or with settled, flocculated particles in the sedimentation basin. If PAC is added prior to sedimentation, residual PAC can be removed by the filter. A few utilities add PAC to the top of the filter, and it remains in contact with water for the length of the filtration cycle, which allows more of its equilibrium capacity to be used. However, some concern exists about PAC passing through the filter. More recently, PAC has been evaluated for use with membrane reactors. There are several advantages to this approach, but concerns still exist about the effect of PAC on membrane integrity.

The further upstream in the treatment train that the PAC is added, the longer the total contact time and the higher fraction of the total adsorption capacity is used (i.e., total τ approaches t_{eq}). However, this increased capacity resulting from upstream addition is offset by the PAC exposure to the competing DOM. DOM has been shown to be rapidly removed during coagulation, so adding PAC after rapid mixing would result in a lower influent DOM that competes with the target compound, leading to improved performance.

One important issue is the interaction of PAC with other water treatment chemicals. Several studies have shown that chlorine and other oxidants can significantly decrease the PAC adsorption capacity for target compounds (Summers et al. 2010). Thus, the addition of PAC at the rapid mixer or other locations when a disinfectant is applied will lead to the use of much higher PAC doses. Most studies have shown that PAC incorporation into floc particles has had little negative effect on the kinetics of the target compound uptake. However, at very high coagulant doses, slower rates have been reported (i.e., longer t_{eq}). Interference from coagulants may be an issue for the high doses used in floc blanket systems like those used for lime softening, but the longer residence times in these reactors is often enough to overcome the slower kinetics.

Bench-Scale Evaluation of PAC

Graham et al. (2000) developed a bench-scale protocol for PAC selection and optimization. The protocol uses standard jar test procedures with an emphasis on simulating full-scale conditions. Figure 2-5 illustrates typical results for two utilities. The procedure takes advantage of initial concentration not affecting the fractional removal. Thus, the utility can run the test at one target compound initial concentration and generalize the removal to other initial concentrations. The protocol allows for the selection of the most economical PAC and the cost to achieve the treatment goal, not just the PAC with the lowest dose or the PAC with the lowest cost per pound. The protocol also allows the point of PAC application to be evaluated.

References

Graham, M.R., I. Najm, M. Simpson, B. MacLeod, R.S. Summers, and L. Cummings. 2000. *Optimization of powdered activated carbon application for geosmin and MIB removal.* Report 90782. Denver, Colo.: American Water Works Association Research Foundation.

Summers, R.S., D.R.U. Knappe, and V.L. Snoeyink. 2010. Adsorption of organic compounds by activated carbon. In *Water quality and treatment,* 6th ed. J.K. Edzwald, ed. New York: McGraw-Hill and American Water Works Association.

Chapter 3

Granular Activated Carbon Systems

Granular activated carbon (GAC) can be an effective treatment approach for organic contaminants of interest that are regularly found in the source waters at concentrations of concern. GAC is typically contacted with the contaminated water in a fixed-bed contactor. The process is not a steady-state process, and the effluent concentration increases with time. Once the target effluent concentration is reached, the adsorber operation stops and the GAC is removed from the contactor. The GAC is replaced with fresh or reactivated GAC, and the spent GAC is normally reactivated, either on-site or off-site. Because the GAC in a fixed bed reactor is in equilibrium with the influent contaminant concentration, it can remove more of the contaminant per mass of activated carbon applied compared to PAC in a CFSTR (chapter 2). However, the high relative capital cost associated with GAC systems reduces its attractiveness on an annualized cost basis.

The most common GAC uses are to control groundwater contaminants, taste-and-odor compounds, pesticides, and organic DBP precursors (i.e., DOM). Some utilities that treat surface waters impacted by municipal and industrial wastewater discharges also use GAC in their filters to serve as a buffer for the emergency control of spills and seasonal taste-and-odor episodes.

This chapter discusses the influence of fixed-bed contactor parameters and the compound adsorbability on GAC performance.

Fixed-Bed Plug Flow Reactor Performance

GAC applied in a fixed-bed contactor can be best be modeled as plug flow reactor (PFR). Because the number of adsorption sites is finite, the GAC eventually becomes exhausted, and the effluent concentration increases with time, creating an unsteady-state condition. PFRs are very efficient reactors from a chemical conversion perspective because reactions occur at the influent concentration level, or in the case of GAC, adsorption occurs at the influent concentration level.

Figure 3-1 shows a diagram of a fixed-bed GAC contactor where V is the GAC bed volume, L is the GAC depth, and A is the bed cross-section surface area. The volumetric flow, Q, is applied to the contactor in a down flow direction where C_0 is the influent liquid-phase adsorbate concentration and C is the effluent liquid-phase adsorbate concentration. Several important design and operating parameters affect GAC performance. An important characteristic is the approach velocity, v, or hydraulic loading rate, HLR, which is defined as the flow rate divided by the cross-section area (Eq 3-1).

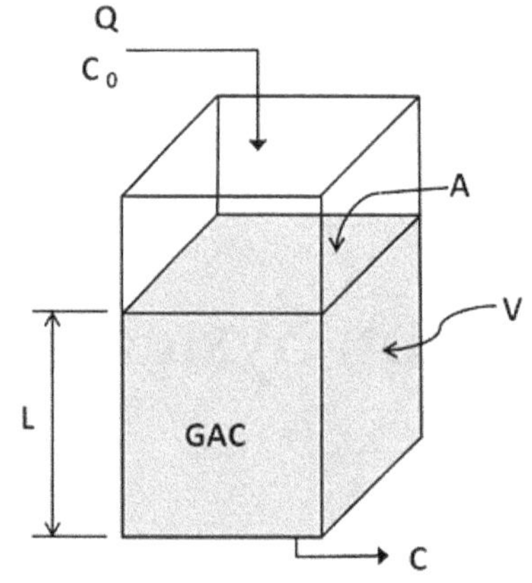

Figure 3-1. Fixed-bed GAC contactor

$$v = HLR = Q / A \qquad \text{(Eq 3-1)}$$

The ratio of the GAC volume, V, to the volumetric flow rate, Q, is termed the *empty bed contact time* (EBCT), which can also be defined as the depth divided by the bed velocity (Eq 3-2).

$$EBCT = V / Q = L / v \qquad \text{(Eq 3-2)}$$

The EBCT is not the true fluid residence time, τ, as it does not account for the reactor volume occupied by the GAC; GAC occupies approximately 50 percent of the bed, so the EBCT is about twice as large as the actual fluid residence time.

Over time, the adsorption sites become occupied, and the adsorbate appears in the effluent once exhausted. For single-solute systems under ideal PFR conditions, the adsorbate "breaks through" completely, with the effluent concentration instantaneously increasing from zero to that of the influent, which is shown in Figure 3-2a as a dashed line under a constant influent concentration condition. Under actual conditions encountered in practice, the breakthrough is spread out due to mass transfer resistance, with the effluent concentration gradually increasing relative to the ideal PFR breakthrough until the GAC is exhausted and the effluent concentration asymptotically approaches the influent concentration. In practice, GAC adsorbers are operated until the effluent concentration approaches a target concentration, C_{max}. At that point, the GAC must be replaced to maintain continued adsorption treatment.

If the influent concentration is not constant, the compound adsorbed to GAC will re-equilibrate to the liquid-phase concentration. For example, if the influent concentration is a pulse, as shown in Figure 3-2b, a reversibly adsorbed compound will desorb from the GAC surface and appear in the effluent, even though no compound remains in the influent. Theoretically, the total mass adsorbed during the pulse will eventually desorb.

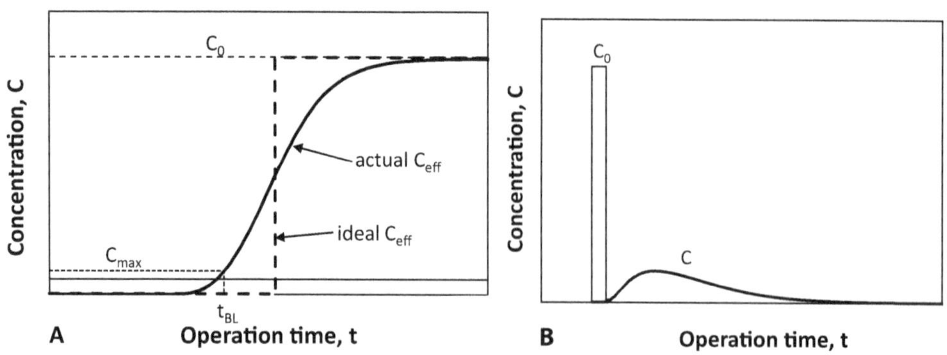

Figure 3-2. Unsteady-state PFR effluent concentration history in a fixed-bed GAC contactor for a single solute: (a) constant influent concentration and (b) pulse influent concentration

Figure 3-3 shows the concentration profile in a GAC adsorber. For an ideal PFR, an adsorption front is created. The GAC above the front is at equilibrium with the influent concentration, and no active net adsorption is occurring. Below the adsorption front, the liquid-phase concentration is zero and no adsorption is occurring. In this ideal case, adsorption only occurs at the depth of the adsorption front. As more adsorbate loads onto the column, more GAC is exhausted, forcing the adsorption front deeper into the bed with time. When the adsorption front reaches the end of the contactor, the compound breaks through instantaneously, as shown in Figure 3-2a for an ideal case.

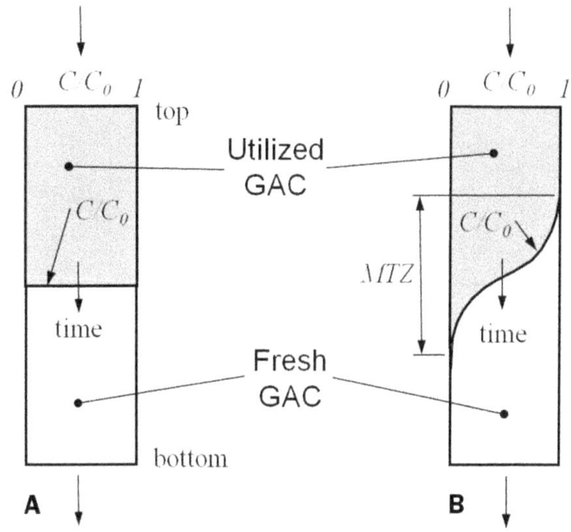

Figure 3-3. Unsteady-state PFR concentration profile in a fixed-bed GAC contactor for a single solute: (a) ideal adsorption front and (b) mass transfer zone (MTZ)

Under actual conditions, dispersion and mass transfer kinetics spread out the contaminant breakthrough. This creates a mass transfer zone (MTZ), which is a spreading of the adsorption front (Figure 3-3b). The MTZ has a defined length that also depends on the adsorption capacity of the GAC for the specific adsorbate. Above the MTZ, the GAC is at equilibrium with the influent concentration and no active net adsorption is occurring. Below the MTZ, the liquid-phase concentration is zero and no adsorption is occurring. Adsorption is actively taking place in the mass transfer zone. As more adsorbate loads onto the column, more of the GAC is exhausted, forcing the MTZ deeper into the bed with time. When the leading edge of the MTZ reaches the end of the contactor, the compound begins to break through at a low concentration. With time and continuous loading of the adsorbate into the contactor, the MTZ moves through the end of the contactor, yielding a gradual breakthrough as shown in Figure 3-2a.

Factors Affecting GAC Performance

Several factors influence the breakthrough pattern shape and hence performance of GAC in a PFR. The spreading of the breakthrough pattern occurs because of (1) dispersion in the fixed bed caused by mixing of the fluid as it passes around the GAC and (2) external and internal mass transfer kinetics. Mass transfer kinetics were discussed in chapter 1 and are affected by the properties of the adsorbent (e.g., the particle diameter, d_{AC}) and the properties of the adsorbate (e.g., the diffusion coefficient, D). Adsorption kinetic effects are illustrated in Figure 3-4a. Note that the ultimate amount adsorbed by the GAC is not affected by the spreading of the MTZ.

The adsorbability of the compound, as measured by the Freundlich adsorption parameters K_F and 1/n, and also the contactor EBCT affect the breakthrough

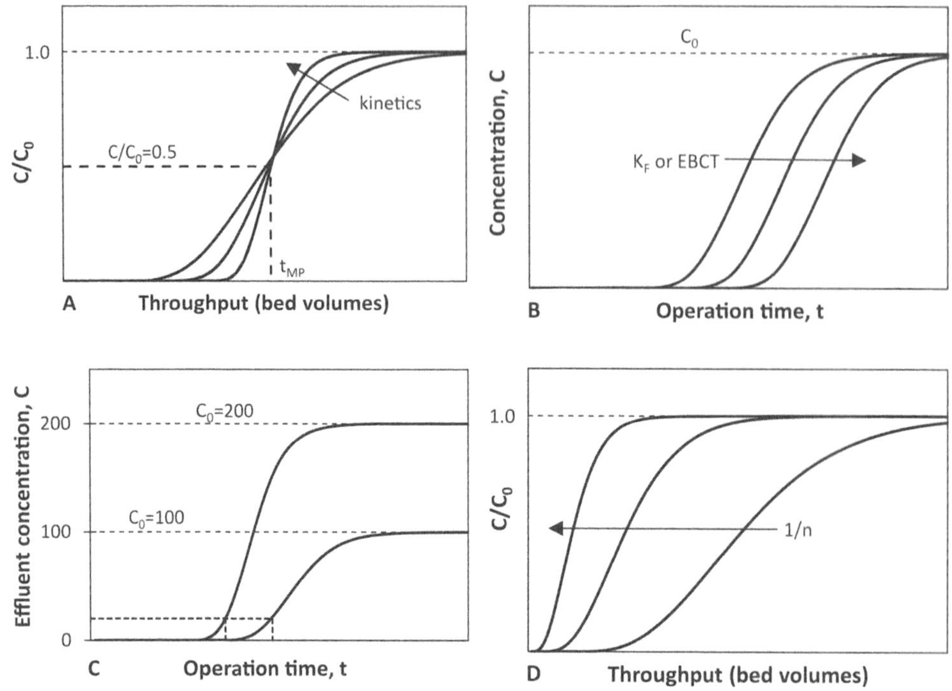

Figure 3-4. Unsteady-state PFR performance for a GAC fixed-bed contactor as affected by (a) dispersion and mass transfer kinetics, (b) K_F and EBCT, (c) C_0, and (d) increasing C_0 at different 1/n values

behavior as illustrated in Figure 3-4b, where the steady-state effluent concentration is shown as a function of increasing the parameter value. As the overall adsorbability of the GAC-adsorbate pair increases, as measured by an increase in the Freundlich K_F value, the PFR effluent concentration is lower at any given time (except after complete exhaustion). Thus, a GAC with a higher K_F value for a given contaminant will yield lower effluent concentrations or, for a given GAC, the contaminant with the higher K_F value will have a lower effluent concentration. When the EBCT increases either by increasing the depth of the contactor or decreasing the flow rate, more GAC is available to remove the compound relative to that applied, thus improving the performance at any given time.

Increasing the influent concentration, C_0, increases the compound loading on the GAC, resulting in earlier mass concentration breakthrough, as shown in Figure 3-4c. By normalizing the effluent concentration to the influent concentration, the breakthrough can be expressed on a fractional concentration basis. The effect of increasing C_0 on the breakthrough in this form is now a function of the Freundlich adsorption parameter 1/n (Figure 3-4d). If the 1/n value is 1.0, the capacity increases directly with increases in C_0 with no effect on the breakthrough. If the 1/n value is less than 1.0, which is common for most compounds of concern in drinking water treatment, then an increase in C_0 causes earlier fractional concentration breakthrough. If the 1/n value is greater than 1.0, which is not common in drinking water treatment, then an increase in C_0 causes later fractional concentration breakthrough.

The time to the approximate midpoint, t_{MP}, of the breakthrough curve, as illustrated in Figure 3-4a, can be estimated from the adsorbate and adsorber properties using Eq 3-3, where q_0 is the solid-phase concentration in equilibrium with the influent concentration, C_0, and M_{GAC} is the mass of GAC in the bed and ρ_{Bed} is the packed bed density of the GAC, $\rho_{Bed} = M_{GAC} / V$.

$$t_{MP} = \frac{M_{GAC}}{Q} \times \frac{q_0}{C_0} \qquad \text{(Eq 3-3)}$$

Assuming that the Freundlich equation represents the equilibrium state, the time to the approximate midpoint can be estimated using Eq 3-4.

$$t_{MP} = \frac{M_{GAC}}{Q} \times \frac{K_F C_0^{1/n}}{C_0} = \rho_{Bed} \times \frac{L}{v} \times \frac{K_F C_0^{1/n}}{C_0} = \rho_{Bed} \times \text{EBCT} \times \frac{K_F C_0^{1/n}}{C_0}$$

(Eq 3-4)

In practice, the difficulty with this approach is finding estimates of the Freundlich parameters K_F and 1/n that are accurate in the presence of competing compounds such as DOM.

The operation time of an adsorber can be normalized to account for the volume of GAC in the contactor such that the breakthrough patterns of adsorbers with different EBCTs can be more directly compared on a throughput basis in bed volumes using Eq 3-5.

throughput (bed volumes) = volume of water treated/GAC volume =
operation time / EBCT

(Eq 3-5)

Theoretically, using throughput to express the operation of the adsorber will normalize for EBCT effect shown in Figure 3-4b and yield the results shown in Figure 3-5a for EBCTs of 10 and 20 min. However, as will be shown in part 2, several factors affect the results leading to earlier breakthroughs for organic micropollutants at longer EBCTs and for DOM at shorter EBCTs, as depicted in Figure 3-5b.

The throughput in bed volumes to the approximate midpoint of the breakthrough curve can be estimated using Eq 3-6.

$$\text{throughput } (BV)_{MP} = \rho_{Bed} \times \frac{q_0}{C_0} = \rho_{Bed} \times \frac{K_F C_0^{1/n}}{C_0}$$

(Eq 3-6)

Another term used to normalize the operation time relative to the mass of GAC is the specific throughput, which is the volume of water treated per mass of GAC. The inverse of the specific throughput is called *GAC use rate*, is more commonly used, and is discussed in chapter 4.

A mass balance can also be used to estimate q_0 if breakthrough data are available by integrating the area above and to the left of the breakthrough curve, either on a concentration versus time format, e.g. Figure 3-2, or a fractional concentration versus throughput in bed volumes (BV) format, e.g. Figure 3-4 (Eq 3-7).

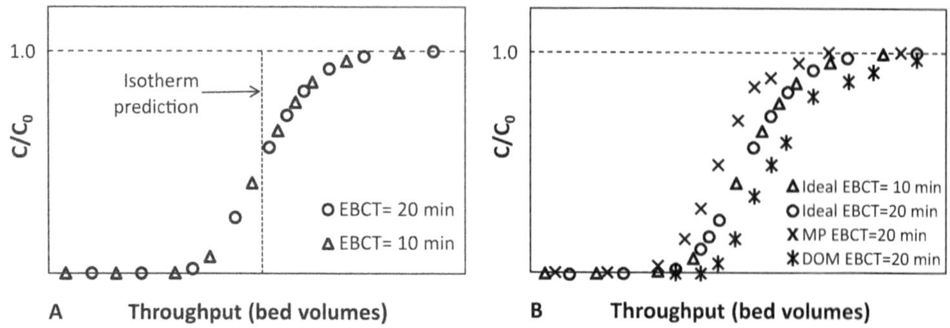

Figure 3-5. GAC breakthrough expressed on a throughput basis for two EBCTs: (a) ideal results and (b) practical results for organic micropollutants and DOM

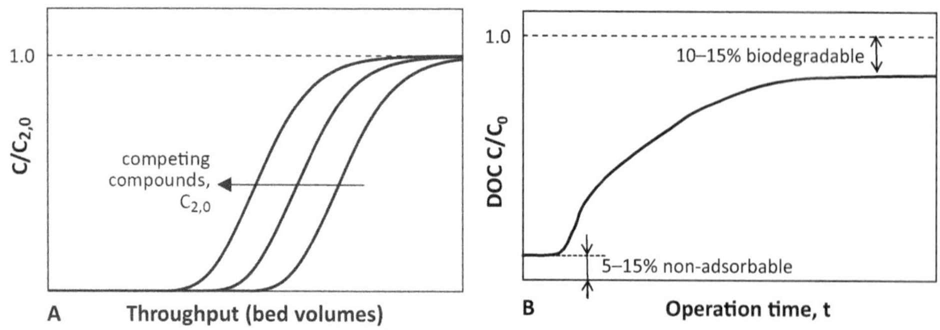

Figure 3-6. Breakthrough behavior of multisolute systems: (a) effect on target compound breakthrough and (b) DOM breakthrough as measured by DOC

$$q_0 = \frac{1}{\rho_{Bed} \times EBCT} \times \int (C_0 - C(t))\, dt = \frac{C_0}{\rho_{Bed}} \times \int (1 - C/C_0(BV))\, dBV$$

(Eq 3-7)

However, the value of q_0 is only applicable to the conditions of the GAC run used to develop the breakthrough curve, i.e., specific C_0 of the target compound and in multisolute systems to the same mixture of compounds found in that water.

Multisolute systems, like the adsorption of a target compound in the presence of DOM, add additional complexity to GAC treatment because of competition for adsorption sites. The influent concentration and the adsorbability of the solutes competing with the target compound affect the breakthrough behavior. In all cases, the presence of competing solutes decreases the adsorbability of the target compound resulting in earlier breakthrough, as shown in Figure 3-6a. In addition to directly competing with the target compound for adsorption sites, the large-molecular-weight DOM can both block pores and irreversibly "foul" the activated carbon surface, thus reducing the available surface area for target compound adsorption. Therefore,

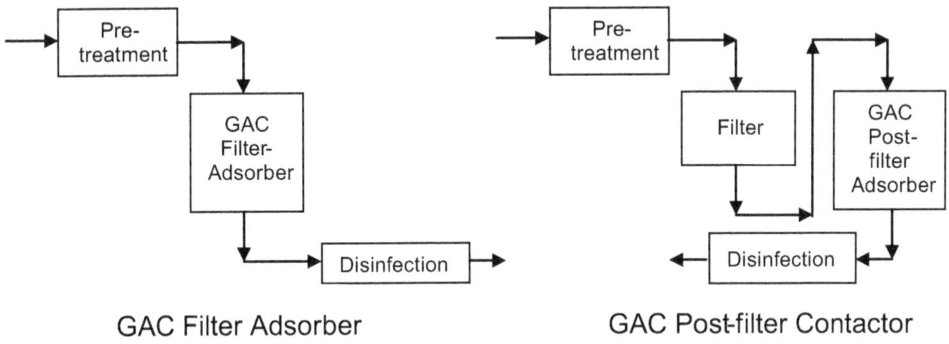

Figure 3-7. Common GAC application points

an increase in the initial concentration of a competing substance, $C_{2,0}$, will cause the effluent concentration of target compound to increase in a manner similar to a increase in the target compound initial concentration; the finite number of adsorption sites are occupied and more appears in the effluent.

DOM is a heterogeneous mixture of organic compounds and is normally dominated by hydrophobic organic acids such as fulvic and humic acids, and it also contains hydrophilic compounds that are weakly adsorbable or nonadsorbable. In addition, a fraction of the DOM is biodegradable. These DOM properties affect the breakthrough behavior, as illustrated in Figure 3-6b. Typically 5 to 15 percent of the DOM is nonadsorbable, resulting in immediate breakthrough. After a period of time that is affected by the adsorber EBCT and the nature of the DOM, initial breakthrough of the weakly adsorbing compounds begins to take place. The DOM concentration in the effluent gradually increases with time and eventually reaches a plateau. The DOM breakthrough is spread out over a longer time period than that expected for a single solute because it is a mixture of compounds with a range of adsorbability and the mass transfer kinetics are slower (Figure 3-4a). The pseudo steady-state plateau achieved is a function of the biodegradable fraction, typically 10 to 15 percent, and the amount of slow adsorption of the large DOM molecules into the small micropores. If the influent water is pretreated with ozone, then the biodegradable fraction of the DOM can increase to approximately 20 to 25 percent.

GAC Application

Common application points of GAC contactors within a treatment process are shown in Figure 3-7. The most common point of GAC application for surface waters is after the sedimentation basin either as media in the filter bed, termed *filter adsorber*, or after the media filters as a *post-filter contactor*. GAC can also be used in filter adsorbers in a direct filtration plant. The GAC used in a filter adsorber should have characteristics that make it an effective filter media. The EBCTs at design flow are typically less than 10 min in a filter adsorber application and 10 min or greater in post-filter contactor application. For groundwater applications, well water may be applied directly to the GAC or after another treatment process, e.g., air stripping or lime softening. Points of GAC application are discussed in more detail in chapter 4.

Pretreatment prior to GAC application can have an important effect on the GAC performance for removing the target compound. Pretreatment that removes competing compounds has a positive effect. For example, removing DOM and lowering the pH by coagulation increases the run time of GAC adsorbers for DOM.

Pilot- and Bench-Scale Evaluation of GAC

Several protocols have been developed for the pilot-scale evaluation of GAC that facilitate the comparison of GAC type and EBCT (Summers et al. 1992; Summers et al. 1996). Rapid small-scale column tests (RSSCTs) have become an effective bench-scale method for evaluating GAC types, EBCTs, and pretreatment effects for the removal of DOM. RSSCTs have been shown to well simulate full-scale GAC run times for DOM (Crittenden et al. 1991; Summers et al. 1996). However, they do not assess the amount of long-term biodegradation expected. These tests are discussed in chapter 5. Application of RSSCTs for the accurate prediction of run times for taste-and-odor compounds and organic micropollutants is still under development, and the limitations are discussed in chapters 6 and 7, respectively.

References

Crittenden, J.C., P.S. Reddy, H. Arora, J. Trynoski, D.W. Hand, D.L. Perram, and R.S. Summers. 1991. Prediction of GAC performance using rapid small-scale column tests. *Journal AWWA* 83(1):77–87.

Summers, R.S., L. Cummings, J. DeMarco, D. Hartman, D. Metz, E.W. Howe, B. MaCleod B., and M. Simpson. 1992. *Standardized protocol for the evaluation of GAC*, 90615-5/92. Denver, Colo.: AWWA [now Water] Research Foundation.

Summers, R. S., S. Hooper, and S. Hong. 1996. GAC precursor removal studies. In *ICR manual for bench- and pilot-scale treatment studies*. EPA 814-B-96-003. Cincinnati, Ohio: USEPA, Office of Water.

PART 2

ADSORPTION APPLICATIONS

Activated carbon can be implemented in several locations throughout a water treatment plant, depending on the contaminants of interest and the desired removal. Understanding the advantages and disadvantages associated with each of the potential application approaches is critical to implementing a successful treatment process that achieves the utility's goals.

The objective of part 2 of this book is to provide an understanding of the various approaches for applying activated carbon within a water treatment plant. First, chapter 4 discusses the master planning process, with particular attention on activated carbon applications and modes of activated carbon operation, including the factors affecting treatment efficacy for each application type. Throughout chapter 4 and supplemented in appendix B, numerous case studies are presented to demonstrate how water providers have planned for and implemented activated carbon. In the remaining chapters, specific information is presented to assist in a facility's planning for removal of the three major classes of contaminants: dissolved organic matter (DOM, in chapter 5), taste- and odor-causing (T&O) compounds (chapter 6), and organic micropollutants (chapter 7).

This section is intended to assist utility personnel and design engineers in the planning process in deciding whether or not activated carbon is a viable and appropriate technology to meet their water quality challenges and in considering the factors affecting design and long-term performance of the systems. Once a decision to implement a specific mode of GAC has been made, the reader is directed to part 3 of the book for concepts related to detailed design of each activated carbon application.

Chapter 4

Activated Carbon Application Approaches

Activated carbon is used in water treatment plants in several ways. Activated carbon may be applied as powdered activated carbon (PAC), which is held in suspension through the treatment process until the removal point; or as granular activated carbon (GAC), which is placed in a fixed-bed reactor. GAC can be applied as a post-filter contactor, a filter adsorber, or a GAC cap. GAC accumulates attached biomass, and after long periods, biotreatment is the primary treatment mechanism instead of adsorption. This mode of operation is often called *biologically enhanced activated carbon* (BAC). Finally, GAC can also be applied in the vapor phase to remove volatile organic compounds (VOCs) following air stripping.

Numerous drinking water utilities across the country are in the process of evaluating the use of activated carbon technologies in their systems; others have already installed activated carbon or in the process of doing so. This chapter provides information on utilizing activated carbon in water treatment systems. First, this chapter will explore the master planning process, with particular attention on activated carbon applications. Next, the activated carbon application modes are presented in more detail, including the factors affecting treatment efficacy for each activated carbon application type. Throughout the chapter, several case studies illustrate the practical experience of utilities, including the decision-making processes they went through to arrive at a decision to improve water quality through the use of activated carbon.

Water Quality Master Planning Approaches

Developing a water quality master plan enables a utility to consider whether activated carbon is appropriate for a treatment facility now or in the future. Sequential steps identified below are typically followed during the development of a master plan. In several of these steps, the decision about the use of activated carbon technology is involved. Following the master planning approach, case studies illustrating the planning process undertaken by five utilities as they considered activated carbon treatment are presented:

- Greater Cincinnati Water Works
- City of Phoenix
- Birmingham, Ala., Water Works Board
- City of Baltimore, Md.
- City of San Diego, Calif.

Water Quality Master Planning Steps

Step 1: Develop Water Quality Goals

The most critical component of water quality master planning is to develop water quality goals. Although some variations occur in how utilities establish water quality goals, three main components include:

- Current and future regulatory environment
- Aesthetic concerns
- Utility's desire to treat for unregulated contaminants

Current and Future Regulatory Environment. The primary driver behind most utilities' water quality goals is the need to comply with existing and forthcoming regulations. As discussed in this book's introduction, existing regulations that may increase the need for activated carbon include the D/DBP Rule (Stage 1 and Stage 2), the Surface Water Treatment Rule, and aspects of the Safe Drinking Water Act that include synthetic organic compounds (SOCs) and VOCs. Dedicated GAC filters can also provide additional removal credits for pathogens such as *Cryptosporidium*. However, considering only the existing regulations can be shortsighted and may leave a utility unprepared to achieve compliance in the future. Based on UCMR and CCL, future contaminants of concern may include MTBE, perchlorate, pesticides, and endocrine-disrupting compounds. While predicting the future regulatory environment is a challenging proposition, all indicators point to more stringent water quality standards and the need to remove broad classes of contaminants in the future. Since such is the case, utilities need to remain informed about regulatory developments under the national and local regulatory agency perspectives.

Aesthetic Concerns. The aesthetic properties of water are very concerning to customers. Some aesthetic parameters, such as odor, color, and total dissolved solids (TDS), have secondary standards. The desire to maintain a high level of consumer confidence has led many utilities to implement additional treatment for T&O episodes. Public concern about endocrine-disrupting compounds and pharmaceuticals in drinking water is now causing utilities to consider additional treatment. Activated carbon is one of the best technologies to remove T&O compounds including MIB and geosmin as well as a wide array of trace organic contaminants.

Utility's Desire to Treat for Unregulated Contaminants. As discussed in part 1 of this book, activated carbon is very good at broad-based removal of several classes of contaminants, including natural organic matter, T&O compounds, and micropollutants. Because the future regulatory environment is uncertain, implementing technologies with a high probability of capturing and removing a broad spectrum of organic contaminants positions the utility very well for future regulatory compliance. In addition, by adding activated carbon as part of a multibarrier treatment strategy, the utility may gain the trust and respect of its consumers for surpassing the federal and state water quality standards.

Depending on the utility's water quality goals, the need for utilizing advanced technologies such as activated carbon may become an obvious choice. For example, if a utility is forward-thinking and is interested in improving water quality beyond what is minimally required by applicable regulations, it may emphasize removal of contaminants and their precursors. This treatment philosophy may result in the selection of activated carbon as a present or future treatment technology in its water treatment portfolio.

Step 2: Review Historical Data and Identify Gaps

Reviewing historical data is critical to determining whether the utility with its existing infrastructure is capable of achieving the water quality goals developed in Step 1. This step involves gathering and analyzing data to describe quality, quantity, and availability of the water resources for the raw water, treated water, and distributed water. The water quality parameters of interest will be specific to the goals established in Step 1 and may include the following:

- Total organic carbon (TOC) or dissolved organic carbon (DOC)
- T&O compounds
- Color
- Micropollutants
- VOCs
- SOCs
- Perchlorate
- Turbidity

By comparing the utility's current water quality performance with the desired goals, the need for additional treatment will be evident. A statistical analysis including percentile calculations and a time-series analysis should be used in combination to determine the extent and frequency of treatment needed. For example, if a spring run-off event causes high TOC but it is only present for a few weeks each year, it may be beneficial to implement a treatment system that can be bypassed when it is not needed. Typically, at least a five-year period for historical data should be used in developing the statistical evaluation of water quality; however, in the event adequate data are not available, a more conservative approach of erring toward worse water quality ranges is used for master planning considerations. When sufficient data records are available, master planning often uses 90th or 95th percentile water quality for developing its plans.

In addition to water quality data gaps, it is also important to understand and close gaps in the treatment technologies, future demands, and future water supply portfolio. By studying the current treatment processes, it can be determined whether they are effective or if additional measures are necessary. Comparing future demands to current demands helps determine whether or not additional treatment capacity is needed or if facility upgrades that address water quality are sufficient. Lastly, the components of the future water supply portfolio will determine if the raw water quality is anticipated to change and how the utility's treatment strategy will need to respond to that change.

Step 3: Identify Alternatives

In this step, multiple viable alternatives are developed to move the utility from its current state of operation to the desired state of operation in which the utility will be able to meet the water quality goals set in Step 1 and close the gaps identified in Step 2. Depending on the utility's preferences toward desired water quality and treatment philosophy (e.g., to merely meet water quality regulation versus further improving the quality of delivered water), some or all of the alternatives identified in this step may include technologies such as PAC or GAC.

Step 4: Evaluate Alternatives

In this step, each of the alternatives identified in Step 3 are analyzed in the context of the expected water quality for the utility and the context of the water quality goals defined in Steps 1 and 2. Alternatives are evaluated with respect to the utility's decision drivers, which often include treatment and distribution costs, ease of operations, amenability to future water quality challenges, stakeholder acceptability, site constraints at treatment facilities, and carbon footprint.

The evaluation can be performed in various ways, including desktop evaluation utilizing models for predicting water quality, resource availability, and distribution capability; bench-scale evaluation using jar testing methods for PAC or RSSCTs (rapid small-scale column tests) for GAC; or pilot-scale evaluation using small (in the range of 2- to 5-gpm) flow-through treatment units that simulate the current or anticipated future treatment configuration. Larger-scale testing is sometimes performed for technologies that require additional operability and operational experience prior to accepting a treatment methodology. Typically, desktop evaluations are performed for all viable alternatives identified in Step 3, and only a subset of more promising treatment configurations are evaluated in bench- and/or pilot-scale testing. Due to the larger cost of conducting bench-scale testing compared with desktop evaluation, only a handful of potential treatment configurations are tested in bench-scale setups. Pilot-scale testing involves higher financial commitment compared with bench-scale testing. As a result, pilot testing, if necessary, is often conducted on a very selective basis on two or three potential treatment configurations. Larger-scale testing using parts of the full-scale plants is complex and expensive. As a result, this type of testing is rarely performed and only undertaken by a water utility to gain confidence on the selected treatment configuration. It is common for water utilities to include GAC or PAC treatment processes during desk-top, bench-scale, or pilot-scale evaluation, particularly if the water quality objectives include improving compliance with disinfection by-product regulations and improving aesthetic quality of the delivered water.

Step 5: Select Best Approach

Based on the evaluation completed in Step 4, the best alternative for water quality improvements is selected. Many approaches for selecting the best alternative are practiced by engineers, including tabulations of pros and cons for each alternative and more systematic approaches in which all alternatives are meticulously scored (often an average of scoring by multiple individuals) using selection criteria and relative weights of the criteria. This latter approach typically results in a more quantitative selection of the best alternative and proper documentation of the decision process. Commercial software is available to aid the process of selecting the best alternative.

Step 6: Develop a Road Map for Implementation

Once the best treatment configuration is decided through the five steps described above, the final step for preparing a master plan is to prepare a systemwide road map for the water utility to achieve the water quality goals selected at the beginning of the process. In order to develop the road map, key decision points need to be identified, and the issues that have to be sorted out at these decision points need to be clearly defined. It could be useful to use a scenario-planning approach where the planning is connected to the future evolution of the utility's customer base. A scenario-planning approach often results in a master plan that offers the flexibility to adapt to changing situations around water distribution, water quality, and financial acceptability.

Utility Experiences with Water Quality Master Planning

Greater Cincinnati Water Works

In December 1976, when Richard Miller became director of the Cincinnati Water Works (now the Greater Cincinnati Water Works, GCWW; Case Study No. 1), he spearheaded the creation and implementation of a vision that would provide GCWW customers with the quality of water they desired at a price they could afford and would be willing to pay.

At the time, the presence of total trihalomethanes (TTHMs) in drinking water was a significant concern, as was another issue: the vulnerability of water from the Ohio River, the system's primary raw water source, to a broad spectrum of contaminants from upriver discharges from chemical industries and river transportation of coal, grain, and chemical products. The health effects of many of these chemicals, at concentrations present either individually or synergistically, had not yet been established. In addition, there was a large spill of carbon tetrachloride, a probable human carcinogen, into the Ohio River system upstream of the city's water intake in 1977, heightening the concern over potential health effects.

Thus began a journey of investigations, studies, pilot plant simulations using GAC, cost analyses, and decisions by the City Council. These efforts culminated in the design and construction of the nation's largest post-filtration GAC treatment system designed to remove a broad spectrum of natural and synthetic organic contaminants in the raw water supply.

Today, the wisdom of taking the bold action to provide a post-filtration GAC system continues to be demonstrated by the multiple benefits being achieved. The system brings peace of mind to the management and operating staff and has provided the foundation for building trust with the city's elected officials, local media, and system customers. It is also a foundation for the utility's current vision, "The Greater Cincinnati Water Works will be the standard for excellence in the water utility industry," and its mission, "To provide our customers with a plentiful supply of the highest quality water and outstanding services in a financially responsible manner."

GCWW's decision to proceed with GAC. In August of 1977, GCWW, with the aid of a USEPA grant, began one of the most comprehensive research projects ever undertaken by a water utility. Its primary objective was to determine whether GAC was a feasible method for removing trace organics from Ohio River water while treating it for use as drinking water (Westerhoff and Miller 1986)

Although the use of GAC in the United States was not new, it had been limited largely to T&O control in potable water system filter beds, polishing in wastewater reclamation operations, and certain industrial processes. To be considered feasible for Cincinnati's purposes, the facility would have to remove trace organics to a desirable level at a cost acceptable to customers without affecting the treatment already provided by the existing plant. Both deep-bed contactors and conventional depth gravity filters were evaluated for use with GAC, along with on-site reactivation of the spent media. A further objective of the studies was to develop plant design, operating parameters, and cost estimates for full-scale plant conversion to GAC treatment.

Four years later, after thousands of samples had been analyzed, GCWW confirmed that the more than 200 synthetic organic chemicals it found in the river on an almost continuous basis could effectively be removed to very low levels by GAC treatment. By using a demonstration plant to process flows up to 4 mgd, it also proved that a GAC facility could be operated successfully and that GAC could be reactivated on-site efficiently and cost effectively with no adverse effect on air quality, wastewater discharges, or worker environment. Finally, the study proved that the use of GAC was the best technology available to remove the broad spectrum of organic contaminants in the Ohio River at Cincinnati.

Based on the results of these studies, GCWW recommended that the City Council proceed with a GAC treatment system with on-site reactivation as an addition to its existing treatment processes. In June 1982, the council authorized GCWW to proceed with the engineering design phase for the project.

GCWW GAC facilities design. Work by the consulting firm of Malcolm Pirnie Inc. on final design of the 175-mgd facility began in early 1984 and was completed a year later in January 1985. In the design, flow is diverted from the existing rapid sand filters to the GAC contactors. The post-filter arrangement provides the opportunity for significant TOC removal in the existing filters ahead of the GAC contactors. To minimize media transport pipe runs, reduce media losses, and centralize operations, the design consists of 12 large gravity downflow GAC contactors, each holding about 720,000 lb of GAC, located on either side of a core facility. The contactor size and number were determined on the basis of economic analyses and the decision to provide an empty bed contact time (EBCT) of 15 min at the design flow rate. A unique underdrain system was selected for the contactors to avoid the need to use any media but GAC. The system consists of a pipe network of wedgewire screen construction, fabricated entirely of stainless steel and designed to provide complete accessibility for maintenance. Six GAC storage tanks located in the core facility minimize media transport distance and ensure the least attrition of the GAC, which is transported in slurry form for reactivation and replacement.

The post-filtration GAC facilities with on-site GAC reactivation were placed in service in 1992. The GAC reactivation system is also located in the core facility. The reactivation system consists of two multihearth furnaces with afterburners and wet scrubbers, each designed to reactivate 50,000 lb/day of GAC. The afterburners increase flue gas temperature to 2,400 °F for 2 sec to destroy potential contaminants. A cyclone removes particulates from the off-gases. Flow from the GAC contactors is chlorinated on its way to clearwells that provide contact time for disinfection.

Multiple benefits of GAC treatment experienced at GCWW. In the spring of 2008, more than 16 years after the GAC facilities were placed in service, former GCWW director Richard Miller, then-director David Rager, superintendent of water quality and treatment Debbie Metz, assistant supervisor of water quality and treatment David Hartman, and supervisor of Treatment Richard Pohlman were interviewed to evaluate system performance and to identify the benefits being achieved.

The philosophy behind the use of GAC in drinking water treatment for Cincinnati was threefold:

- It is better to remove contaminants by adsorption with GAC instead of adding chemicals that might result in unintended consequences.
- Science is continually identifying additional chemicals in the drinking water supply, often in minute concentrations. While there may be lack of sufficient evidence regarding the specific health risks of these chemicals, removing them is considered advantageous.
- Periodic surveys provide a valuable method to evaluate customer attitudes toward water quality, service, and costs.

One of the primary objectives for installing the GAC treatment system was to provide an effective barrier for periodic spills in the Ohio River of chemicals potentially hazardous to human health. While this objective has been achieved, numerous other significant benefits have been realized, including:

1. *Peace of mind.* On top of the list of key benefits is the "peace of mind" experienced by the system leaders and operators who are responsible for protecting the public health of their drinking water customers. As David Hartman expressed when interviewed, "We can't analyze every bit of water" being delivered into the distribution system, "but we have confidence in the knowledge, from thousands of analyses over many years, that the GAC would provide a significant barrier for removal of organic contaminants that might be present in the raw water supply."
2. *High level of public trust.* GCWW has used its GAC treatment approach for the protection of public health as a foundation for building a high level of confidence and trust among elected officials, the media, and customers. According to director Rager, this foundation has enabled GCWW to successfully communicate the value of water and service being provided and has supported a financially sound rate structure to ensure sustainable operations.

   A survey in spring 2006 of more than 1,000 local residents conducted by the University of Cincinnati Institute for Public Policy found that almost 90 percent of customers said that the water provided by GCWW is considered safe to drink, and almost three-quarters felt that GCWW uses state-of-the-art technology.
3. *Reducing TOC and DBPs.* Through the process of adsorption, GAC reduced levels of TOC and hence, DBP precursors, ultimately reducing the formation of DBPs during chlorine disinfection in the water treatment process.

   GCWW recognized that its conventional treatment alone would not permit compliance with then-existing DBP regulations and, in particular, with regulations that were on the horizon. Its work had demonstrated that GAC would adsorb portions of the total organics and DBP precursor material in the water.

   The post-filtration GAC treatment system achieves a more than 50 percent TOC reduction, maintaining TOC less than 1 mg/L. As a result, the GAC system can be operated to achieve compliance with drinking water regulations.

   Figure 4-1 illustrates the benefit of GAC treatment on TTHM formation over a one-year period with the free chlorine disinfection. Similar reductions are achieved with other DBPs.
4. *Savings from reduction in use of chlorine.* By reducing organics in the treated water, the GAC treatment system substantially reduces the amount of chlorine required in the system. Experience indicates a reduction of about 60 percent, which represents a savings of approximately $200,000 in 2008 annual operating costs.
5. *Ability to economically expand service area.* The reduction in TOC and DBP formation potential resulting from GAC treatment enables the utility to more easily maintain the free chlorine residual in the transmission and distribution system, even with long travel (detention) times. Thus, GCWW can expand its service area without the need for additional chlorine booster stations.

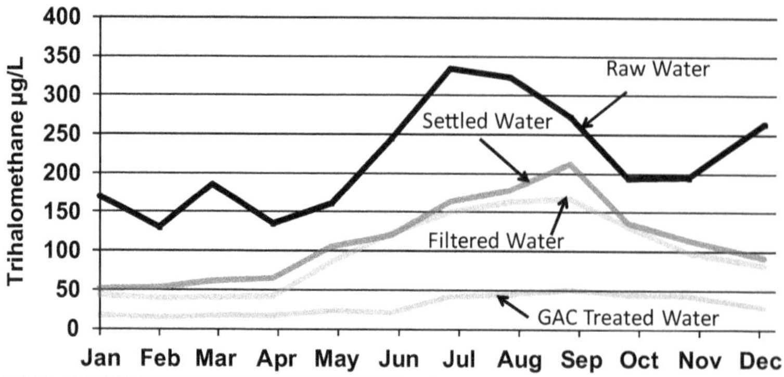

Figure 4-1. Richard G. Miller Treatment Plant TTHM concentration through the treatment process

6. *Elimination of the need to use chloramines.* The GCWW's choice is to use free chlorine as its distribution system disinfectant rather than chloramines. It is noteworthy that hundreds of DBPs can be formed with different characteristics and health effects during chlorination and chloramination. Some chloraminated DBPs (e.g., NDMA) are believed to be more carcinogenic than chlorinated DBPs (e.g., TTHMs). GCWW can successfully use free chlorine as its disinfectant while maintaining low levels of DBP formation because of the reduction in precursors afforded by the GAC treatment.
7. *Compliance with Lead and Copper Rule.* Although higher pH in drinking water results in decreased levels of lead and copper at the customer's tap, a higher pH also increases rate of TTHM formation. With the reduction of DBP precursors by the GAC treatment, GCWW can keep the pH sufficiently high to comply with the lead and copper regulations while minimizing TTHM formation.
8. *Improved control of taste and odor.* A significant benefit of GAC treatment is its effectiveness at reducing tastes and odors in water. Tastes and odors are generally the result of three sources: MIB, geosmin, and high concentrations of chlorine. The GAC system, with downstream chlorination, permits biological activity in the sand filters. Biotreatment results in sustained removal of 80 percent of the MIB and 50 percent of the geosmin. The lower use of chlorine discussed in Item 4 above results in reduced complaints of chlorinous tastes and odors. The "odor threshold" prior to GAC ranges on a scale from 1 to 9 (averages between 2 and 3). After GAC and prior to addition of chlorine, it is close to zero.
9. *Improved turbidity and* Cryptosporidium *removal.* As the effluent from the plant's filters pass through the 11.5-ft-deep GAC contactor, additional particulates are removed, resulting in further turbidity reduction, as shown in Figure 4-2, and improved microbial protection through removal of *Cryptosporidium*.
10. *Facilitating biologically active rapid media filters ahead of treatment.* The relocation of the point of chlorine disinfection downstream of the filters and GAC contactors results in biological activity within the filters. This has two benefits. First, it facilitates some removal of TOC before the water is applied to the GAC. Second, it increases the efficiency of the GAC because reducing the organic loading on the GAC extends its useful life.

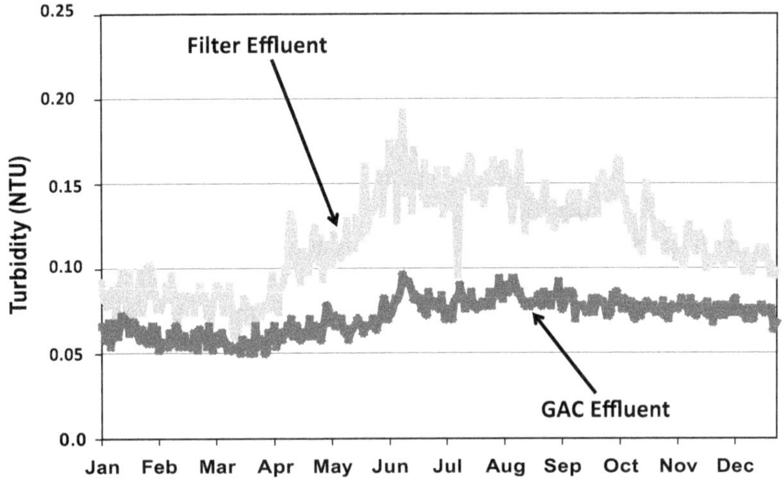

Figure 4-2. Richard G. Miller Treatment Plant turbidity removal through GAC

Pharmaceuticals and personal-care products in water supply. Vast quantities of daily used pharmaceuticals, including prescription, over-the-counter, and veterinary drugs, as well as personal-care products, are entering many of the country's drinking water supplies, raising questions about the potential human health effects of long-term exposure to minute concentrations of these chemicals. The GAC treatment process used in Cincinnati is very effective at removing the trace levels of these compounds.

Much has been written concerning this issue by prominent researchers, including Shane Snyder (University of Arizona), Jörg Drewes (Colorado School of Mines), Paul Westerhoff (Arizona State University), David Sedlak (UC Berkeley), and others. There are many ongoing efforts to determine whether pharmaceutical chemicals pose an actual health threat at the concentrations being found in drinking water.

In March 2008, a news story, "AP Probe Finds Drugs in Drinking Water," was highlighted on *Good Morning America* and CNN, describing how drug residues had been found in a number of the country's raw water supplies. Shortly afterward, many local media reporters interviewed their water system managers and featured reports on their own newscasts. Most satisfying to Cincinnati political and GCWW leaders was the very positive coverage GCWW received on the topic.

Local ABC TV Channel 9 ran a number of reports on the AP release and the Cincinnati drinking water. It reported that "none of these compounds [drug residues] get to you because of three words: granular activated carbon." The investigative reporter interviewed GCWW director David Rager, who was able to report that "the only compound the AP study found in the treated Cincinnati drinking water was minute quantities of caffeine." Another reporter on local CBS TV Channel 12 offered in his report, "the carbon filtration system we have is considered the gold standard nationally and around the world." The investigative reporter reinforced the decision made some two decades earlier of the value of paying the higher costs associated with the GAC treatment, stating "everybody now thinks it's worthwhile."

Regardless of the outcome of current ongoing research evaluating the potential health implications of very small concentrations of pharmaceuticals in drinking water, the positive media coverage that GCWW received is an example of the trust and confidence gained by providing the GAC barrier to remove a wide spectrum of organics from its water supply. GCWW continues to evaluate the effectiveness of GAC for these trace organic contaminants through its involvement in several research studies (e.g., Water Research Foundation Project 4235).

Perfluorooctanoic acid (PFOA) in the water supply at GCWW. Perfluorooctanoic acid (PFOA) is a synthetic chemical that does not occur naturally in the environment. PFOA, sometimes also called C8, is used to produce fluoropolymers, substances with special properties used in manufacturing and industrial applications. Fluoropolymers impart properties such as fire resistance and oil, stain, grease, and water repellency, and are used in hundreds of ways in nearly every industry segment. Consumer products made with fluoropolymers include nonstick cookware and breathable all-weather clothing.

PFOA is very persistent in the environment and is found to remain in the human body for long periods of time. Some studies have indicated that PFOA can cause adverse developmental effects in laboratory animals. Currently, the Centers for Disease Control and Prevention are including PFOA in their upcoming National Reports on Human Exposure to Environmental Chemicals. The USEPA is also investigating whether PFOA poses a risk to human health and the environment. The agency has worked with upriver industries along the Ohio River, and the concentration of PFOA now being observed in the plant's raw water has decreased considerably over time.

Testing done at GCWW indicates that the GAC treatment process removes a significant quantity of PFOA from the filtered water. Testing data from 2007 showed an average of 16 ng/L of PFOA entering the plant and below the detection limit of 5 ng/L PFOA after GAC treatment.

Costs of GCWW's GAC facilities. The 175-mgd post-filtration GAC facilities with on-site GAC reactivation at Cincinnati were placed in service in January 1992 at a cost of approximately $63.9 million (1992 dollars), including construction and engineering. The average estimated cost of amortized construction and operation was estimated at 6 cents per day or $5.40 per quarter, for the average household, single-family customer.

Now, after 16 years of operation, the actual 2008 cost of treating with GAC was less than 20 cents per 1,000 gallons or, expressed another way, about $5.00 per quarter for a single-family household for both debt service and operations, compared with the quarterly average cost for a single-family household with a ⅝-in. meter of $53.83. It confirms that the GCWW is achieving their original vision of providing customers with very high–quality water at a price they can afford and are willing to pay.

The reactivation of GAC is a significant part of operation costs. GCWW has been able to lower the originally anticipated reactivation costs for the following reasons:

- The per-pound cost for new GAC declined for a few years. However, that trend has changed recently, and the most recent contract has a 50 percent increase in cost per pound.
- The reactivation system can be optimized to take advantage of the seasonal cycle of natural gas costs. Reactivation during the summer months and operating using stored reactivated GAC in the winter months reduces reactivation costs.
- The number of reactivations per year decreased. Originally it was expected that each of the 12 contactors would be reactivated twice each year for a total

of 24 reactivations. Through improving the efficiency of the GAC adsorption process, it has been determined that the contactors can be operated for much longer periods of time, reducing the reactivations from 24 to 16 per year.
- Careful management of the movement of GAC through the plant has reduced the GAC loss. It was projected that 10 percent of the GAC would be lost through reactivation, but loss has been reduced to 7 percent.
- Automating the reactivation process has reduced labor costs. Originally it was planned to have three staff involved in the reactivation process. Through automation with the supervisory control and data acquisition (SCADA) system, staffing has been reduced to one person.

GCWW conclusions. In Cincinnati, nearly two decades of experience have demonstrated how one system approached the integrated issues of protecting public health, meeting or surpassing government water quality regulations, satisfying customer expectations, and securing community trust and confidence. GCWW's experience has demonstrated the effectiveness and viability of the use of post-filtration GAC treatment to provide a barrier for a high-level removal of organic compounds from the water supply.

City of Phoenix, Ariz.

Like many communities in the United States, the city of Phoenix (Phoenix) sought prudent solutions to address the Stage 2 D/DBP Rule. Phoenix is served by five water treatment plants.

Plant	Design Capacity	Raw Water Source
24th Street WTP	140 mgd	Salt River Project Arizona Canal
Deer Valley WTP	100 mgd	Salt River Project Arizona Canal
Lake Pleasant WTP*	80 mgd	Central Arizona Project Canal
Union Hills WTP†	160 mgd	Central Arizona Project Canal
Val Vista WTP‡	220 mgd§	Salt River Project South Canal

*conventional treatment (high-rate clarification with sand ballasted flocculation and granular media filtration) with a bypass for direct filtration, chlorine dioxide, post-filter GAC contactor, and ultraviolet (UV) disinfection.

†direct filtration plant

‡co-owned by the Cities of Phoenix and Mesa

§130 mgd allocated to the City of Phoenix.

Phoenix updated water quality master plan. In anticipation of the Stage 2D/DBP Rule, Phoenix conducted testing and evaluated treatment alternatives to develop a water quality compliance strategy and a capital improvements program (CIP). The Water Quality Master Plan (WQMP) and subsequent updates (Case Study No. 2) were prepared to identify a water quality strategy that focused on removing DBP precursors prior to disinfection. In addition to controlling DBP formation, particularly total trihalomethanes (TTHMs), this approach also provided for an improved barrier against taste-and-odor compounds and organic micropollutants. The most current update to the Water Quality Master Plan for four of Phoenix's Water Treatment plants (Val Vista, 24th Street, Deer Valley, and Union Hills) was completed in July 2009.

The technologies selected for DBP precursor removal included enhanced coagulation, biological filtration, and adsorption by GAC. During the development of these plans, it was accepted that a more robust way to use GAC is in post-filter contactors, where GAC is used only for the purpose of DBP precursor and other trace organics removal by adsorption. However, in an effort to develop a lower-cost approach for GAC use at the water treatment plants, Phoenix adopted a strategy that uses GAC in filter adsorbers, in which GAC serves the dual purpose of filtering particulate contaminants and removing DBP precursors.

Subsequent to adopting the filter adsorber treatment strategy, the Salt River Project Watershed experienced some unusual water quality conditions due to extensive wildfires in the watershed area and other climate-related reasons. The change in source water quality resulted in higher levels of TOC. Due to this elevated TOC, concerns were raised about the practicality of GAC filter adsorbers because the elevated TOC would require frequent replacement and reactivation of GAC. Consequently, Phoenix conducted an extensive bench-scale evaluation of various alternatives for extending the life of GAC, including PAC, chlorine dioxide, magnetic ion exchange, manganese dioxide, softening, and dissolved iron. The results, which are summarized in their *GAC Road Map* (Phoenix 2008), found that PAC and ClO_2 were the most promising techniques for extending GAC life while balancing capital and operating costs.

Full-scale testing of both technologies was conducted to confirm their effectiveness for TTHM reduction in the distribution system. Full-scale testing results showed that while ClO_2 was very effective for reducing TTHM formation for one of the water sources (Salt River water), it had little to no effect on TTHM formation in one of the other sources (Verde River water). Because the water source is controlled by the Salt River Project, Phoenix cannot predict which source it will be receiving at any given time. Since such is the case, using ClO_2 at 3 mg/L was not sufficiently reliable to be a long-term operating strategy.

While the design of facilities and demonstration testing were progressing, the Phoenix water department staff continued to evaluate the complexity of operating filter adsorbers and explored other options based on their own evaluations as well as the experiences of other utilities. As a result, the following preferences emerged:

- Post-filter contactors (due to the ability to bypass the contactors during part of the year)
- Extensive optimization of distribution system operation (to reduce water age)
- Localized treatment in areas with elevated DBPs (e.g., tank aeration, distribution system GAC, or pipe looping)

At the conclusion of the full-scale demonstration testing, a cost analysis was done to help set the water quality strategy for Phoenix. Changes in the watershed water quality and the rising cost of GAC led Phoenix to reevaluate potential DBP compliance strategies including the use of GAC (as filter adsorbers, post-filter contactors, and BAC), ClO_2, chloramines, PAC, and localized treatment in the distribution system. The cost analysis included several variables: raw water TOC conditions, presence or absence of various treatment technologies at each water treatment plant (WTP), unit chemical costs, source water mix, distribution system water age, and water treatment plant production scenarios. The evaluation concluded that the following combination of treatment technologies would position Phoenix well for surpassing all

TTHM goals under typical and historically challenging (e.g., high TOC) raw water conditions:

- GAC post-filter contactors at Val Vista WTP
- GAC filter adsorbers at Deer Valley WTP
- Biologically active GAC filters at 24th Street and Union Hills WTPs
- PAC at all WTPs to reduce TOC peaks
- GAC post-filter contactors at Lake Pleasant WTP

In addition, Phoenix is planning to use chlorine dioxide as the pre-oxidant to limit TTHM formation within the WTPs and is also planning to optimize the distribution system (e.g., tank levels, looping to reduce dead ends, and pressure zone interconnects) to reduce water age and minimize DBP formation once water leaves the treatment plants. By locating the activated carbon facilities prior to disinfection, the city will avoid converting its secondary disinfection from free chlorine to chloramination, thus avoiding the risk of nitrogenous DBPs and nitrification within its warm-water distribution system. The activated carbon facilities offer numerous additional benefits, including providing a proactive barrier against compounds that cause T&O and for other adsorbable contaminants that are present or may be present in its raw water supplies in the future.

Birmingham, Ala., Water Works Board

The Water Works Board of the City of Birmingham (BWWB) operates four water treatment plants (Case Study No. 3):

Plant	Design Capacity	Raw Water Source
H. Y. Carson Filter Plant	24 mgd	Mulberry River/Inland Lake
Putnam Station Filter Plant	24 mgd	Mulberry River
Shades Mountain Filter Plant	80 mgd	Cahaba River (Lake Purdy)
Western Filter Plant	60 mgd	Mulberry/Sipsey River

Combined, the four plants have a total treatment capacity of 188 mgd. The system has 50 finished-water storage tanks (74.5 mil gal total capacity) and 3,800 mi of water main. Primary and secondary disinfection at all four plants is achieved using free chlorine.

Several years ago, the BWWB undertook the development of a Road Map, or Water Quality Master Plan, to improve water quality and plan for complying with the Stage 2 D/DBP Rule. The framework of this Road Map is shown in Figure 4-3.

Step 1: System evaluation, goal setting, and alternative analysis. The process of developing the Road Map to the Stage 2 D/DBP Compliance Plan began with a complete systemwide evaluation of raw water supplies, the four water treatment plants, the distribution system, and the 50 water storage tanks.

Although BWWB consistently provides high-quality drinking water that meets or exceeds current state and federal regulations, historical water quality data indicated that the in-place DBP practices were not sufficient to meet the internal BWWB performance goals or regulatory requirements for the Stage 2 D/DBP Rule that became effective in 2012.

Internal water quality performance goals for DBPs have been established by the BWWB. The present goals are to be at or below the USEPA Stage 2 D/DBP

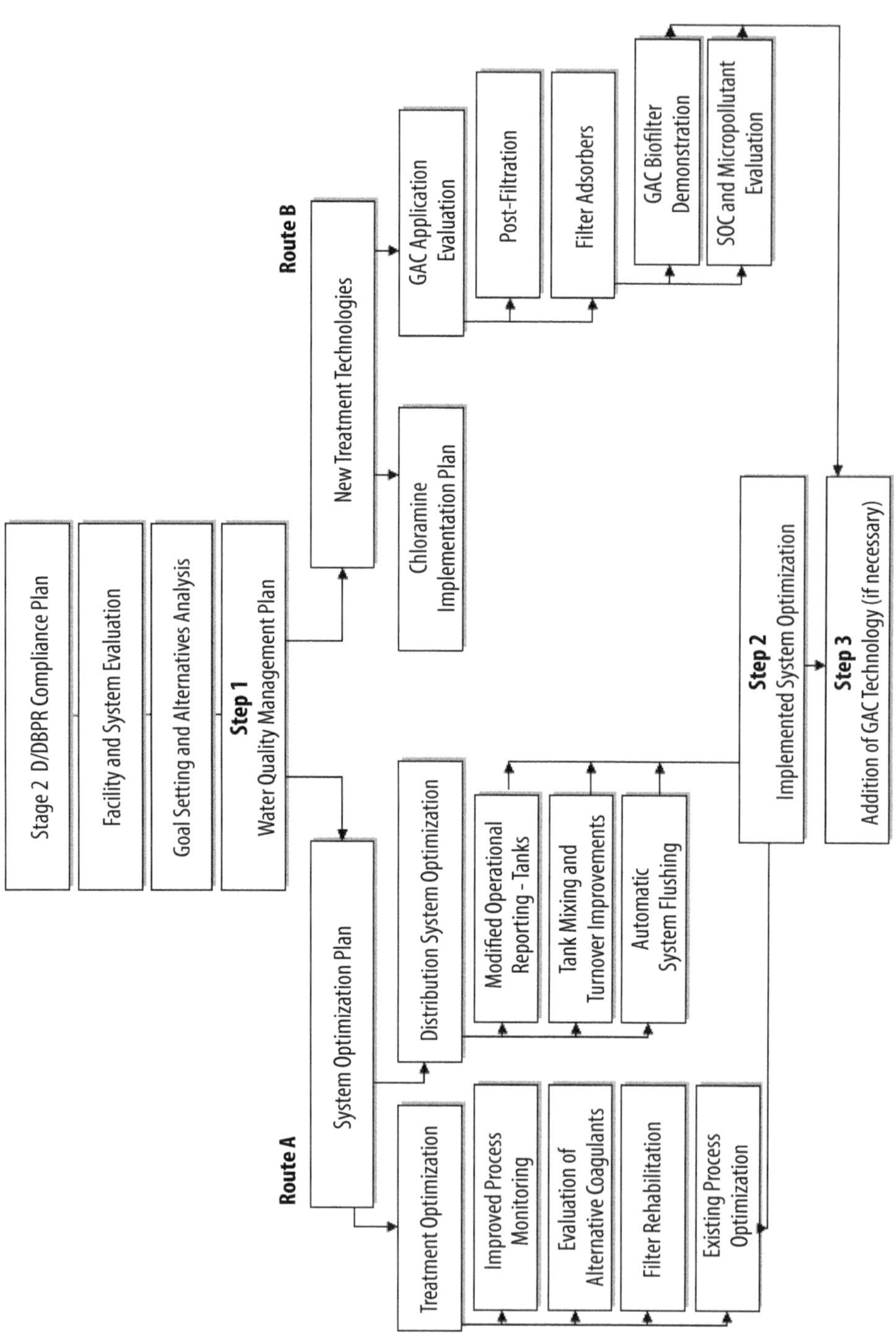

Figure 4-3. BWWB Road Map to improved water quality

maximum contaminant levels (MCL) for TTHMs and the HAA5 for every sample collected at each sampling location. These goals are somewhat more restrictive than Stage 2 compliance criteria of being at or below the MCLs for each sampling location on an annual average basis.

A number of treatment strategies for reducing DBPs were evaluated, including alternative disinfectants (chloramines, ozone, UV light) and improving DBP precursor removal prior to disinfection. GAC was identified as the most promising precursor reduction option while retaining the use of free chlorine as the disinfectant. GAC adsorption was recognized as a best available technology (BAT) for DBP precursor removal, and it was determined that it may be the best way to achieve compliance with the BWWB's internal water quality goals and the Stage 2 D/DBP Rule. Preliminary design criteria, preliminary site layouts, and budget-level cost estimates were developed for the use of GAC.

The outcome of these Step 1 efforts led to the development of the Road Map of a Water Quality Management Plan to proceed to Step 2 on two concurrent routes:

- Route A: System optimization plan
- Route B: Incorporation of new treatment technologies at the treatment facilities

Step 2: Implementing system optimization. In Step 1, it was determined that further studies would be necessary to determine the extent to which system optimization alone would permit BWWB to cost effectively meet its established internal water quality goals and comply with the Stage 2 D/DBP Rule, without the use of a GAC technology. If system optimization is effective, Step 3 would not be necessary.

Step 3: Addition of GAC technology (if necessary). If Step 2 is insufficient to permit BWWB to reliably and cost effectively achieve its internal water quality goals and comply with the Stage 2 D/DBP Rule, then in addition to system optimization a GAC treatment technology would need to be added to the treatment process at one or more of the water treatment plants.

Road Map Route A: System Optimization Plan. This route expands on the preliminary investigations and pilot studies performed under Step 1 in two areas: the water treatment systems and the distribution system.

BWWB has replaced the filter underdrains at the Putnam Station Filter Plant with stainless-steel, low-profile underdrains and has installed a filter media of anthracite coal and sand. Similar work is planned for the other three plants. The rehabilitation will facilitate future media replacement with sand and GAC (state regulations prohibit the use of GAC alone as monomedia), converting the units into filter adsorbers with the following GAC EBCTs at design flows:

Plant	GAC EBCT
Shades Mountain Filter Plant	7.1 min
Western Filter Plant	5 to 7.8 min
H. Y. Carson Filter Plant	6.6 min
Putnam Station Filter Plant	3.2 min

In addition to the filter rehabilitation work, online TOC monitors have been installed on filtered water. TOC monitoring will assist plant-scale evaluation of alternative coagulants, coagulant aids, and filter aids to optimize the removal of DBP precursors and provide effective filter-backwashing operations.

Concurrent with the treatment system improvements, work is proceeding on distribution system optimization. Optimization consists of efforts to eliminate the very long detention time in water storage tanks using a combination of operational actions to improve water turnover in the tanks and the installation of internal mixing in selected tanks. To reduce long detention times in selected sections of the distribution system, a program of manual and automatic flushing operations is being implemented.

Road Map Route B: New Treatment Technologies. Based on the results of Step 1, it was determined that the most effective alternative disinfection process would be chloramination and that GAC could effectively reduce DBP precursors.

BWWB made a policy decision to discontinue further work on chloramination because of concerns over the effect the change might have on corrosion control in the distribution system as well as the presence of nitrogenous DBPs (N-DBPs) and the possibly higher public health risk.

The pilot studies during Step 1 confirmed the effectiveness of GAC in adsorbing DBP precursors. Based on the significantly higher construction and operating costs associated with post-filtration GAC, it was temporarily suspended because it appeared that adequate removals would be achieved using GAC filter adsorbers. The pilot studies also indicated that some of the raw water supplies would permit the GAC filter adsorbers to be biologically active, without the added cost of the use of ozone, extending the intervals between GAC replacements and thus reducing operation costs.

GAC Biofilter Demonstration Plant. In order to explore the use of GAC biofilters on a larger scale than pilot plant, a demonstration plant was constructed at the Shades Mountain Filter Plant and operated through the spring of 2009. BWWB worked with Malcolm Pirnie, the University of Colorado, and the University of Alabama on the GAC demonstration plant, which resulted in valuable information on GAC biofilters for the water industry (Chowdhury et al. 2010).

The GAC demonstration plant consists of four 40-ft^2 filter boxes with a dual media of sand supporting the GAC. They are operated at a filtration rate of 2.8 gpm/ft^2 to match the Shades Mountain Filter Plant design rate. Three of the filter adsorbers provide an EBCT of 7 min while the fourth provides a 9-min EBCT. Of the three 7-min EBCT units, one is subject to water with a chlorine residual in the influent and backwash, one has a chlorine residual only in the backwash, and one has no chlorine residual in either influent or backwash. Operations were evaluated based on effectiveness of DBP reduction (as measured by UV absorbance at 254 nm), TOC removal, turbidity removal, run time, head loss development, and particle count reduction.

The process is unique in that ozone is not being applied ahead of the filter adsorbers to enhance biological activity as is the case with most other biofilter operations. Avoiding the use of ozone results in both capital and operating cost savings and leverages the use of existing treatment infrastructure. The results from the demonstration plant allowed evaluation, on a reasonable scale, of the effectiveness and efficiency of using GAC biofilters to reduce DBP precursors, permitting BWWB, along with its system optimization, to achieve its internal water quality goals and comply with the Stage 2 D/DBP Rule.

Analysis conducted during the demonstration plant operation provided important information of the capability of the GAC biofilters to remove micropollutants, pesticides, pharmaceutical compounds, and other synthetic organic chemicals that may be present in the filter influent.

City of Baltimore, Md.

The Susquehanna River drains a large watershed that encompasses many agricultural and industrial zones as well as numerous wastewater treatment plant discharges, which are all potential sources of contamination. These can contribute to higher raw water turbidities, heightened TOC levels, and elevated concentrations of nitrates, pesticides, and micropollutants such as endocrine disrupting compounds (EDCs) and pharmaceutically active compounds (PhACs). The city's existing conventional treatment plants have generally not performed well in treating Susquehanna River water, but the new 120-mgd Fullerton Plant will be designed to include the advanced treatment technologies needed to overcome these treatment challenges.

A pilot study has been completed to evaluate short-listed technologies and define preliminary design criteria for each process. These criteria were used to develop realistic concept-level cost estimates of the most appropriate treatment train for the new Fullerton Plant.

Process Studies. Prior to the pilot study, a series of bench-scale tests were conducted to help answer key questions and thereby refine the pilot testing program. Bench-scale testing included jar tests to compare various coagulants (including alum, ferric chloride, ferric sulfate, and polyaluminum chloride) and coagulant-aid polymers with respect to turbidity and TOC removal. Jar tests were also used to evaluate the performance of various PAC products for removing T&O compounds. These preliminary studies helped to streamline the pilot testing effort and thereby save time and money.

The complex 250-gpm pilot plant evaluated the following processes:

- Iron and manganese removal using potassium permanganate and chlorine dioxide
- T&O control using activated carbon (PAC and GAC) and ozone
- Enhanced coagulation using alum, ferric chloride, ferric sulfate, polyaluminum chloride (PACl), and pH control for enhanced TOC removal
- Flocculation/sedimentation (solids removal) with inclined plate settlers and sand ballasted flocculation (ACTIFLO®)
- Filtration using low-pressure membranes (microfiltration [MF] and ultrafiltration [UF]) and granular media (various media configurations)
- Disinfection/pathogen inactivation with chlorine and UV light
- Disinfection by-product control (TOC removal) with post-filter GAC contactors, GAC filter adsorbers, and biologically active GAC filters
- Destruction of micropollutants (EDCs, PhACs) using UV + peroxide (advanced oxidation) and ozone + peroxide (peroxone)

The Fullerton Water Treatment Plant. The Fullerton Water Filtration Plant will be the most significant addition to Baltimore's regional water supply since the development of the Susquehanna River supply in the 1960s. Certain parts of the existing water system, especially the tunnel between Loch Raven Reservoir and the Montebello Filtration Plants, cannot be taken out of service for extended maintenance or repair without curtailing demand. The Fullerton Plant will be designed with utmost flexibility to address these issues.

First, it will be capable of treating raw water from either the Susquehanna River or the Loch Raven Reservoir supply. A proactive raw water management plan will be used to balance the water supplies. This plan will minimize treatment costs and avoid using the river during times of poor water quality.

Second, the Fullerton Plant will be master planned with the option to build an additional 120-mgd of oxidation, PAC contact, and clarification capacity, to "pre-treat" Susquehanna River water prior to sending it to the Montebello Plants and make it more comparable to raw water from Loch Raven Reservoir, for which the Montebello Plants were designed.

As part of the plant development, a one-year pilot study was used to demonstrate the recommended treatment scheme, providing a "proof of concept" and specific design criteria to be used in developing budget-level cost estimates and conceptual site layouts.

The Fullerton treatment train was derived as a result of a detailed process evaluation. At the start of the project, a benchmark treatment train was developed to represent the preferred treatment processes as envisioned by city of Baltimore and Baltimore County staff and by a consultant team. A subsequent alternatives analysis was performed to validate the benchmark treatment train and determine if additional processes should be pilot tested. A matrix of noneconomic decision criteria was used to evaluate different treatment train options on the basis of regulatory drivers, industry trends, and client preferences.

Early in the project, Baltimore made a decision not to use chloramines for DBP control. Pilot testing showed that coagulation alone could not be relied upon for adequate removal of DBP precursors; therefore, advanced treatment technologies were considered (including ozone and biological filtration, nanofiltration, and magnetic ion exchange [MIEX®] resin). Ultimately GAC was recommended because it can also reduce T&O levels to below human detection and remove a variety of micropollutants (including certain endocrine disruptors, pharmaceutically active compounds, and personal care products).

Both GAC filter absorbers and post-filter contactors were evaluated as part of the Fullerton pilot study. Testing showed that both filter adsorbers and post-filter contactors would be suitable for reducing finished water TOC concentrations to a level low enough for Stage 2 D/DBP Rule compliance. Ultimately the choice of post-filtration GAC contactors was selected by Baltimore because of its desire to incorporate low-pressure membrane filtration, which precluded the use of filter adsorbers.

City of San Diego, Calif.

The Water Quality Master Plan for the city of San Diego, Calif. (Case Study No. 4), was prepared in August 1990, also in response to the state and federal regulatory changes resulting from the 1986 Amendments to the Safe Drinking Water Act. Here, the primary concerns were the need to develop a long-term treatment strategy for disinfection and control of DBPs at the city's Alvarado and Miramar water treatment plants.

Based on the evaluation of water quality requirements, assessment of existing plant performance, pilot-scale treatability testing results, and economic analysis, a two-phase water quality strategy was developed for San Diego's surface water treatment plants.

Phase 1 included:

- Conventional treatment using coagulation, flocculation, and sedimentation, consistent with the city's existing process.

- Ozone to provide primary disinfection and reduce the formation of undesirable DBPs with hydrogen peroxide added to later stages of ozone contact to assist in oxidizing T&O compounds.
- GAC caps as a replacement of the anthracite medium in conventional filters to stabilize biodegradable organics generated during ozonation and to provide an additional barrier for T&O compounds.
- Chloramines for secondary disinfection, similar to existing city practice.

Phase 2 of the strategy, to be implemented if limits for organic chemicals and DBPs become more stringent than anticipated after initial regulatory implementation, includes post-filter GAC contactors for substantial DOM removal. This alternative would also provide a significant barrier for intermittent T&O episodes.

Application Modes for Activated Carbon

As explained in the introduction to part 2, activated carbon technologies can be incorporated in several ways in water treatment plants. The two main classifications of treatment technologies involving activated carbon are the use of PAC, which is generally preferred for intermittent operation, and that of GAC, placed in fixed-bed filters or contactors, which is typically used for prolonged operation.

Several treatment modes follow, each with their own site-specific limitations and design considerations. Following the descriptions of each application type are case studies that illustrate real-world applications of activated carbon technology.

PAC. PAC is contacted as slurry and removed by a combination of settling and filtration.

- City of Columbus, Ohio
- Consolidated Mutual Water Company, Colorado

Post-filter contactors. Contactors are placed in series with media or membrane filters, contaminants are removed via adsorption, and GAC is replaced on an as-needed basis.

- City of Scottsdale, Ariz.: Central Arizona Project WTP
- City of Scottsdale, Ariz.: Chaparral WTP
- City of Phoenix, Ariz.: Lake Pleasant WTP

Filter adsorbers. The filter media is exclusively GAC, and media is replaced on an as-needed basis to retain the adsorptive capacity of the GAC.

- City of Glendale, Ariz.
- Town of Gilbert, Ariz.

GAC caps. A portion of anthracite or sand media filter is replaced with GAC.

Biologically enhanced activated carbon (BAC). The GAC is not replaced on a regular basis; instead, contaminant removal relies on biological processes and not adsorption processes.

- City of Peoria, Ariz.

GAC pressure vessels. GAC contactors in the form of pressure vessels are typically implemented by smaller systems, particularly wellhead groundwater systems. Similar to post-filter contactors, contaminants are removed via adsorption, and GAC is replaced on an as-needed basis.

Table 4-1. Activated carbon applications

		GAC				
Water Quality Goal	PAC	Post-Filter Contactor	Filter Adsorber	GAC Caps	BAC	Vapor Phase
TOC/DOC Removal	X	X	X	X	X	
SOCs and VOCs		X	X	X	X	X
Taste and Odor	X	X	X		X	
Color	X	X	X		X	
Micropollutants	X	X	X			

- Borough of Fair Lawn, N.J.
- Dover Township, NJ.
- Flowing Wells Water District, Tucson, Ariz.
- Metro Water South Shannon Wellhead Treatment, Tucson, Ariz.

Vapor-phase adsorption. Air stripping is first used to remove volatile compounds from water, and then the air is treated with GAC to adsorb the contaminants.

- City of Scottsdale, Ariz., Central Ground Water Treatment Facility

Table 4-1 shows the applicability of various types of activated carbon technologies for achieving compliance with different water quality goals.

PAC Use to Achieve Water Quality Goals

PAC is used for many treatment issues including removing T&O compounds, trace organics, and DBP precursors (Najm et al. 1991, Graham et al. 2000). The effectiveness of PAC adsorption for these compounds varies. Many herbicides and pesticides have been shown to adsorb strongly to PAC, while the adsorption of synthetic organic compounds can be variable. Similarly, removal of DBP precursors by PAC has "varied from poor to very good" (Najm et al. 1991). Typically, the PAC is added on an as-needed basis in conventional water treatment plants. Although capital cost is small, operating costs can be significant.

PAC systems might prove to be an effective and economical treatment method for facilities with the following characteristics:

- Periodic or seasonal issues with T&O or trace organic compounds at low to moderate concentrations.
- A potential need to remove organic contaminants from the water on short notice, such as emergency response to a spill upstream of the water treatment plant.
- A solids handling facility that has sufficient treatment capability to handle the additional load resulting from PAC use.
- Limiting conditions, such as a lack of space or funding for plant expansion, where installing new or retrofitted GAC treatment units is not appropriate.

Factors that affect PAC adsorption include the type of PAC being used, PAC dose, interaction with other chemicals, mixing, and contact time.

The effectiveness of using PAC for water treatment greatly depends on the type of PAC used and the characteristics of the source water. Many PAC products are currently available, but it is important to note that they are not identical. Even PAC products that are made from the same source material can have significantly different iodine numbers and particle size distributions (Table 4-2); therefore, one PAC might

Table 4-2. Comparison of several types of powdered activated carbon

PAC Brand	Manufacturer	Made From	Specified Minimum Iodine Number (mg/g)	Particle Size Distribution: % Less Than 100 Mesh	% Less Than 200 Mesh	% Less Than 325 Mesh
Hydrodarco B	Norit Americas	Lignite coal	570	99.9	99.8	96
Hydrodarco 0	Norit Americas	Lignite coal	550	99.9	99.8	96
Hydrodarco W	Norit Americas	Lignite coal	525	99.9	99.8	96
AquaPAC 500	Siemens	Bituminous coal	500	99	95	90
AquaPAC 800	Siemens	Bituminous coal	800	99	95	90
Norit PAC 20B	Norit Americas	Bituminous coal	925	99	n/a	80
WPH	Calgon Carbon	Bituminous coal	800	99	95	90
WPH-M	Calgon Carbon	Bituminous coal	500	99	95	90
AquaNuchar	MeadWestvaco	Wood	900	99	95	90
Nuchar MWC	MeadWestvaco	Wood	900	95	85	65
PACarb	Cameron Carbon	Wood	550	n/a	n/a	90
PACarb Plus	Cameron Carbon	Wood	800	n/a	n/a	90

be more effective than another for treating a given source water. In addition, some PAC products are specially designed for removal of a certain contaminant, such as DBP precursors (Anderson et al. 1981).

PAC adsorption can be affected by other constituents in the water being treated, including:

- Chlorine, potassium permanganate, or other oxidants that react with the PAC
- Coagulants
- Lime
- Background organics that compete for adsorption sites

Mixing and contact time also affect PAC performance as discussed in chapter 2. Adequate mixing should be provided at the PAC addition point to decrease the external mass transfer limitations. In bench-scale tests, Lalezary et al. (1986) found that vigorous mixing resulted in the most efficient adsorption of T&O compounds on PAC. Similarly, sufficient contact time should be provided to allow compounds in the water to adsorb to the PAC.

PAC Application Point

The location of the PAC addition within the treatment process is critical for effective use of the carbon. The application point should be located (Najm et al. 1991) such that:

- The PAC has effective mixing and contact with the water.
- A long enough contact time is provided to allow adsorption to occur.
- Interaction of other water treatment chemicals with the PAC that might detrimentally affect the adsorptive capacity is minimized.
- The finished water quality is not negatively affected by the addition of PAC.

As shown in Figure 4-4, common locations for adding PAC are the plant intake, rapid mix, and filter influent (Najm et al. 1991). PAC may also be added in a separate

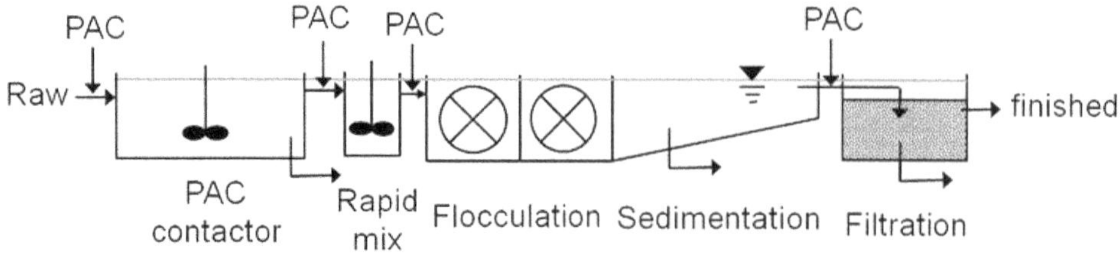

Figure 4-4. PAC addition and removal points

PAC slurry contactor. Some facilities apply PAC at multiple addition points (Graham et al. 2000, Brady 2005). The advantages and disadvantages of each addition point are discussed below.

Plant Intake. Adding PAC at the plant intake provides the longest possible contact time and may avoid interferences with other chemical additions. However, it is possible that some substances that would normally be removed during coagulation will be adsorbed onto the PAC, which could increase the amount of PAC required for treatment (Najm et al. 1991). Although uncommon, PAC may be added directly to a raw water reservoir or presedimentation facility. Adding PAC to a reservoir or presedimentation facility is generally inefficient because of the poor mixing conditions, and the accumulation of PAC at the bottom of the reservoir/basin can lead to increased dredging/removal requirements.

Rapid Mix. The most common PAC addition point is at the rapid mix (Graham et al. 2000). Adding PAC at the rapid mix can provide good mixing and contact time. Disadvantages of this addition point include the possibility of coagulants interfering with adsorption, competition from compounds that would otherwise be removed by conventional treatment (coagulation, flocculation, and sedimentation), and a contact time that is potentially too short to remove some contaminants (Najm et al. 1991). Several studies (Ho and Newcombe 2005, Graham and Summers 1996) have shown that even though floc formation decreases the overall dispersion of PAC, generally there is no notable decrease in adsorption of T&O compounds at coagulant doses less than 40 mg/L. The contact time of the PAC in this application is congruent with the flocculation period, and this results in sufficient exposure time for effective T&O compound adsorption in most cases.

Filter Inlet. PAC can also be added at the filter inlet. PAC addition at the filter inlet minimizes the competitive adsorption effect from organics that are removed by coagulation, but it reduces the contact time. Another disadvantage of this addition point is that adding PAC to the filters will also increase the head loss through the filter; in fact, the PAC dose might be limited by the allowable head loss through the filter (Najm et al. 1991). There is also a possibility that some of the PAC might break through the filter into the clearwell or distribution system (Najm et al. 1991). PAC is removed from the filters during backwashing (Brady 2005).

Slurry Contactor. A dedicated PAC slurry contactor may be installed upstream of the rapid mix chamber (Najm et al. 1991). Here, the PAC is mixed with the water prior to coagulation and flocculation, so mixing conditions and contact time can be optimized for efficient use of the PAC and interference from other chemical additions is avoided. A new contact chamber and mixer would need to be installed for this option, which could make it expensive compared with the other addition points. Facilities equipped with presedimentation or equalization processes may be able to provide opportunities to retrofit PAC addition at lower cost. Competition from compounds that would be removed by coagulation could affect the adsorption efficiency (Najm et al. 1991, Graham et al. 2000).

PAC Removal Points

PAC must be removed from the process stream before the treated water enters the distribution system. A logical place for PAC removal is the sedimentation basins, where the PAC settles as part of the coagulated floc. Hydraulic conditions during sedimentation may vary along with PAC particle size and density, thus the PAC product should be evaluated to verify its settling performance. Removal of PAC that escapes sedimentation occurs in the filter beds. If PAC is applied just upstream of the filters, then all of the PAC must be removed in the filters. As previously mentioned, relying on the filters to remove all the PAC can lead to head loss problems and PAC breakthrough. PAC removal can be a limiting factor in inline and contact filtration plants. Any PAC added to the process will eventually be processed through the solids handling facility. The effect on the solids handling process, including excessive wear on the pumping, piping, and mechanical dewatering equipment and the cost of residuals disposal, should be considered.

PAC Performance Evaluation

The goal of a PAC performance evaluation is to determine whether PAC will be an economical solution to meet the treated water quality goals. As such, PAC performance evaluation ultimately consists of determining the PAC dose that should be applied to meet the treatment objective. Graham et al. (2000) developed a protocol for bench-scale PAC performance evaluation. Their protocol makes a comparison of the lowest PAC dose with the treatment objective—and also takes cost into consideration, resulting in cost per volume treated as an endpoint for comparison. Several parameters determine the necessary PAC dose: contact time and mixing, PAC product, the influent concentration of the target compound, the background water quality, and the treatment objective. Each of these variables will be discussed more thoroughly in chapters 5, 6, and 7, organized by the primary treatment objective. Chapter 5 discusses activated carbon for removal of dissolved organic matter, chapter 6 discusses activated carbon for removal of T&O compounds, and chapter 7 discusses activated carbon removal of organic contaminants. Graham et al. (2000) developed a protocol for bench-scale PAC performance evaluation.

Regulatory Guidance on PAC Design

The USEPA sets regulations for the entire United States, and individual states are permitted to adopt their own regulations provided that they are at least as stringent as the national regulations. Therefore, local requirements can vary. The design engineer should consult with the local regulatory agency having jurisdiction to determine local approval and enforcement practices.

One of the better-known sets of regulations in the United States is known as the "Ten States Standards." The formal title is *Recommended Standards for Water Works* (GLUMRB 2007). The Ten States Standards were developed by ten partner states (and Ontario, Canada) to guide water treatment project implementation. The standards are frequently referred to by the partner states as well as other states for design of drinking water treatment facilities.

Ten States Standards refer to PAC treatment in chapter 4, "Filtration." Section 4.9.4 specifically states the following about PAC:

> "4.9.4 Powdered activated carbon
>
> a. Powdered activated carbon should be added as early as possible in the treatment process to provide maximum contact time. Flexibility to allow the addition of carbon at several points is preferred. Activated carbon should not be applied near the point of chlorine or other oxidant application.
> b. The carbon can be added as premixed slurry or by means of a dry-feed machine as long as the carbon is properly wetted.
> c. Continuous agitation or re-suspension equipment shall be provided to keep the carbon from depositing in the slurry storage tank.
> d. Provision shall be made for adequate dust control.
> e. The required rate of feed of carbon in a water treatment plant depends upon the tastes and/or odors involved, but provision should be made for adding from 0.1 milligrams per liter to at [most] 40 milligrams per liter.
> f. Powdered activated carbon shall be handled as a potentially combustible material. It should be stored in a building or compartment as nearly fireproof as possible. Other chemicals should not be stored in the same compartment. A separate room should be provided for carbon feed installations. Carbon feeder rooms should be equipped with explosion-proof electrical outlets, lights, and motors." (GLUMRB 2007)

Utility Experiences Using PAC

The following examples provide case studies that illustrate how two utilities are using PAC.

City of Columbus, Ohio. The city of Columbus uses PAC at its Hap Cremean Water Treatment Plant to remove atrazine (Case Study No. 5). This plant treats 130 mgd of raw water from the Hoover Reservoir using a conventional two-stage lime softening process with recarbonation and sand filtration. Facilities for feeding PAC at a maximum dose of 50 mg/L were placed in service in 1998. Benefits of using PAC include reduction of T&O compounds, removal of SOCs and VOCs that might result from a potential chemical spill into Big Walnut Creek, and reduction of DBP precursors. Flow from Big Walnut Creek feeds the Hoover Reservoir.

The PAC facility includes six 40,000-gal tanks for storing PAC slurry. Six metering pumps are provided to feed PAC slurry to three separate feed points in each plant. Feed points are provided at the raw water intake, the first-stage rapid mix (prior to the sedimentation basins), and the lime channel upstream of the softening basins. The metering pumps are sized to feed a maximum of 50 mg/L at the maximum treatment rate of 130 mgd.

Consolidated Mutual Water Company, Colo. (CMWC). The Maple Grove Water Treatment Plant is owned and operated by the Consolidated Mutual Water Company (Case Study No. 6), a not-for-profit private utility that provides treated water to residents in Denver's western suburbs, including portions of Lakewood and Wheat Ridge. Originally constructed in the 1950s, the facility was recently expanded and upgraded. The plant treats raw water diverted from Clear Creek and stored in nearby Maple Grove Reservoir. Clear Creek water is characterized by moderate alkalinity and low turbidity, is susceptible to turbidity spikes during spring runoff events, and also contains high levels of dissolved manganese due to mining activities in the watershed. Storage in the Maple Grove Reservoir has led to algal blooms and seasonal T&O episodes at the facility. The Maple Grove Water Treatment Plant employs a membrane treatment process train consisting of PAC addition, oxidation, coagulation, flocculation, sedimentation, and ultrafiltration. The facility has a capacity of 15 mgd and space for expansion to 18 mgd.

The PAC system at the Maple Grove plant is designed for seasonal operation during T&O events, which generally occur in the summer. PAC is stored in 50-lb bags delivered on pallets of 40 bags. The operator manually empties the bags into the PAC feeder, which is equipped with dust collection equipment to minimize fine particles escaping the feeding unit. PAC is paced volumetrically based on raw water flow into the plant. The equipment is capable of feeding up to 200 lb/hr of PAC and is capable of 30:1 turndown. The average PAC dose is 20 mg/L, but the design range is 5 mg/L to 40 mg/L. PAC is fed upstream of the PAC contactor, a 170,000-gal tank that provides 15 min of contact time between the PAC and the raw water before chlorine dioxide is added to oxidize dissolved manganese prior to flocculation. When PAC is not in use, the contactor can be bypassed.

GAC Use to Achieve Water Quality Goals

GAC contactors can be classified by the following characteristics: (1) driving force: gravity versus pressure; (2) flow direction: downflow versus upflow; (3) configuration: parallel versus series; and (4) position: filter adsorber versus post-filter contactors. GAC may be used in pressure or gravity contactors. Pressure filters enclose the GAC and can be operated over a wide range of flow rates because of the wide variations in pressure drop that can be used. An advantage of these filters is that they can be prefabricated and shipped to the site. A disadvantage is that the GAC cannot be easily visually observed. Gravity contactors are better suited to systems (1) when wide variations in flow rate are not desirable because of the need to remove turbidity, (2) when large pressure drops are undesirable because of their effect on operation costs, (3) when there is available hydraulic head in the plant, and (4) when visual observation is needed to monitor the condition of the GAC. For many systems, the decision between pressure or gravity contactors is made on the basis of cost. Medium- and large-sized systems normally use gravity contactors, and smaller groundwater systems typically use pressure filters.

Water may be applied to GAC either downflow or upflow, and upflow columns may be either packed bed or expanded bed. Downflow columns are the most common and seem best suited for drinking water treatment. McCarty et al. (1979) found that in tertiary wastewater treatment carbon fines were produced during packed bed upflow operation and were not produced during downflow operation.

GAC Post-filter Contactors

Post-filter contactors are concrete basins, steel vessels, or fiberglass vessels filled with GAC that are located downstream of the filters (Figure 4-5). Post-filter GAC systems normally include multiple contactors, and these are usually operated in a staged

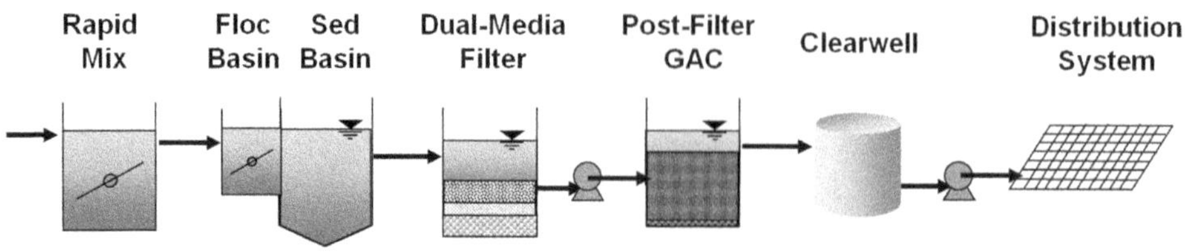

Figure 4-5. Post-filter contactor schematic

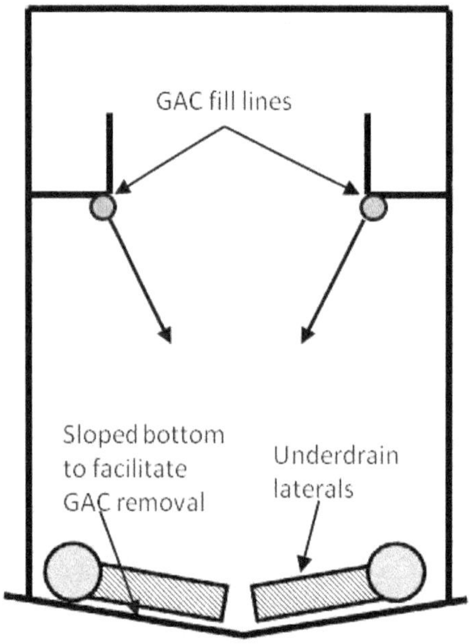

Figure 4-6. Schematic cross-section of a gravity post-filter contactor

parallel mode (described later in this chapter), where each contactor is at a different point in its life cycle.

Figure 4-6 shows a schematic cross-section of a post-filter contactor. The purpose of the post-filter contactor GAC treatment unit is adsorption rather than filtration. Because particles have been removed in the upstream filtration process, backwashing to alleviate excessive head loss is typically not needed. However, a form of backwashing often termed *fluffing* is still recommended to reduce the bed compression that develops over time.

The following design features are typical of a post-filter contactor facility:

- Stainless-steel underdrains, typically installed on the contactor floor.
- Small-mesh-size GAC (typically 12×40) in a deep bed design up to 12 ft deep.
- Water backwash capability.
- Wall washdown system to assist in removing spent GAC from the contactor.
- GAC transfer systems for filling and emptying a post-filter contactor, including eductor(s), service water, and a loading/unloading station.
- Means of measuring bed depth/bed expansion.
- Means of obtaining water samples at various GAC depths.
- Means of isolating a contactor from service, using gates or valves.
- Means of controlling flow to each contactor.

Utility Experiences Using Post-Filter Contactors

City of Scottsdale, Ariz., Central Arizona Project WTP. The City of Scottsdale operates a 75-mgd Central Arizona Project (CAP) Water Treatment Plant (Case Study No. 7). Water from this plant can travel more than five days to the farthest reaches of the distribution system. GAC was selected to reduce DBPs, for its ability to reduce the potential for biological regrowth, and to eliminate nitrification by avoiding chloramination in the warm-weather distribution system. The decision to use post-filtration GAC treatment to provide the best available technology and highest quality water for Scottsdale was based on extensive input from elected and appointed officials and public forums that were held by the city.

Construction cost of the GAC facility, completed in September 2008, was $45 million. This facility put Scottsdale ahead of the curve in responding to the requirements of the Stage 2 D/DBP Rule and substantially eliminates the potential for seasonal T&O episodes. The CAP Water Treatment Plant is Scottsdale's largest facility, and combined with the Chaparral Water Treatment Plant using post-filter contactors, Scottsdale supplies high-quality GAC treated water to nearly all of its customers.

Because the facility would not fit within the existing plant hydraulic profile, a pump station had to be included. A diversion structure allows the facility to be partially or totally bypassed as water quality dictates, reducing operations and maintenance costs for GAC replacement. The pump station combines diversion, backwash, and service water pumping systems in one location, utilizing water treated by the plant's filtration facilities.

City of Scottsdale, Ariz., Chaparral Water Treatment Plant. The City of Scottsdale operates the Chaparral Water Treatment Plant (Case Study No. 7) to treat its allocation of Salt River water from the Arizona Canal. The plant began producing potable water in March 2006. The total capacity of the plant is 30 mgd with a firm capacity of 27 mgd. The process train at the Chaparral Water Treatment Plant consists of the raw water strainer gallery, followed by raw water chemical addition and mixing (ferric sulfate is added for adsorption of arsenic), membrane filtration, and post-filter GAC contactors. Following the GAC contactors, the water is disinfected using on-site generation of sodium hypochlorite and stored in the finished-water storage reservoir prior to being pumped into the distribution system.

The GAC contactors are used to reduce TOC and to improve T&O by adsorbing MIB and geosmin. The gravity flow GAC contactors consist of a single deep layer of GAC resting on an underdrain system contained in concrete box structures. Ten contactors operate and one contactor is out of service at any given time. The GAC contactors are arranged in a single row with access along one side. Each GAC contactor is 17×34 ft and has a capacity of 3 mgd. The design loading rate of each contactor is 3.65 gpm/ft^2, and the empty bed contact time at design flow is 17.1 min. The GAC bed depth in each contactor is 8.1 ft, and the media has a mesh size of 12×40.

City of Phoenix, Ariz., Lake Pleasant Water Treatment Plant. Lake Pleasant Water Treatment Plant (Case Study No. 8) has an initial nominal design capacity of 80 mgd, and it is master planned to be expandable to 320 mgd. Raw water is pumped from the Waddell Canal to ballasted flocculation units for initial clarification. The facilities include four basins each consisting of flash mixers, an injection zone, a flocculation zone, and a sedimentation zone. From there, flow is ozonated to reduce DBPs and TOC and to facilitate the filters' operation as biologically active, deep-bed monomedia. The filtered water flows to post-filter GAC contactors. Collectively, these processes provide a robust and multibarrier approach to remove DBPs, TOC, geosmin, and MIB.

The GAC contactors are open concrete basins in an enclosed building with provisions for backwash and contactor to waste. The eight gravity flow contactors (each 42×10×8 ft) have 8 ft of GAC and provide an EBCT of 14.1 min with all eight in service. At 82-mgd flow, seven contactors in service provide 12.4 min of EBCT. Reactivation of spent GAC is on-site in a separate gas-fired rotary kiln. Off gas from the reactivation process is scrubbed to comply with air quality regulations.

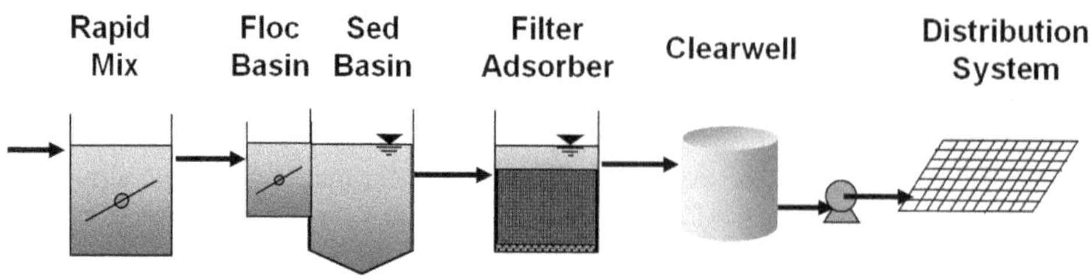

Figure 4-7. Filter adsorber schematic

GAC Filter Adsorbers

GAC filter adsorbers are a relatively simple way of implementing GAC treatment, often without the need for the construction of new structures. The filter adsorber approach (Figure 4-7) involves replacing all of the granular media in a conventional filter with GAC, resulting in the GAC serving as a filter media as well as an adsorbent. The GAC can be any depth but may be limited by the existing filter dimensions.

Figure 4-8 shows a schematic cross-section of a filter adsorber. This option allows optimal use of plant footprint and capital funds (especially for a retrofit project) because both organic compounds and particles are removed without the need for separate adsorption facilities. The following design features are typical of a filter adsorber facility:

- Floor-mounted underdrains.
- Larger-mesh-size GAC (typically 8×20) in a deep-bed design (less media depth than a post-filter contactor).
- Water backwash capability supplemented by surface wash and/or capability for air scour.
- Wall washdown system.
- GAC transfer systems for filling and emptying a filter adsorber, including eductor(s), service water, and a loading/unloading station.
- Means of measuring bed depth/bed expansion.
- Means of obtaining water samples at various GAC depths.
- Means of isolating a contactor from service, using gates or valves.
- Means of controlling flow to each filter.

Utility Experiences Using GAC Filter Adsorbers

City of Glendale, Ariz., Cholla and Oasis Water Treatment Plants. The Cholla Water Treatment Plant (Case Study No. 9) was originally commissioned in 1979 and currently treats a maximum flow of approximately 30 mgd. Its service area includes primarily central and southern Glendale. The Cholla WTP treats water delivered from the Salt River Project Arizona Canal, which is a highly variable blend of water from several sources including the Salt River, the Verde River, the Central Arizona Project aqueduct, and groundwater. The plant employs a conventional water treatment process train consisting of presedimentation followed by coagulation, sedimentation, and filtration. Prior to 2005, the filters were dual-media filters with anthracite and sand. In 2005, a Filter Retrofit and Enhancement Project was undertaken, which included converting the existing dual-media filters to GAC filter adsorbers for the removal of

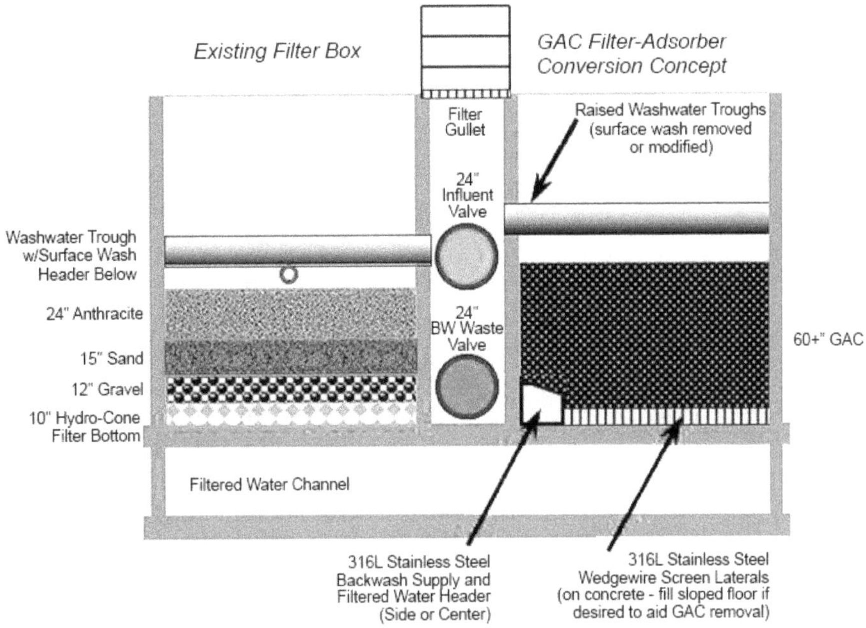

Figure 4-8. Schematic cross-section of a gravity filter adsorber

TOC. Following filtration, free chlorine is used as the primary and secondary disinfectant in the finished water reservoirs and in the distribution system.

Implementation of GAC at the Cholla Water Treatment Plant, and also at the newly constructed Oasis Water Treatment Plant, provides Glendale the means to stay in compliance with the Stage 2 D/DBP Rule and supply its citizens with high-quality water.

Operations staff has experienced improved treatment performance through optimization of enhanced coagulation and chlorination processes, including reduced chemical usage. The use of GAC has already resulted in a significant decrease in DBPs within the distribution system. GAC has also eliminated most customer taste-and-order complaints.

Town of Gilbert, Ariz., North Water Treatment Plant. Prior to 1996, the Town of Gilbert relied solely upon groundwater wells to meet the potable water demand. In 1996, its first surface water treatment plant, the North Water Treatment Plant, came online. The plant was originally constructed to treat 15 mgd, and in 2002, the capacity was expanded to treat up to 40 mgd. The plant was originally a conventional water treatment plant, but during the 25-mgd expansion, ozonation and biological filtration were included in the process train. Ozone is applied to the raw water between the presedimentation and flocculation/sedimentation processes. The clarified water passes through the biologically active GAC filters. No chlorine is added to the water prior to filtration. Following filtration, the water is disinfected with chlorine and is held in the finished water reservoir prior to being discharged to the distribution system.

As mentioned above, filters at the North Water Treatment Plant were converted from dual-media to GAC during the 2002 expansion. Today, 18 GAC biologically active filter adsorbers operate at variable declining rate mode. Each GAC filter adsorber

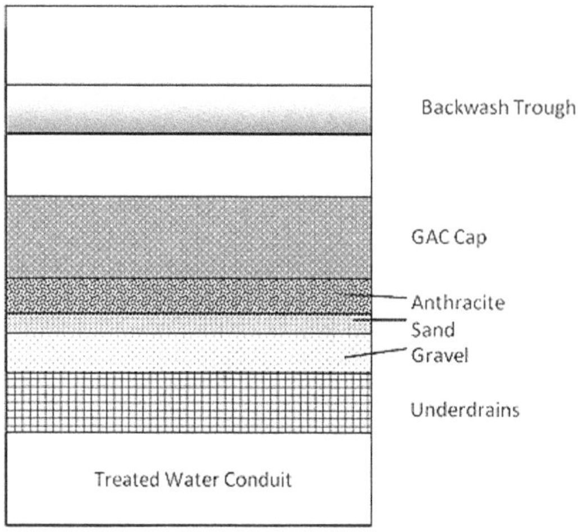

Figure 4-9. Schematic cross-section of a GAC cap

is 16 ft wide and 24 ft long, thus providing a total surface area of 384 ft^2. The filter media is made up of 42 in. of GAC over 6 in. of sand.

GAC Caps

Figure 4-9 shows a schematic cross-section of a GAC cap installation. GAC caps are an economical, easy-to-install option to provide adsorption capability in an existing filter. As the term implies, GAC is added on top of, or to cap, an existing filter. The GAC layer can be any depth but may be limited by the existing filter dimensions. In some states, GAC filter adsorbers without a layer of sand are discouraged or not allowed. Thus, the use of GAC filter caps with a layer of sand is the only approach allowed for GAC as a filter media.

The GAC cap as a GAC treatment technique has limitations. First, unless the GAC is matched with the existing media for uniform backwashing as described by Kawamura (2000), either the existing media will not be thoroughly cleaned or the GAC will be washed out of the filter during backwashing. Second, the limited space available to install a GAC cap on top of existing filter media will provide limited adsorption capability. These limitations make GAC caps a treatment technique that may be implemented as an interim measure to improve water quality.

The following design features are typical of a GAC cap installation:

- Utilization of existing filter box (including underdrains, troughs, media) and backwash system.
- Larger-mesh-size GAC (for example, 8×20).
- Provisions to install or remove the GAC cap.

Biologically Enhanced Activated Carbon

Biological processes such as slow sand filtration and bank filtration have a long history of use in drinking water treatment. In recent decades, rapid biologically active filters (BAF) achieved increasing acceptance for treatment of potable water (Servais et al. 2005). In a filter or packed column, indigenous microorganism will accumulate on the surface of media, like sand or GAC, as long as a disinfectant residual is not present in the effluent and if a biodegradable substrate, organic or inorganic, is available. Although a filter will operate biologically whenever no disinfectant residual is present throughout the depth of the filter, a major impetus to biological filtration was given by the increased use of ozonation, which increases the concentration of the easily biodegradable fraction of dissolved organic matter (DOM) measured as TOC. This process is often termed *biofiltration*, and when GAC is the media, the term *biologically enhanced activated carbon* (BAC) has been used (Figure 4-10). Because chlorine reacts with GAC, even if chlorine is in the influent, it is very likely not in the effluent; thus all applications

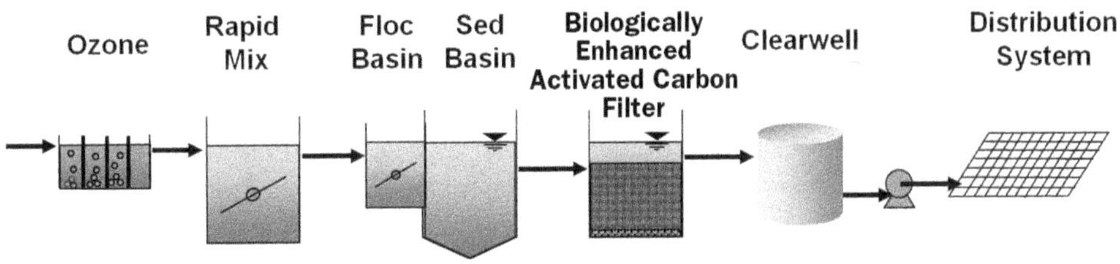

Figure 4-10. BAC application schematic

of GAC in water treatment have a biomass growth and biological removal component.

Figure 4-11 shows a schematic cross-section of a BAC filter. When ozone is included in a treatment train, BAC filters are frequently used to reduce concentrations of ozone by-products and biodegradable dissolved organic carbon (BDOC) (Hozalski et al. 1995 Wang et al. 1995). BAFs also have been shown to reduce chlorine demand, DBP formation (due to the reduction of DBP precursor concentrations), and assimilable organic carbon (AOC) (Servais et al. 2005). BAFs with GAC are not operated for adsorption because the adsorption capacity is exhausted long before the GAC is replaced. It should be noted that a filter adsorber operated past the point of exhaustion of the GAC adsorptive capacity may continue to operate as a BAC filter.

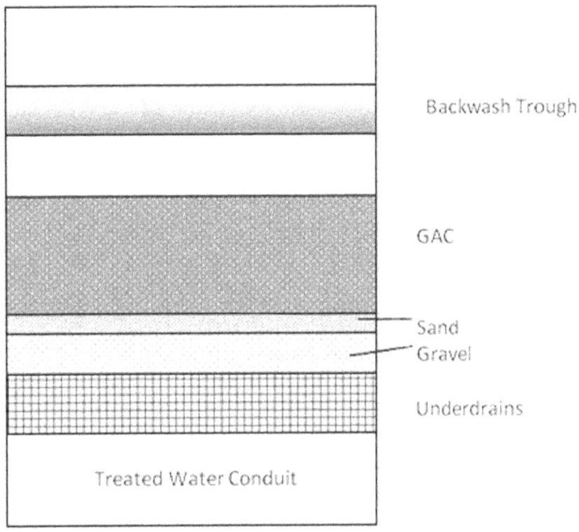

Figure 4-11. Schematic cross-section of a gravity biologically enhanced activated carbon filter

The following design features are typical of a BAC facility:

- Floor-mounted underdrains.
- Larger-mesh-size GAC (e.g., 8×20) in a deep-bed design (less than a post-filter contactor), which may include a sand layer between the media retention layer and the GAC for enhanced particle removal/polishing.
- Water backwash capability supplemented by air scour.
- Means of measuring bed depth/bed expansion.

Utility Experiences With Biologically Enhanced Activated Carbon

City of Peoria, Ariz., Greenway Water Treatment Plant. The Greenway Water Treatment Plant in Peoria, Ariz., is an ozonation/conventional water treatment plant with a design capacity of 16 mgd (Case Study No. 10). Raw water from the Arizona Canal enters the presedimentation basin, which is designed to remove large settleable solids. The plant recycle stream is also returned prior to the presedimentation basin. From there, the water is ozonated to reduce DBPs and TOC. Following ozonation the flow goes through two trains of three-stage flocculation/sedimentation basins. The settled water is filtered through GAC filters, pH adjusted, disinfected with chlorine, and stored in the finished water reservoir prior to being pumped into the distribution system.

In 2007, Peoria established its Stage 2 D/DBP Rule compliance strategy based on using GAC filter adsorbers with GAC replacement frequency as required to achieve internal water quality goals.

GAC Pressure Vessels

When smaller systems implement GAC, pressure vessels are typically being used rather than constructing concrete basins that are open to the atmosphere. Groundwater systems can contain a variety of synthetic and naturally occurring organic chemicals including pesticides, volatile organic compounds, endocrine-disrupting compounds, pharmaceuticals and personal care products, and petroleum-related compounds, along with other industrial organic compounds causing tastes and odors. Although not usually targeting TOC, the removal mechanisms and activated carbon replacement strategies are similar to post-filter contactors or filter adsorbers.

Utility Experiences Using GAC Pressure Vessels

The Borough of Fair Lawn, N.J. The Borough of Fair Lawn, N.J. (Case Study No. 11), worked with the New Jersey Department of Environmental Protection (NJDEP) to develop the first treatment facility in the state specifically designed to demonstrate the ability of GAC to remove a broad spectrum of unregulated contaminants from groundwater supply sources.

This full-scale demonstration project will help the NJDEP promote the use of GAC elsewhere in the state as a cost-effective and efficient means of removing contaminants from potable water sources that could potentially pose a long-term threat to human health. NJDEP had two main goals for this project: (1) to use this facility to gather data on performance of GAC for unregulated organic contaminant removal and (2) to use it as an example to demonstrate the viability, community acceptability, functionality, cost, and operability of these types of facilities.

In the proposed facility, which was designed in 2008–2009, there are two GAC treatment trains. Each train has two contactors that can be operated in series or parallel to provide 20 min of EBCT at 900 gpm. The skid-mounted GAC contactors will be sized and configured to provide operational flexibility by allowing the contactors to work in series or parallel and with different flow rates and thus EBCTs. The GAC treatment system will be installed downstream of an existing air-stripping tower and upstream of the finished-water storage tanks.

Dover Township, N.J. This GAC adsorption facility in Dover Township, N.J. (Case Study No. 12), uses pre-engineered GAC contactors. Treated water is recharged to groundwater. The facilities were placed in service in 1997.

Flowing Wells Water District, Tucson, Ariz. In this project, liquid-phase GAC treatment is used by the Flowing Wells Irrigation District in Tucson (Case Study No. 13) to treat a volatile organic compound (VOC) contamination plume.

Design of the treatment system was completed in 2005, and the facilities were placed in service in December 2006. Liquid-phase GAC for VOC treatment was integrated with granular iron media adsorption for arsenic treatment. Two 12-ft-diameter liquid-phase GAC vessels operate in series to treat 900 gpm of flow from Wells 70 and 75 prior to arsenic adsorption. GAC treatment is also mitigating radon and DBP formation.

Metro Water South Shannon Wellhead Treatment, Tucson, Ariz. In 2004, a design was completed to replace a multistage diffused bubble aeration system with two liquid-phase, 12-ft diameter GAC contactors operating in series at a flow rate of 750 gpm. The system was placed in service in July 2006, and the results from recent sampling at the South Shannon Well (Case Study No. 14) indicate that the treated water VOC goal was met.

Vapor-Phase Adsorption

Emissions of contaminated air from water treatment technologies can result in air quality issues or permit noncompliance. In instances where emission rates exceed air quality regulations or where health and safety concerns exist, vapor-phase GAC treatment systems may be used to mitigate the release of contaminants into the atmosphere. Examples of water treatment technologies that may require vapor-phase GAC treatment include air stripping and multistage diffused bubble aeration (MSDBA). Air stripping is recognized by the USEPA as a best available technology for removing VOCs (USEPA 1989) from water, and MSDBA is also used for some applications. Consequently, numerous examples exist of using air stripping to remove VOCs from contaminated groundwater sources and using the treated groundwater as a potable supply.

Utility Experience Using Vapor-Phase Adsorption

City of Scottsdale, Ariz., Central Ground Water Treatment Facility (CGTF). The Scottsdale CGTF was originally designed in 1992. It is an air-stripping facility for removal of trichloroethylene (TCE) from groundwater. Water is delivered to the CGTF from four off-site wells. Air from the stripping process is conditioned and treated by vapor-phase GAC. Water from the facility is used for potable purposes and is introduced to the distribution system via the Site 80 reservoir and booster station. The maximum capacity of the plant is 13.6 mgd. The facility operates under USEPA mandate and must maintain a year-round average capacity of 9 mgd.

The plant has three packed column air-stripping towers, and each tower has a dedicated vapor-phase GAC contactor. Each vapor-phase GAC contactor can hold 20,000 lb of GAC, has a GAC bed area of 113 ft^2 and depth of 3 ft, and provides a contact time of 2–3 sec. GAC is replaced annually.

GAC Operation Strategies

Single-stage contactors are often used for small groundwater systems, but if more than one unit is required, more favorable GAC use rates (chapter 1) can be achieved by arranging the units either in series or in parallel, as shown in Figure 4-12. GAC in a single-stage adsorber should be replaced sometime before breakthrough exceeds the treatment objective. At this point, not all of the adsorption capacity of the GAC has been utilized, so the GAC use rate will be higher than if further breakthrough had occurred.

Adsorbers in Series. Columns may be arranged in series so that the mass transfer zone (MTZ) is contained within the lag column after the lead column has been saturated with the influent concentration. When the GAC is replaced in the lead column,

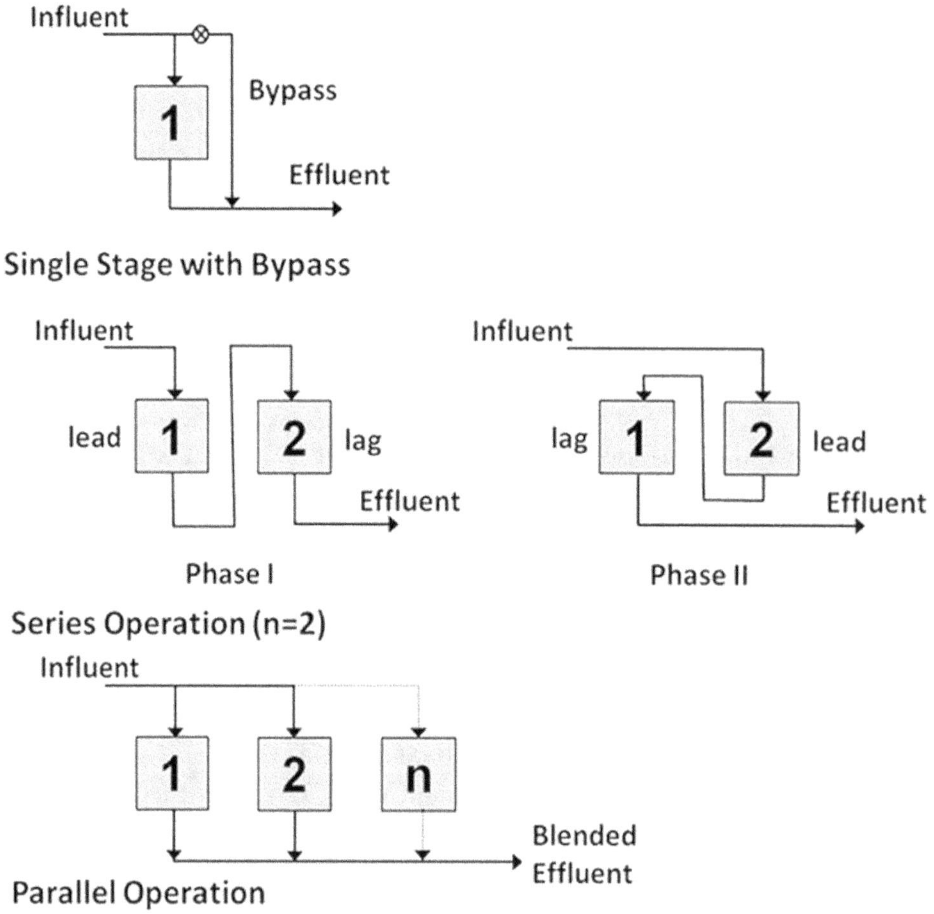

Figure 4-12. Adsorber operation schematics

the flow is redirected so that it goes through the freshest GAC last. In series operation, lower GAC use rates are achieved than with single-stage contactors. Series configuration is best for when the effluent criteria is very low compared with the influent concentration (Wiesner et al. 1987). However, for many SOCs at low concentrations, the adsorption capacity decreases when the GAC is exposed to the background DOM for prolonged time; thus the capacity in the second column in series may be lower than the first (Jarvie et al. 2005). The increased cost of plumbing also counters the cost–benefit of reduced GAC use rate, especially when more than two columns must be used in series.

Analysis of absorbers in series requires single-adsorber breakthrough curve at the total EBCT (lead + lag EBCT). The lead column performance can be estimated by converting the total adsorber breakthrough curve into half, if breakthrough data are not available for both EBCTs. Using the total EBCT to predict performance is preferred over using the lead column EBCT, because the long-term effects of DOM preloading will be better represented in the longer column. The performance of both the lead and total EBCTs is plotted as concentration versus operation time as shown in Figure 4-13. For maximum bed life, the lead column is replaced when the lag column reaches the treatment objective (47 days in Figure 4-13). The mass of adsorbate adsorbed by the lead column is represented by Area A and can be calculated by integrating the area above the breakthrough curve to the influent. The amount of adsorbate allowed in the effluent up to the treatment objective is represented by Area B. Area C represents the additional mass of adsorbate that is adsorbed on the lag column that will continue to be used. Area D represents the adsorption capacity of the GAC that is not used in the lead adsorber. While the EBCT of the lead column can be specified to contain the entire MTZ, often it is not economical to provide so much more EBCT to capture so little mass.

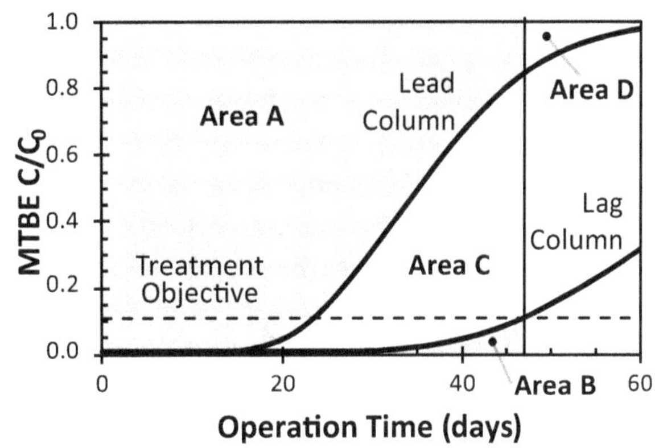

Figure 4-13. Operation of two adsorbers in series for the removal of MTBE

The improvement in performance of adsorbers in series over a single-stage adsorber occurs because the mass of contaminant represented by Areas A and B is removed with only half the mass of the GAC media. The single-stage adsorber would remove the mass of contaminant represented by Areas A, B, and C with replacement of the entire GAC media. Thus, because Area A plus Area B is greater than Area C, the GAC use rate will improve with adsorbers in series.

Adsorbers in Staged Parallel Mode. When GAC adsorbers are operated in staged parallel mode, they can also be used to decrease the GAC use rate as compared with a single-stage contactor (Westrick and Cohen 1976, Roberts and Summers 1982). Because only the composite flow must meet the effluent quality goal and because the effluent from each of the units is blended, each unit can be operated until it is producing water with an effluent concentration in excess of the treated water goal. Dvorak and Maher (1999) showed that the GAC use rate for DOM removal decreased asymptotically with increasing number of adsorbers in parallel with the largest decrease occurring when moving from a single adsorber to two adsorbers in parallel.

Staged parallel performance can be estimated if the single-adsorber breakthrough curve is known. Each unit in parallel provides the full EBCT, but it only treats a portion of the flow. Thus, each unit will exhibit the single-adsorber breakthrough curve, but each adsorber is replaced at different times, offsetting the breakthrough curves. The combined effluent is calculated by averaging the effluent concentration of each adsorber (assuming all are treating an equal amount of the flow). When the adsorber facility is first put into service, all the units are fresh, and overall performance is identical to the single-adsorber breakthrough curve as shown for DOC in Figure 4-14 because all the adsorbers are performing identically. Once the treatment objective is reached (e.g., 1.4 mg/L DOC in Figure 4-14) the first adsorber is replaced, and its breakthrough curve is reset to time = 0. The combined effluent is monitored until the treatment objective is reached again and the next adsorber is reset to time = 0. The cycle repeats until steady-state time to adsorber replacements is achieved.

Figure 4-14 shows analysis results for two, four, and eight adsorbers in parallel to a TOC treatment objective of 1.4 mg/L. In this example, single-adsorber bed life is

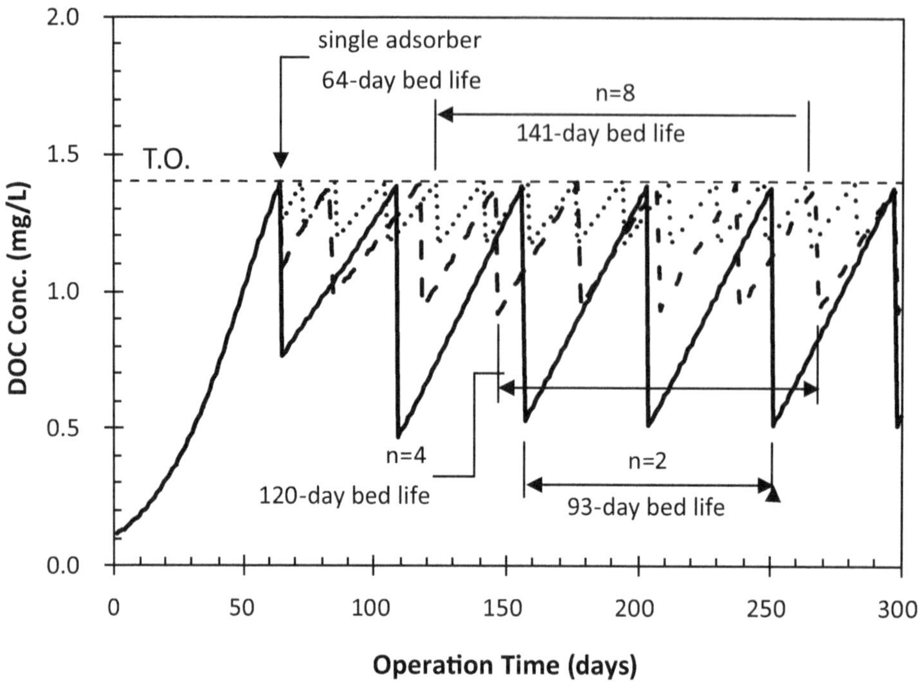

Figure 4-14. Combined adsorber effluent versus time profile for adsorbers operated in staged parallel mode (solid and dashed lines represent the DOC profile when contactor operation is staggered and off-set by various number of days)

64 days. Two adsorbers in staged parallel mode increase the bed life to 93 days, four adsorbers increase bed life to 120 days, and eight adsorbers to 141 days. Going from one to two adsorbers increases bed life by 45 percent, whereas doubling the number of adsorbers again to four increases the bed life by 29 percent, and finally doubling the adsorbers to eight increases bed life by 18 percent. Thus, there will be a point of diminishing returns with adding more adsorbers in staged parallel mode.

Also, the time between bed replacements depends on the number of adsorbers. For example, each of the two adsorbers in parallel has a bed life of 93 days, meaning a bed replacement must be performed every 46 days; whereas each of eight adsorbers have a bed life of 141 days, but a bed replacement must be made about every 18 days. The distribution of labor costs should be included in the economic analysis.

Adding adsorbers in staged parallel mode increased bed life by managing the effluent concentration nearer to the treatment objective, increasing the total mass of contaminant into the distribution system. Such a strategy may not be appropriate for all contaminants.

Denning and Dvorak (2009) used the pore-surface diffusion model to extensively evaluate the performance of lead-lag series operation, blended parallel operation, and bypass operation, which is similar to parallel operation with a column at exhaustion. They found that when high levels of removal were required, $C/C_0 = 0.05$, the lead-lag system was most efficient. For a less stringent criterion, $C/C_0 = 0.25$, the lead-lag series and staged parallel operation yielded similar results.

Bypass Operation. Post-filter GAC contactors in single-stage mode or series mode can also be outfitted with a bypass pipeline. The flow rate in the bypass pipeline varies so the effluent water quality always meets the treatment objective with the maximum bypass flow rate. GAC bed life increases because the dissolved organic carbon in the bypassed flow does not compete for adsorption sites, but like staged parallel operation, the concentration of the contaminant is managed near the treatment objective, which may not be appropriate. Filter adsorbers cannot use bypass lines because all the water must be filtered for turbidity and pathogen removal. Bypassing is not practiced with staged parallel operation because operating a filter after complete breakthrough is essentially the same as bypassing without the need for any untreated flow, and if the compound of interest is biologically degradable, some removal may continue, even after adsorption capacity is exhausted.

Denning and Dvorak (2009) found bypassing to be an effective approach for reducing costs. However, bypassing is generally not encouraged because the controlling treatment objectives are generally the most weakly adsorbing compounds. Thus, bypassing flow releases more strongly adsorbing compounds downstream whereas operating units in staged parallel mode would provide nearly the same level of performance for the target organic while removing more strongly adsorbed compounds. Bypassing flow also reduces the effectiveness of the post-filter contactor in providing redundant filtration for the removal of pathogens and particulates.

Chlorine and GAC. Allowing chlorine residuals in the GAC adsorber is typically avoided. Free chlorine is reduced to chloride in a reaction at the GAC surface. The reaction is very fast and occurs in the first few minutes of EBCT. Continuous exposure to chlorine results in degradation of the GAC structure and will produce more fines and GAC losses over time. The chlorine may also react with organics adsorbed on the GAC surface creating potentially toxic by-products that may desorb into the finished water. In many applications, chlorinated backwash water is used to control undesirable biological growth, while minimizing the exposure of chlorine to the GAC.

Combined chlorine also reacts with GAC in manner similar to free chlorine; however, because the ammonia is not destroyed, free ammonia will be released into the system. Free ammonia may affect downstream chlorination chemistry and possibly result in nitrification in the adsorber.

GAC Performance Evaluation

GAC performance evaluation is undertaken in order to determine whether the treatment objective(s) can be economically achieved with GAC. Ultimately, the GAC use rate determines the economics of adsorption and is considered the figure of merit for GAC performance evaluation. Like most GAC processes, determining the GAC use rate for target organics depends on EBCT, the GAC product, the influent concentration of the target organic, the background water quality, and the treatment objective. Evaluating GAC is typically much more complex than evaluating PAC because PAC operates at steady state and GAC is an unsteady-state process. More complete guidance on GAC evaluation is given in the following chapters organized by treatment objective. Chapter 5 is specific to DOM control, chapter 6 discusses T&O episode control, and chapter 7 presents GAC for target organic contaminant control.

Logistic Model of Breakthrough Curves. Occasionally, it is useful to be able to simulate a breakthrough curve with a simple mathematical model. The logistic curve is a three-parameter model that meets the parameters in Eq 4-1:

$$C(t) = C_0 + \frac{A}{1 + Be^{-Dt}} \qquad \text{(Eq 4-1)}$$

Where

$C(t)$ = the concentration at time t
C_0 = the initial concentration
A, B, and D = logistic curve fit parameters
t = time (can be in days or bed volumes)

The logistic curve fit parameters can be determined using software functions such as the Solver in Microsoft Excel®or the «fit curve» function of SigmaPlot.® The equation allows for a nonadsorbable fraction and can therefore be used to model TOC as well as single solutes. However, the parameter values cannot be used to predict specific contaminant or TOC breakthrough at other influent TOC concentrations because the parameter values change with influent TOC concentration.

The logistic curve can be integrated to quickly estimate the best performance of adsorbers in staged parallel mode (Chowdhury et al. 1996). The integral of Eq 4-1 is the blended curve for an infinite number of treatment units, and it is shown in Eq 4-2.

$$C_{blend} = C_0 + A + \frac{A}{Dt}\ln\left(\frac{1 + Be^{-Dt}}{1 + B}\right) \qquad \text{(Eq 4-2)}$$

where C_{blend} is the blended GAC effluent concentration and the other variables are as defined in Eq 4-1.

Comparison of Gravity GAC Treatment Technologies. All four types of gravity GAC treatment systems (post-filter contactors, filter adsorbers, GAC caps, and biologically enhanced activated carbon) are typically installed in concrete box–style structures that contain a certain quantity of GAC or a combination of other media and GAC. During normal operation, water enters the structures via channels located at the upstream end of the GAC bed. Hydraulic head prior to the GAC unit process drives flow through the media. Treated water exits at the downstream end of the GAC bed to enter the treated water piping.

Gravity GAC treatment systems should be backwashed periodically to remove particles that have accumulated within the bed, reduce the head loss through the media (filter adsorbers, BAC, GAC caps), or alleviate compaction of the media (post-filter contactors). During backwashing, treated water is pumped back through the underdrains and is collected in the troughs. The backwash water is typically conveyed to a backwash equalization basin where it can be recycled to the head of the plant or conveyed to the solids handling process.

Table 4-3 provides a brief comparison for some characteristics of the gravity GAC technologies. Variations will occur, but the information below summarizes typical approaches to each technology. Further details are included in the remainder of part 2 of this book.

Table 4-3. Comparison of gravity GAC technologies

	Post-filter contactors	Filter adsorbers	GAC caps	Biologically enhanced activated carbon
Primary treatment mechanism	Adsorption	Adsorption	Adsorption	Biological treatment and some adsorption
Relative frequency* of GAC replacement	High	High	Low	Low
GAC removal considered in design	Present	Present	Absent	Absent
Preoxidation typically used?	No	No	Varies	Yes – often ozone

* The frequency of replacement is a strong function of organic loading; the information in this row reflects frequency under similar organic loading conditions.

Regulatory Guidance on GAC Design

Ten States Standards (GLUMRB 2007) refer to GAC treatment technologies in two places. The first is in chapter 4, "Filtration": GAC is one of the media types allowed for filtration under paragraph 4.2.1.6. Subparagraph d.4 specifically states the following about GAC:

> "4. Granular activated carbon (GAC) Granular activated carbon as a single media may be considered for filtration only after pilot or full-scale testing and with prior approval of the reviewing authority. The design shall include the following:
>
> a. The media must meet the basic specifications for filter media as given in Section 4.2.1.6.a through c.
>
> b. There must be provisions for a free chlorine residual and adequate contact time in the water following the filters and prior to distribution (See 4.3.2.d and 4.3.3).
>
> c. There must be means for periodic treatment of filter material for control of bacterial and other growth.
>
> d. Provisions must be made for frequent replacement or regeneration."

Paragraphs 4.2.1.6.a through c state that,

> "The media shall be clean silica sand or other natural or synthetic media free from detrimental chemical or bacterial contaminants, approved by the reviewing authority, and having the following characteristics:
>
> a. A total depth of not less than 24 inches and generally not more than 30 inches,
>
> b. A uniformity coefficient of the smallest material not greater than 1.65,
>
> c. A minimum of 12 inches of media with an effective size range no greater than 0.45 mm to 0.55 mm."

The second place GAC is mentioned in Ten States Standards is in Section 4.9, "Taste and Odor Control." Subparagraph 5 states the following about GAC:

> "Replacement of anthracite with GAC may be considered as a control measure for geosmin and methyl isoborneol (MIB) T&Os from algae blooms. Demonstration studies may be required by the reviewing authority." (GLUMRB 2007)

Review of Ten States Standards yields conditions of GAC use for filtration and T&O removal. Specific standards are provided in the absence of other information, and the standards are flexible enough that testing data can be provided to allow variations from the guidance provided.

Regulatory Guidance for Pressure Contactors. Guidance for pressure contactors will be specific to each regulatory agency. The state of California policy provides an example of how GAC pressure systems are regulated when used to remediate contaminated groundwater sources for potable use. California public drinking water systems are regulated by the Department of Health and Safety (DHS) under Element 4 of the DHS 97-005 Policy. This policy states that the DHS will evaluate the effectiveness and reliability of treatment based on the following items:

- Use of best available technology (BAT).
- Incorporation of reliability features consistent with the risks posed by the type and degree of contamination.
- Provision of optimized treatment to the lowest concentrations feasible at all times.
- Treatment of the entire flow from an impaired source.
- Provision of multibarrier treatment, if warranted.

GAC is considered the BAT for many organics, but treatment selection should be confirmed with the DHS prior to beginning detailed design. The treatment process must include reliability features that are consistent with the type and degree of contamination (California DHS 1997). For example, the use of multiple vessels in series would increase reliability by providing intermediate water quality sampling points between vessels. Further evaluation is required to determine the appropriate reliability features for the treatment trains under consideration.

Also, the GAC facility must treat the entire flow as the DHS does not allow bypassing a portion of impaired water, in conjunction with blending, as a means of water quality compliance under Policy 97-005 guidelines.

References

Anderson, M.C., R.C. Butler, F.J. Holdren, and B.H. Kornegay. 1981. Controlling trihalomethanes with powdered activated carbons. *Journal AWWA* 73(8):432-439.

Brady, R.D. 2005. Activated carbon process. In *Water treatment plant design*. 4th ed. New York: McGraw-Hill, American Water Works Association, and American Society of Civil Engineers.

California Department of Health and Safety (DHS). 1997. Policy guideline for direct domestic use of extremely impaired sources. DHS Policy Memo 97-005.

Chowdhury, Z.K., G. Solarik, D.M. Owens, S.M. Hooper, and R.S. Summers. 1996. NOM removal by GAC adsorption: implications of blending. In *Proc. Amer. Water Works Assoc. Conference, Toronto, Ont., Canada*. Denver, Colo.: AWWA.

Chowdhury, Z., A. Traviglia, J. Carter, T. Brown, R.S. Summers, C. Corwin, T. Zearley, M. Thurman, I. Ferrara, J. Olson, R. Thacker, and P. Barron. 2010. *Cost-effective regulatory compliance with GAC biofilters*. Denver, Colo.: Water Research Foundation.

Denning, P.C., and B.I. Dvorak. 2009. Maximizing sorbent life: comparison of columns in parallel, lead-lag series, and with bypass blending. *Water Environment Research* 81(2):206–216.

Dvorak, B.I., and M.K. Maher. 1999. GAC contactor design for NOM removal: implications of EBCT and blending. *Journal of Environmental Engineering* ASCE 81(2):206.

Great Lakes–Upper Mississippi River Board of State and Provincial Public Health and Environmental Managers (GLUMRB). 2007. *Recommended standards for water works (known as Ten States Standards)*. Albany, N.Y.: Health Education Services.

Graham, M., and R.S. Summers. 1996. The role of floc formation and presence on adsorption by powdered activated carbon. In *Proc. of the AWWA Annual Conference and Exposition, Toronto, Ont., Canada.* Denver, Colo.: AWWA.

Graham, M.R., I. Najm, M. Simpson, B. MacLeod, R.S. Summers, and L. Cummings. 2000. *Optimization of powdered activated carbon application for geosmin and MIB removal.* AWWA RF report 90782. Denver, Colo.: Water Research Foundation.

Ho, L., and G. Newcombe. 2005. Effect of NOM, turbidity and floc size on the PAC adsorption of MIB during alum coagulation. *Wat. Res.* 39(15):3668–3674.

Hozalski, R.M., S. Goel, and E.J. Bouwer. 1995. TOC removal in biological filters. [87(12):45–54.] *Journal AWWA*

Jarvie, M.E., D.W. Hand, S. Bhuvendralingam, et al. 2005. Simulating the performance of fixed bed granular activated carbon adsorbers: removal of synthetic organic chemicals in the presence of background organic matter. *Wat. Res.* 39, 2407–2421.

Kawamura, S. 2000. *Integrated design and operation of water treatment facilities.* 2nd ed. New York: Wiley.

Lalezary, S., M. Pirbazari, and M.J. McGuire. 1986. Evaluating activated carbons for removing low concentrations of taste- and odor-producing organics. *Journal AWWA* 78(11):76–82.

McCarty, P.L., D. Argo, and M. Reinhard. 1979. Operational experiences with activated carbon adsorbers at Water Factory 21. *Journal AWWA* 71(11):683–689.

Najm, I.N., V.L. Snoeyink, B.W. Lykins Jr., and J.Q. Adams. 1991. Using powdered activated carbon: a critical review. *Journal AWWA* 83(1):65–76.

Phoenix, City of, Water Services Department (Phoenix). 2008. *Water quality master plan update amendment, Vol. 7: GAC road map.* Prepared for the City of Phoenix by Carollo Engineers, Malcolm Pirnie, and Narasimhan Consulting Services. http://www.westcas.com/PDF/WESTCAS-October_28_2011-Hayes_final.pdf. 25–31.

Roberts, P.V., and R.S. Summers. 1982. Granular activated carbon performance for organic carbon removal. *Journal AWWA* 74:113–118.

Servais, P.,M. Prevost, ic matter in drinking water treatment. In *Biodegradable organic matter in drinking water treatment and distribution.*, M. Prevost, P. Laurent, P. Servais, and J.C. Joret, eds. Denver, Colo.: AWWA.

Ten States Standards (see GLUMRB 2007).

USEPA. 1989. *Volatile organic chemicals: are VOCs in your drinking water?* 1989. EPA 5709-89-EEE. USEPA Office of Water.

Wang, J.Z., R.S. Summers, and R.J. Miltner, 1995. Biofiltration performance: part I, Relationship to biomass. *Journal AWWA* 87(12):55–63.

Westerhoff, G., and R. Miller. 1986. Design of the GAC treatment facility at Cincinnati. *Journal AWWA* 78 (4):147–156.

Westrick, J.J., and J.M. Cohen 1976. Comparative effects of chemical pretreatment on carbon adsorption. *Journal Water Pollution Control Federation* 48:323–338.

Wiesner, M.R., J.J. Rook, and F. Fiessinger. 1987. Optimizing the placement of GAC filtration units. *Journal AWWA* 79(12):39–49.

Chapter 5

Dissolved Organic Matter Control

The Disinfectants and Disinfection By-Products (D/DBP) Rule (stages 1 and 2) has provided the impetus for many surface water plants to consider activated carbon as a DBP control strategy. Generally, approaches to reduce DBP formation are preferred over removing DBPs after their formation because of concerns over presently unregulated DBPs that may be formed and not removed during subsequent treatment. In addition, precursors (e.g., DOM) are more adsorbable than DBPs; therefore, if adsorption by activated carbon occurs before chlorination, the DBP formation potential of the water will decrease (Summers et al. 2010).

The use of PAC is typically limited to controlling seasonal DOM peaks. GAC may be used continuously or seasonally with a bypass to optimize use rate. This chapter discusses the use of PAC and GAC for the control of DOM.

DOM Characteristics

DOM is a complex mixture of ill-defined and specific organic compounds of both natural and synthetic origin. In most cases, DOM from natural sources dominates, but when source waters are impacted by upstream human activity, wastewater treatment plant effluent organic matter (EfOM) can become important. EfOM includes background DOM, compounds from human activity, and metabolites from biological wastewater treatment. Building blocks of natural DOM primarily include fulvic acids, humic acids, hydrophilic acids, and carbohydrates. DOM is widely present in natural waters and is derived primarily from decaying vegetative and animal material, as well as from soil. The type and nature of DOM vary widely, which makes it difficult to effectively remove from water with a single-unit process. For example, larger-molecular-weight DOM is effectively removed by coagulation, while oxidation can be effective in mineralizing (to a certain extent) some of the lower-molecular-weight DOM fractions. The heterogeneous nature of both DOM and activated carbon also affects the DOM adsorptivity. Large-molecular-weight DOM can be excluded from much of the internal surface area of activated carbon, while polar DOM is soluble and not attracted to the activated carbon surface. Even with these limitations, activated carbon can remove more than 80 percent of the DOM; however, the activated carbon use rate required to do so is likely cost-prohibitive. Depending on the influent conditions, most notably influent DOC (or TOC) concentration, activated carbon can effectively remove 30 to 60 percent of DOM that remains after coagulation.

PAC for Removal of Dissolved Organic Matter

PAC Applications

Najm et al. (1991) provided a comprehensive review of the PAC adsorption literature and concluded that although one of the main advantages of PAC is low capital cost, relatively high PAC doses are required to achieve substantial removal of DOM and other DBP precursors. Therefore, the applicability of PAC for organics control is generally limited to plants where only seasonal removal of DOM is needed or where a low percentage of DOM removal is necessary.

PAC Performance Evaluation

Evaluating PAC performance consists of determining the required PAC dose to meet the stated treatment objective. This section discusses the issues related to these PAC evaluations.

Determination of Required Dose

The required PAC dose needed to meet the treatment objective depends on contact time and mixing conditions, the PAC product, the influent DOM, and the DOM treatment objective. These variables are considered in more detail in the following sections.

Contact Time and Mixing. An important factor that affects the removal of DOM by PAC is the time that the DOM-containing water stays in contact with the PAC particle. In a quiescent basin, such as a presedimentation or sedimentation basin, the PAC readily settles out and is no longer effective as an adsorbent. While the contact time should be maximized for best performance, mixing or agitation should also be provided to keep the PAC in suspension. Often, PAC is added at rapid mix, stays in contact with the water through flocculation, and is removed in the sedimentation basin.

PAC Product and Dose. The applied PAC product and dose have perhaps the largest effect on DOC removal. Figure 5-1a shows DOC removal for three PAC products over a range of PAC doses with a constant coagulant dose. The three products (made of different base materials) each perform differently, and the higher applied doses result in more DOC removal. However, under practical dose limitations, less than 40 percent of the total DOC is removed with PAC, whereas most target organic contaminants can be completely removed with reasonable PAC doses. Figure 5-1b also shows DOM removal regardless of PAC type and includes the effect of contact time. The data in Figure 5-1b indicate that most of the DOC removal takes place in the first 60 min. Because PAC and coagulation are capable of removing some of the same DOM fraction, the combined removal from PAC and coagulation is less than PAC alone plus coagulation alone.

Influent DOC Concentration and Treatment Objective. The influent concentration and treatment objective determine the required removal. As shown in Figure 5-1, the DOC removal attainable with PAC at economic doses is limited.

Under certain circumstances, up to 50 percent DOC removal may be attainable with PAC alone; however, removals of less than 30 percent are more common without the use of coagulants.

PAC Evaluation Methods

PAC is typically evaluated with a combination of bench-scale tests. Model calculations may also be applicable during the feasibility phase.

Jar Testing. Jar testing is an efficient method for evaluation of PAC performance for adsorption of DOC. Mixing, contact time, and addition of other chemicals should match the actual process to the maximum extent possible. A dose-response curve is generated by applying a range of PAC doses to the jars (Graham et al. 2000).

SDS-DBP Testing. For most applications where DOC is to be controlled, the actual treatment objective is to control DBP formation. DBP formation is typically measured by a bottle point method called the *simulated distribution system* (SDS) DBP formation test. Source water is collected and treated with one or several methods, or treated water is directly sampled, and then the water is chlorinated under conditions similar to the distribution system (hold time, temperature, pH, and chlorine dose or target residual). After the hold time is reached, the residual chlorine is measured and quenched, and the DBP concentrations are measured. Besides DOC concentration, DBP formation depends on pH and bromide concentration, so these conditions should be carefully controlled (Summers et al. 1996a).

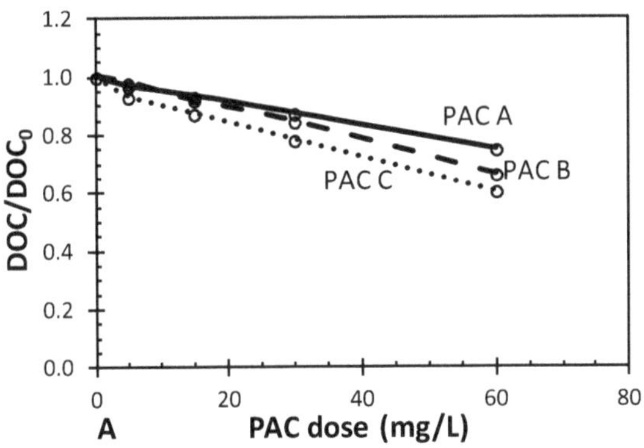

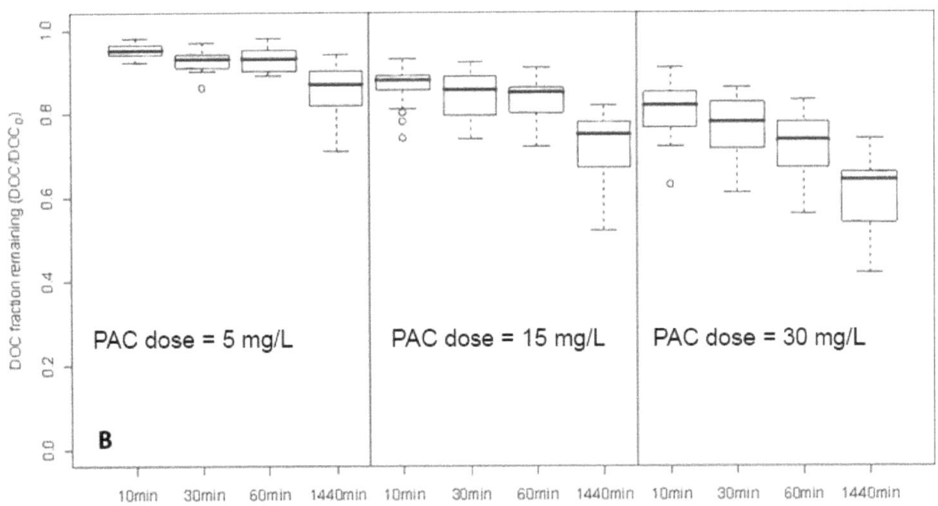

Figure 5-1. (a) Removal of DOC by three PAC products and (b) boxplot of DOC fraction remaining with time regardless of PAC type (total n = 223)

Table 5-1. Fitting parameters for the PAC removal of DOC (n = 173) (Cho and Summers 2007)

PAC base material	β_1	β_2	β_3	β_4	R^2
Bituminous (WPH-Calgon)	0.1561	0.9114	−0.0263	−0.002	0.977
Lignite (HDB- Norit)	0.4078	0.8516	−0.0225	−0.002	0.977
Wood (CASP- Norit)	0.3653	0.8692	−0.0151	−0.0025	0.988

Model Predictions. Cho and Summers (2007) proposed empirical models for DOM adsorption by three PAC types: bituminous, lignite, and wood. The models were developed from the results of five waters (n = 173) with DOC_0 1.3–5.4 mg/L, without coagulation for PAC doses between 5 and 30 mg/L and contact times between 10 and 60 min. The models are based on the form shown in Eq 5-1:

$$DOC = ß_1 + ß_2 \times DOC_0 + ß_3 \times dose + ß_4 \times time \qquad \text{(Eq 5-1)}$$

Where

DOC = DOC concentration after PAC adsorption (mg/L)
DOC_0 = DOC concentration of influent water (mg/L)
dose = PAC dose (mg/L)
time = contact time (min)
$ß_1$ through $ß_4$ = fitting parameters

The values of $ß_1$ through $ß_4$ are presented in Table 5-1 for the three representative PACs. PAC adsorption is not linear with respect to dose and time as modeled in Eq 5-1; however, for DOC over the stated ranges, the models provided a good approximation of observed performance in the waters tested. The model was verified with the results from three other waters (n = 80) with DOC_0 of 2.4 to 3.8 mg/L with an average R^2 of 0.91.

GAC for Removal of Dissolved Organic Matter

GAC Applications

GAC is generally applied as a filter adsorber or post-filter contactor for the adsorption of DOM. GAC caps that rely mainly on adsorption are generally avoided as the EBCTs and resulting adsorption capacity are too low and would require frequent replacements. BAC filters, which rely on biological removal mechanisms rather than adsorption, may remove 10 to 15 percent of the TOC in conventional surface waters and up to 25 percent in ozonated waters on a steady-state basis. Generally, GAC in the adsorption dominated mode is used to meet D/DBP regulations when TOC removals of 20 to 50 percent are needed, which is best achieved with post-filter contactors.

In order to reduce GAC operating expense, some utilities may consider bypassing flow around GAC for a portion of the year when the source water TOC is low. However, this reduces some of the advantages of using GAC, such as multibarrier filtration, reduction of objectionable T&O, adsorption or attenuation of target organic contaminants, and biological removal of additional DOC. Another approach to balance use rate and maintain the advantages of GAC is to schedule GAC replacements before the highest source water TOC of the year. The fresh GAC will remove substantial portions of TOC when concentrations are highest, but as the adsorption capacity of the media is reduced, the GAC can still provide biological removal of TOC and continue to adsorb organic contaminants, remove T&O, and provide extra protection against particulates and pathogens (Corwin and Summers 2012).

GAC Performance Evaluation

Typical DOM breakthrough behavior was introduced in chapter 3, and now it will be examined in further detail. Figure 5-2 shows a typical DOM breakthrough curve, as measured by DOC, with some traits that are commonly observed.

First, there will generally be immediate breakthrough of some DOC, which is the nonadsorbable fraction. The nonadsorbable fraction of DOM is generally composed of low-molecular-weight hydrophilic molecules with a high negative-charge density; this fraction accounts for 10 to 20 percent of the total DOC. After a short time of a constant DOC effluent, DOC breakthrough begins. The first compounds to break through are either very weakly adsorbed compounds or very large compounds that do not have access to the majority of the internal adsorption surface area because of size exclusion. Breakthrough continues as the adsorption capacity is consumed; the continuum of weakly to strongly adsorbed compounds starts to appear in the effluent. A significant change in slope of the effluent DOC curve is often observed around 60 to 75 percent breakthrough. At this point, only relatively strongly adsorbed compounds are still being removed from the influent. The change in slope of the breakthrough curve signals a change in transport, or mass transfer resistance. Thus, the fraction being removed is strongly adsorbed but slow-diffusing material.

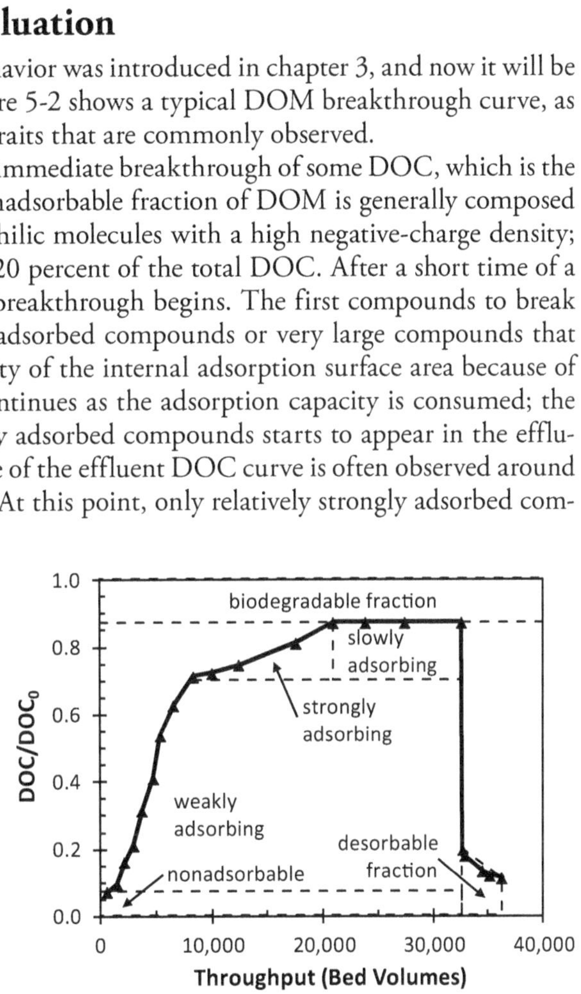

Figure 5-2. Typical DOM breakthrough behavior

The strongly adsorbed phase is quickly followed by the establishment of pseudo-steady state, where DOC removal is nearly constant and may be difficult to distinguish from the strongly adsorbed phase. In this near steady state, the primary removal mechanism is biodegradation, with a small amount of adsorption still occurring. Most anthropogenic organic contaminants are still being removed to various degrees after DOC breakthrough has reached steady state. Finally, if the influent DOC concentration falls to zero, only a small portion of the adsorbed DOC will desorb. In real systems, the DOC will likely not fluctuate enough to ever observe any desorption. It is important to note that nearly all DOM fractions react with chlorine to form some level of DBPs, regardless of molecular weight or other separations based on physiochemical properties (Sontheimer et al. 1988).

When ultraviolet absorbance (UVA) at 254 nm is used as a surrogate for DOM breakthrough instead of DOC, several differences are normally found as the UV-adsorbing fraction of DOM is more strongly adsorbed. First, usually all of the UV-adsorbing DOM is adsorbed, thus no immediate breakthrough occurs. The breakthrough curve shifts to the right indicating later breakthrough of this DOM fraction. The biodegradable fraction of DOM is not well detected by UVA, so that fraction at the end of the breakthrough is small.

Determination of the Use Rate

In this section, the factors affecting GAC breakthrough and use rate for organics control are explored.

EBCT. Theoretically, EBCT does not affect breakthrough of a single adsorbate on a throughput basis. However, an increase in EBCT generally decreases use rate for DOC systems. Figure 5-3 shows a generalization of EBCT behavior. The largest benefits to the use rate by increasing EBCT are observed at low EBCTs. At longer EBCTs, the breakthrough curves and use rate converge until little effect is observed. The cause for the behavior in Figure 5-3 is uncertain but can be explained by how equilibrium is defined. As shown in chapter 1, equilibrium is established after some minimum amount of time. Thus short EBCTs may not have enough residence time to come to the same level of equilibrium as in longer EBCT columns. The effect is most readily observed in DOM because the wide distribution of molecular weights and adsorbabilities accentuates the time needed to reach equilibrium (short to very long mass transfer zones).

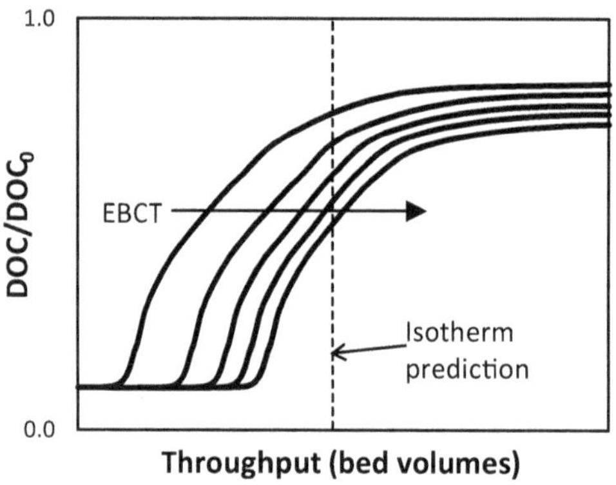

Figure 5-3. Effect of EBCT on DOC breakthrough

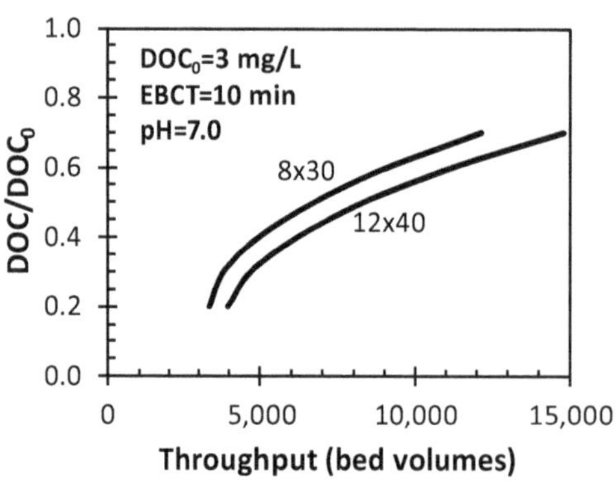

Figure 5-4. Effect of GAC media size on DOC breakthrough

From a practical perspective, longer EBCTs mean more efficient use of the GAC and longer bed life for DOM removal.

GAC Product Selection. As with PAC above, different GAC products may perform differently for DOC removal. The economical choice is often the most available product that is efficient to reactivate. The rapid small-scale column test (RSSCT) is often the best way to evaluate different GAC products under site-specific conditions to determine economics and verify manufacturer's claims (discussed later in this chapter).

GAC performance also depends on media size. Figure 5-4 shows the predicted performance for a 12×40 bituminous GAC and a larger 8×30 product (Zachman and Summers 2010). The smaller GAC media outperform the larger product on a mass of GAC (use rate) basis. The increase in adsorption capacity cannot be explained by the improved adsorption kinetics of the smaller media; however, one explanation for the improved performance in the smaller media is that the activation process can more easily create internal pore structure throughout the entire GAC particle. While smaller GAC particle sizes perform better both in adsorption and particle filtration, smaller GAC particle sizes also result in more clean-bed head loss and faster accumulation of head loss in the filter, thereby limiting the minimum size of the media used.

Influent DOC Concentration. DOC influent concentration (DOC_0) is one of the most important factors affecting DOC breakthrough. Figure 5-5 shows the sensitivity to DOC_0 with breakthrough curves from Zachman and Summers (2010) model predictions, which are presented later in this chapter. Figure 5-5a shows the data on a normalized DOC basis, and Figure 5-5b shows the same data on a DOC concentration basis. The normalized data in Figure 5-5a indicate that the higher the DOC_0, the earlier the breakthrough. In terms of the isotherm, the higher DOC_0 concentration drives a larger concentration to the surface phase; however, this is counteracted by the higher mass loading rate (flux) on the column.

For DBP control, the DOC treatment objective is generally a target DOC concentration. Figure 5-5b shows the wide performance differences between the systems in this case. If a DOC treatment objective is 1.5 mg/L, the DOC_0 of 4 mg/L will have a bed life of about 5,000 BV, the DOC_0 of 3 mg/L will be 8,000 BV, and the DOC_0 of 2 mg/L may last indefinitely. It would not be uncommon for a surface water DOC_0 to vary from 2 to 4 mg/L throughout an annual cycle. The performance difference illustrates the importance of (1) characterizing the DOC_0 annual cycle, (2) carefully choosing a representative DOC_0 for bench-scale testing, and (3) pretreatment at the full scale to reduce the DOC_0 that is loaded onto the GAC.

Background Water Quality. In the case of DOM removal, the goal is to remove the aggregate background organic matter. The influence of DOC_0 was discussed

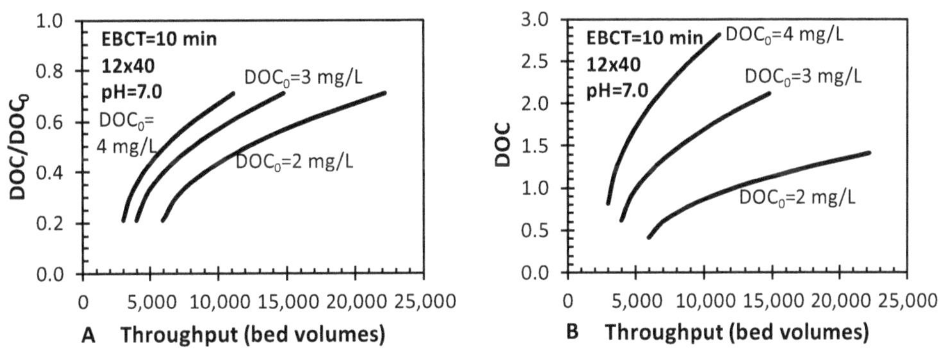

Figure 5-5. Sensitivity of DOC to influent concentration on a normalized DOC basis (a) and a concentration basis (b)

above. However, other factors affect the adsorption of DOC. Primarily, pH determines the charge density of the DOM and affects adsorbability (Summers et al. 2010). As discussed previously, bromide concentration also affects DBP formation. Finally, temperature can affect biotreatment efficiency but generally has little effect on adsorption mechanisms.

Charge density affects adsorption of DOM in two ways. First, a high negative-charge density affects the affinity of the DOM for the GAC surface. Even for GACs with a positive surface charge, the surface quickly becomes negatively charged as DOM accumulates on the surface. The like negative charges repel each other as DOM accumulates on the GAC surface. Secondly, the charge density of DOM affects the liquid diffusion coefficient of the DOM molecule. Molecules with high charge densities have multiple sites that repel each other, and the molecule will spread out to its maximum dimensions, resulting in slower diffusion (Cornel et al. 1986). Alternatively, as the pH shifts near the point-of-zero-charge (pzc) of the molecule, the molecule may coil up to minimum dimensions, allowing faster diffusion.

Figure 5-6 shows the sensitivity of DOC breakthrough to pH from Zachman and Summers (2010) model predictions and shows longer bed life at lower pH. DOM is also preferentially removed at lower pH during coagulation, which may provide impetus for lowering pH for more effective coagulation, GAC adsorption, and chlorine disinfection prior to raising the pH for distribution.

Treatment Objective. Generally, the goal of DOM removal is to control the DBP precursors before chlorination. Therefore, the DOC treatment objective is based on the DBP formation under conditions that simulate those of the distribution system (i.e., SDS testing). Figure 5-7 presents a typical SDS-DBP formation plot after GAC treatment. Figure 5-7 plots the post-GAC DOC concentration on the x-axis versus the TTHM and HAA5 SDS formation on the y-axis. The first behavior to note is that the TTHM and HAA5 formation is linear with DOC, a typical observation, in the DBP concentration range of concern. At the very low DOC concentrations representative of the nonadsorbable fraction, the reactivity to form DBPs is lower. The linear behavior allows the use of DOC as a surrogate for DBP control. Also, the slope of the TTHM formation relationship is higher than that for HAA5. While this is a common observation, it is not universal and is highly dependent on the pH and bromide concentration of the water. The DOC treatment objective, as shown in Figure 5-7, is defined by the DOC concentration where the DBP MCL is exceeded. Some utilities will choose to

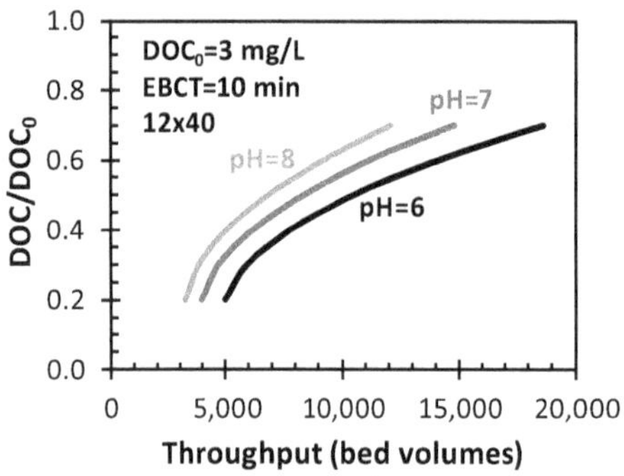

Figure 5-6. Influence of pH on breakthrough of DOM

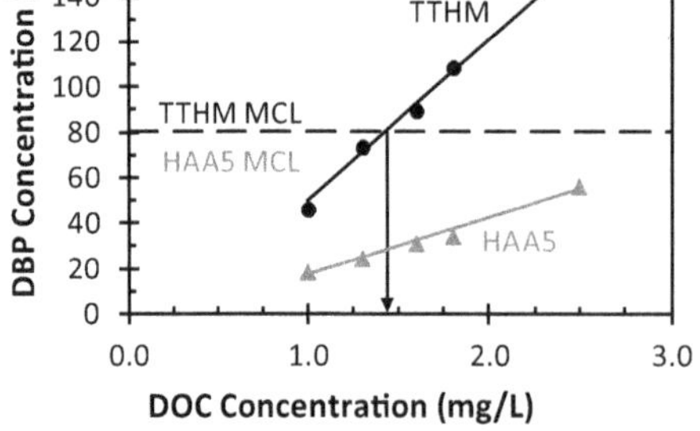

Figure 5-7. Typical post-GAC DBP formation results

apply a factor of safety to the MCL, while others may use the MCL because the regulation is based on an annual running average and they may be well under the MCL for two or three quarters. For the example illustrated in Figure 5-7, the TTHM MCL will be met at a DOC of 1.4 mg/L, and the HAA5 MCL will be met around 2.7 mg/L, higher than the DOC_0 of 2.5 mg/L.

Desorption. Summers and Roberts (1988) showed about 10 percent of the adsorbed DOM to be desorbable. The nondesorbable/nondisplaceable nature of the DOM results in two practical phenomena. First, because desorption is not of concern, chromatographic effects of having higher concentrations of DOM in the effluent over the influent are not of concern. Second, the nondesorbable/nondisplaceable nature of the DOM requires thermal reactivation of the GAC media instead of less intensive regeneration methods.

GAC Evaluation Methods

Methods for evaluating GAC performance for DOM removal are presented in this section. They include pilot-scale columns, bench-sale tests, and predictive models; each has its own strengths and weaknesses.

Pilot-Scale Testing. Pilot testing is still the most reliable way of predicting actual GAC performance. Long-term pilot testing can capture the day-to-day and season-to-season variation in DOC_0 concentration, background DOM composition, and other water quality parameters that can affect adsorption and pretreatment effectiveness. Long-term pilot tests can also capture the seasonal potential for biodegradation of DOM. However, pilot testing takes long periods of time and can be expensive. The duration of pilot tests may lead to a rush in design and startup of the pilot study, which can lead to an unnecessarily large (and expensive) pilot scope or a pilot scope that quickly becomes outdated and inadequate as data are obtained. Pilot testing is subject to changes in source water quality, which is representative of the actual system. However, the fluctuations in source water quality may make analysis of the data difficult to optimize the design. Bench-scale testing with stable influent water quality may be more efficient for optimizing facility design, e.g. GAC selection and EBCT evaluation. Bench-scale testing can then be followed by an efficient pilot program for proof of performance and determining operation and maintenance needs.

Bench-Scale Testing. Bench-scale testing is particularly effective for predicting full-scale adsorption for DOC. The proportional diffusivity RSSCT (PD-RSSCT) has been shown to match adsorption kinetics and capacities of DOC breakthrough profiles for many waters (Crittenden et al. 1991; Summers et al. 1992, 1996b, 1998). PD-RSSCT assumes the intraparticle diffusivity of the DOM changes proportionally to GAC particle size. Any nonconstant diffusivity design, like the PD-RSSCT, prevents satisfying all the scaling equations of the dispersed flow pore and surface diffusion model. Therefore, use of the PD-RSSCT design requires that internal mass transfer, or intraparticle diffusion, control the uptake rate of the adsorbate. DOM has a wide molecular weight distribution, but even the low end is large enough (~500 Da) to ensure that film mass transfer can be neglected and the PD-RSSCT can be used. The effect on the practical design of the PD-RSSCT is that the filter velocity in the small column can be any convenient value that does not result in film mass transfer control. Thus, the filter velocity can be reduced to the practical minimum so the least amount of water is needed as long as the Reynolds number in the small column is greater than or equal to 0.5. Minimizing the filter velocity also shortens the PD-RSSCT column length, which reduces head loss and other operational difficulties.

To determine the EBCT in the small column, Eq 5-2 is used to ensure similitude:

$$EBCT_{SC} = EBCT_{LC}\left|\frac{d_{p,SC}}{d_{p,LC}}\right| = \frac{t_{SC}}{t_{LC}} \qquad \text{(Eq 5-2)}$$

Where

EBCT = empty bed contact time, min
d_p = log-mean GAC particle diameter, mm
t = operation time
SC and LC subscripts refer to small (RSSCT) and large columns, respectively

The small-column EBCT can be converted to a mass of crushed GAC needed for the RSSCT with the apparent density. Alternatively, the small column can be filled to the specified bed volume (media depth). In either method, the density of the crushed media must be known, but in the latter method the difficult task of loading all of the measured media without spillage is avoided.

To ensure film mass transfer does not become significant in the small column, Eq 5-3 is used to check the minimum filter velocity:

$$v_f \geq \frac{0.5\,\varepsilon \times \upsilon}{d_{p,SC}} \qquad \text{(Eq 5-3)}$$

Where

v_f = filter velocity
ε = bed porosity
υ = kinematic viscosity
$d_{p,sc}$ = long-mean GAC particle diameter for small column, mm

Model Predictions. The Zachman and Summers (2010) model is useful in predicting GAC breakthrough curves of coagulated waters in the feasibility phase as a function of DOC_0, pH, EBCT, and bituminous GAC size. Once site-specific breakthrough curves are generated, the model can be used to compare the adsorbability of DOM in a specific water with that in an "average" water.

The Zachman and Summers model is an empirical model of DOC breakthrough developed from 221 RSSCTs in 35 different waters and externally validated by 55 RSSCTs in 14 different waters and 24 field-scale tests in 12 different waters. The database was heavily weighted with coagulated surface waters, and experience shows the models fit these best. Independent variables consist of the DOC influent concentration (1.5 to 11.5 mg/L) and the pH (6.2 to 10). Four models were developed to cover two bituminous GAC sizes (12×40 and 8×30) and two EBCTs (10 and 20 min).

The form of the Zachman and Summers model is shown in Eq 5-4:

$$BV_x = A \times TOC_0^{-1} \times pH^{-1.5} \qquad \text{(Eq 5-4)}$$

Where

BV_x = bed volumes to x = 20, 30, 40, 50, 60, and 70 percent breakthrough
A = fitting parameter from Table 5-2
TOC_0 = influent TOC or DOC concentration in mg/L

Table 5-2. Zachman and Summers (2010) models

Bituminous GAC mesh size	EBCT (min)	Model (x = 20, 30, 40, 50, 60, 70% breakthrough)
12 × 40	10	$A = 196x^2 - 5{,}589x + 252{,}992$
12 × 40	20	$A = 164x^2 - 1{,}938x + 245{,}064$
8 × 30	10	$A = 178x^2 - 6{,}208x + 238{,}321$
8 × 30	20	$A = 202x^2 - 5{,}995x + 261{,}914$

The model is only valid from C/C_0 of 0.2 to 0.7; lower values (below 0.2) depend on the nonadsorbable fraction, and higher values (above 0.7) are strongly affected by biodegradation. Care should be used in extrapolating the model for other EBCTs as the dependence of breakthrough on EBCT is generally not linear. This model was used to generate Figures 5-4, 5-5, and 5-6.

EXAMPLE 5-1: A utility needs help meeting DBP regulations. They have an influent DOC of 3.6 mg/L and currently coagulate the DOC down to 2.5 mg/L. The utility wishes to investigate activated carbon to achieve the treatment objective DOC of 1.7 mg/L needed to meet regulations. They have asked you to perform a desktop evaluation of the following scenarios:

A. A post-filter contactor, 20-min EBCT, 12×40 bituminous GAC

B. A filter adsorber, 10-min EBCT, 8×30 bituminous GAC

C. Bituminous PAC with a 45-min contact time

Assume the addition of coagulant at the same time as the PAC will result in 20 percent additional removal of DOC over the Cho and Summers model (because PAC and coagulation have overlap of DOC that each remove, increasing overall removal, but less than each separately added together). Assume pH = 7 and the bituminous GACs have a density of ρ_{GAC} = 0.48 g/cm^3.

SOLUTION: For the two GAC scenarios, use the Zachman and Summers model to predict the throughput to 1.7 mg/L from the coagulated water DOC of 2.5 mg/L. Results are shown in Figure 5-8.

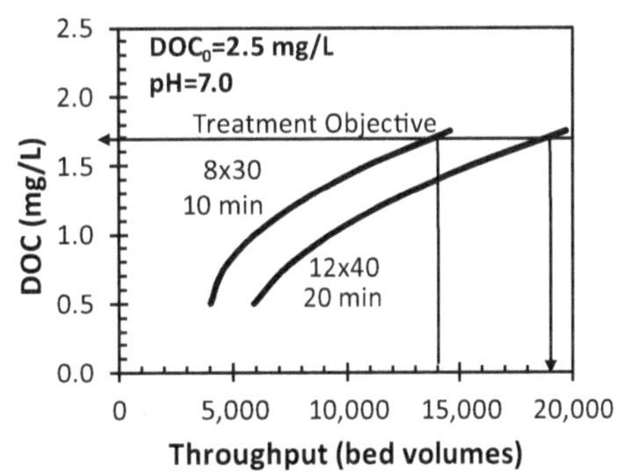

Figure 5-8. Zachman and Summers prediction of GAC performance

The bed life for Scenario A is 19,000 bed volumes and for Scenario B, 14,000 bed volumes. The use rate in milligrams GAC per liter water treated is found by dividing the ρ_{GAC} by the bed volumes treated (Eq 3-7), which can be compared directly with the PAC dose:

$$\text{Scenario A:}\quad \text{Use Rate} = \frac{\rho}{BV} = \frac{0.48\ \frac{g}{cm^3}}{19{,}000} \times \frac{1{,}000\ mg}{g} \times \frac{1{,}000\ cm^3}{L} = 25\ \frac{mg}{L}$$

$$\text{Scenario B:} \quad \text{Use Rate} = \frac{\rho}{BV} = \frac{0.48\ \frac{g}{cm^3}}{14{,}000} \times \frac{1{,}000\ mg}{g} \times \frac{1{,}000\ cm^3}{L} = 34\frac{mg}{L}$$

If beds are operated in staged parallel mode, the use rates should be approximately halved to 12.5 and 17 mg/L, respectively.

SCENARIO C. For the PAC option the treatment objective is 20 percent higher than the actual treatment objective of 1.7 mg/L × 1.2 = 2.0 mg/L (accounting for simultaneous coagulation). The Cho and Summers model is rearranged to solve for dose:

$$DOC_{T.O.} = 0.1561 + 0.9114DOC_0 - 0.0263\text{dose} - 0.002\text{time}$$

$$\text{dose} = \frac{0.1561 + 0.9114DOC_0 - 0.002\text{time} - DOC_{T.O.}}{0.0263}$$

$$\text{dose} = \frac{0.1561 + 0.9114(3.6) - 0.002(45) - 2.0}{0.0263} = 51\ mg/L$$

The dose of 51 mg/L is outside the limits of the model (30 mg/L) and needs to be verified because of the combined coagulation anyway. The level of removal required may be difficult to achieve with PAC. The PAC dose is significantly higher than either GAC use rate for post-filter contactor (Scenario A) or filter adsorber (Scenario B). However, DOC may not be a problem for the whole year. The use rates can be equated to determine the maximum length of the year the DOC can be at 3.5 mg/L for PAC to be an economic solution:

$$\frac{12.5}{51} = 0.25 \times 12\ \text{months} = 3\ \text{months per year}$$

$$\frac{17}{51} = 0.33 \times 12\ \text{months} = 4\ \text{months per year}$$

If the DOC must be controlled for less than 3 months per year, the mass of PAC required will be less than the mass of GAC required for post-filter contactor. Similarly, if the DOC must be controlled for less than 4 months per year, then less mass of PAC will be used than GAC in the filter adsorber.

BAC Performance Evaluation

Biologically enhanced processes have a long history of use in drinking water treatment, in the form of processes such as slow sand filtration and bank filtration. Biofiltration is a modification of conventional filtration in that no sustained disinfectant is applied prior to the filters. In the absence of disinfection, heterotrophic bacteria attach to the medium as biofilm, oxidize biodegradable organic matter (BOM), and use it as a source of carbon (Urfur et al. 1997, Huck et al. 2000, Wang et al. 1995,

Servais et al. 2005). As utilities throughout the country are facing decisions regarding treatment alternatives for achieving compliance with the Stage 2 D/DBP Rule, biofiltration is receiving increased attention.

Factors Affecting BAC Performance

A number of biofiltration reviews have been published (Urfer et al. 1997, Huck et al. 2000, Servais et al. 2005), and the factors affecting the removal of biodegradable organic matter in rapid biological filtration can be summarized as follows:

- Contact time
- Media surface area
- Type of media
- Temperature
- Nature and concentration of the influent BOM
- Biomass disruption (due to backwashing)

Contact Time. Huck et al. (2000) found that removals of biodegradable dissolved organic carbon (BDOC) increased with increasing EBCT, but not proportionally. The effect of EBCT was shown previously in Figure 5-3.

Media Type and Size. A study conducted by the Sweeney (WTP) in Wilmington, N.C., found that lignite and bituminous GAC performed equally well for turbidity removal and biofiltration; however, the lignite GAC provides a number of advantages with respect to lower costs. Lignite GAC is less dense than bituminous GAC; therefore it requires a lower backwash rate for effective backwash. The lower backwash rate translates into less backwash water and lower energy consumptions. Moreover, the lower density also results in fewer pounds of GAC required to fill the filter volume, leading again to lower costs (Najm et al. 2005).

For a given bed depth, the media surface area within the filter for biofilm growth is inversely related to media diameter; thus an inherent tradeoff occurs between removals of biodegradable material (favored by small media) and head loss (favored by larger media).

Temperature. Higher temperatures will promote biodegradation and biomass growth, and thus performance can be significantly affected by temperature. Although temperature is generally not controllable, the lowest temperatures are experienced in winter when the DOM concentration is often lowest and removal may not be a major issue.

Concentration and Composition of BOM. Concentration and composition of DOM may vary throughout the year. While these may also be thought of as out of the control of the operators, oxidation with ozone or advanced oxidation process (AOP) can convert some of the recalcitrant DOM to BOM.

Backwashing. The use of air scour in backwashing (under collapse pulsing conditions; see Amirtharajah et al. 1991) did not affect BOM removals (oxalate, an ozonation by-product), assimilable organic carbon (AOC), and DOC compared with water-only backwashing. These results were obtained using a GAC filter at temperatures above 10 °C without chlorine in the backwash water. These data supported results of modeling conducted as part of the study (Hozalski and Bouwer 2001a, b) that indicated backwashing would have to achieve biomass removals greater than 60 percent to have an effect on performance. Also, in this investigation, the BOM removal capability of the full-scale filters was essentially unaffected by free chlorine in the backwash water.

Practical Aspects of BAC Performance

From a practical point of view, the level of control that can be exerted over the various factors is important. Huck et al. (2000) addressed both the degree of control and effect of the major parameters. For example, the type of media can be specified, but it may only have a moderate influence on process performance, depending on the specific parameter being considered. Alternatively, temperature has a major effect on performance and essentially cannot be controlled. A major factor for chlorination is manifested through its effect on biomass. In the absence of chlorinated backwash water, the effect of the backwashing method may not be large, except when biomass levels in the filter are low. Although the extent of biodegradation is related to the amount of biomass, the latter cannot be independently controlled because it is related to biodegradation through the bacterial yield coefficient. Thus the factors listed above that control the extent of biodegradation also essentially control the amount of biomass. The level of biomass in a BAF can be estimated by measuring phospholipids (Wang et al. 1995) or using other methods.

Results from Huck et al. (2000) provide a useful context for a practically based approach to quantify biofiltration performance. In contrast to many investigations that are conducted at the pilot scale, a substantial portion of this study was conducted on demonstration-scale and full-scale biological filters. The objective was to optimize the performance of biological filters that were also intended for particle removal.

Results from a demonstration-scale plant (Chowdhury et al. 2010) indicated that approximately 10 to 20 percent TOC removal is achievable through biofiltration without preozone for this utility. The effect of EBCT on DOM removal was also seen with a 9-min EBCT filter outperforming one with an EBCT of 7 min. It was also found that chlorine in the backwash water did not affect the bioremoval results. During this study, GAC slightly outperformed anthracite, without chlorinated backwash water, for TOC removal during the steady-state biological removal phases. This TOC removal could yield a 20 to 40 percent reduction in TTHM formation. In order to evaluate the effect of these results on a broader scale, data from the ICR database were used to estimate the number of utilities that use free chlorine for disinfection and have difficulty achieving compliance with the Stage 2 D/DBP Rule. While compliance with the Stage 2 D/DBP Rule is based on locational running annual average, this analysis estimated compliance status by comparing the maximum distribution system sample TTHM concentration for each utility with the MCL. Using these data, and assuming a 30 percent reduction in TTHM levels can be achieved through GAC biofiltration without preozone, it is estimated that more than half of the selected utilities may benefit from retrofitting filters with GAC for biofiltration.

BAC Evaluation Methods

Pilot-Scale Testing. TOC removal performance of a biologically active GAC filter can be very site specific. Because the biological processes greatly depend on the variations in site conditions, only pilot testing provides a true representation of field performance and operating needs. Pilot-testing campaigns need to include an acclimation period and a long evaluation period to capture the effects of temperature and DOC concentration. To achieve a condition when bioremoval dominates (shown previously in Figure 5-2), at least 3 months of acclimation are required, and the GAC must be exhausted or near exhausted, which can take at least 6 months and is function of DOC_0, GAC type, and EBCT.

Bench-Scale Testing. Accurate assessment of BAC removal of BOM is difficult. The maximum amount of removal can be accessed through BDOC tests, but accurate removal simulation at a given EBCT is elusive.

Model Predictions. Various investigators have attempted to model biofiltration; Servais et al. (2005) and Huck and Sozański (2008) have summarized the practical usefulness of the most widely known models. The authors note that the models are all "research models"; i.e., they are relatively complex and cannot be directly used by practitioners because they require parameters that cannot be readily obtained.

In terms of a simpler approach, a first-order empirical model was presented by Huck et al. (1994) for the prediction of approximate performance. Those authors showed that the amount of BOM removed in a given biofilter was directly proportional to the influent concentration. Therefore removals can be approximated as a first-order process, and a biofilter at apparent steady state will essentially achieve a constant percentage removal for a given EBCT and temperature. In addition to applying to the removal of AOC and BDOC, the relationship was shown to hold for TTHM formation potential (TTHMFP) and chlorine demand. Wang et al. (1995) have found that first-order models with respect to influent concentration and biomass to be effective for TOC. Results from the study conducted by Huck et al. (2000) suggested that the biomass level as measured by the phospholipid method only becomes important for performance if it drops below a certain level.

References

Amirtharajah, A., N. McNally, G. Page, and J. McLeod. 1991. *Optimum backwash of dual media filters and GAC filter-adsorbers with air scour.* Denver, Colo.: Water Research Foundation.

Cho, H., and R.S. Summers. 2007. Powdered activated carbon for controlling total organic carbon: Application and model development. In *Proc. AWWA Water Quality Technology Conference, Charlotte, N.C.* Denver, Colo.: AWWA.

Chowdhury, Z., A. Traviglia, J. Carter, T. Brown, R.S. Summers, C. Corwin, T. Zearley, M. Thurman, I. Ferrara, J. Olson, R. Thacker, and P. Barron. 2010. *Cost-effective regulatory compliance with GAC biofilters.* Denver, Colo.: Water Research Foundation.

Cornel, P.K., R.S. Summers, and P.V. Roberts. 1986. Diffusion of humic acid in dilute aqueous solution. *Jour. Colloid Interface Sci.* 110(1):149.

Corwin, C.J., and R.S. Summers. 2012. Controlling trace organic contaminants with granular activated carbon adsorption. *Journal AWWA* 104(1):E36–E47. doi.org/10.5942/jawwa.2012.104.0004.

Crittenden, J.C., P.S. Reddy, H. Arora, J. Trynoski, D.W. Hand, D.L. Perram, and R.S. Summers. 1991. Predicting GAC performance with rapid small-scale column tests. *Journal AWWA* 83(1):77–87.

Graham, M.R., I. Najm, M.R. Simpson, B. MacLeod, R.S. Summers, and L. Cummings. 2000. *Optimization of powdered activated carbon application for geosmin and MIB removal.* 909. Denver, Colo.: AWWA [now Water] Research Foundation.

Hozalski, R.M., and E.J. Bouwer. 2001a. Non-steady state simulation of BOM removal in drinking water biofilters: model development. *Water Res.* 35(1):198–210.

Hozalski, R.M., and E.J. Bouwer. 2001b. Non-steady state simulation of BOM removal in drinking water biofilters: applications and full-scale validation. *Water Res.* 35(1):211–223.

Huck, P.M., S. Zhang, and M.J. Mitton. 1994. Kinetic and empirical models for the removal of biodegradable organic matter in drinking water treatment. Presented at International Symposium on Biodegradable Organic Matter in Drinking Water, Montreal, Quebec.

Huck, P.M., B.M. Coffey, A. Amirtharajah, and E.J. Bouwer. 2000. *Optimizing filtration in biological filters.* Report No. 90793. Denver, Colo.: AWWA [now Water] Research Foundation and American Water Works Association.

Huck, P.M., and M.M. Sozański. 2008. Biological filtration for membrane pre-treatment and other applications: Towards the development of a practically-oriented performance parameter. *Jour. Water Supply Res. Technol.–Aqua* 57(4):203–224.

Najm, I.N., V.L. Snoeyink, B.W. Lykins, Jr., and J.Q. Adam 1991. Using powdered activated carbon: A critical review. *Journal AWWA* 83(1):65–76.

Najm, I., M. Kennedy, and W. Naylor. 2005. Lignite versus bituminous GAC for biofiltration: a case study. *Journal AWWA* 97(1):94-101.

Servais, P., M. Prevost, P. Laurent, J.C. Joret, R.S. Summers, B. Hamsch, and C. Ventresque. 2005. Biodegradable organic matter in drinking water treatment. In *Biodegradable organic matter in drinking water treatment and distribution.* M. Prevost, P. Laurent, P. Servais, and J.C. Joret, eds. Denver, Colo.: AWWA.

Sontheimer, H., J.C. Crittenden, and R.S. Summers. 1988. *Activated carbon for water treatment.* Karlsruhe, Germany: DVGW-Forschungsstelle am Engler-Bunte Institut der Universitat Karlsruhe.

Summers, R.S., and P.V. Roberts. 1988. Activated carbon adsorption of humic substances. I. Heterodisperse mixtures and desorption. *J. Colloid and Interface Sci.* 122(2):367–381.

Summers, R.S., L. Cummings, J. DeMarco, D. Hartman, D. Metz, E.W. Howe, B. MaCleod, and M. Simpson. 1992. *Standardized protocol for the evaluation of GAC.* 90615-5/92. Denver, Colo.: AWWA [now Water] Research Foundation.

Summers, R.S., S. Hooper, H. Shukairy, G. Solarik, and D.M. Owen. 1996a. Assessing DBP yield: uniform formation conditions. *Journal AWWA* 88(6):80–93.

Summers, R.S., S.M. Hooper, and S. Hong. 1996b. *ICR manual for bench- and-pilot-scale treatment studies, part 2: granular activated carbon precursor removal studies.* EPA 814-B-96-003. Washington, D.C.: Office of Water, US Environmental Protection Agency.

Summers, R.S., G. Solarik, S.M. Hooper, Z.K. Chowdhury, and D.M. Owen. 1998. *Removal of DBP precursors by GAC adsorption.* 816. Denver, Colo.: AWWA [now Water] Research Foundation.

Summers, R.S., D.R.U. Knappe, and V.L. Snoeyink. 2010. Adsorption of organic compounds by activated carbon. In *Water quality and treatment.* 6th ed. New York: AWWA and McGraw-Hill.

Urfer, D., P. Huck, S.D.J. Booth, and B.M. Coffey. 1997. Biological filtration for BOM and particle removal: a critical review. *Journal AWWA* 89(12):83–98.

Wang, J.Z., R.S. Summers, and R.J. Miltner. 1995. Biofiltration performance: Part I, relationship to biomass. *Journal AWWA* 87(12):55–63.

Zachman, B.A., and R.S. Summers. 2010. Modeling TOC breakthrough in granular activated carbon adsorbers. *Jour. Environ. Engnrg.* 136(2):205–210.

Chapter 6

Taste-and-Odor Control

Taste-and-odor (T&O) control has long been an important aspect of drinking water treatment. Although many naturally occurring T&O-causing compounds pose no known health threats, their presence can greatly undermine public confidence in a water supply. The principal causes of objectionable tastes and odors in drinking water supplies include algae and algal metabolites, various microorganisms, as well as industrial and domestic wastes. Mallevialle and Suffet (1987) discuss the origins, sources, and causes of tastes and odors in detail.

One of the most common T&O problems reported by utilities is the presence of earthy or musty odors and flavors caused by 2-methylisoborneol (MIB) and geosmin (Suffet et al. 1999, Young et al. 1996). These organic compounds are microbial by-products that are easily detected by human olfactory senses. In fact, many people will notice an earthy or musty smell and flavor if MIB concentrations exceed 7 to 10 ng/L (parts per trillion) or if geosmin concentrations exceed 4 ng/L (Ashitani et al. 1988, Chen et al. 1997, Chudyk et al. 1979). Thus, for water systems faced with MIB and geosmin, effective T&O control can only be achieved by utilizing a technology that can reduce MIB and geosmin concentrations to extremely low levels. As mentioned above, T&O in drinking water can come from a variety of sources (not just MIB and geosmin), but a treatment system that effectively eliminates MIB and goesmin is normally well prepared for dealing with other types taste or odor compounds.

Various studies have shown that conventional treatment processes such as aeration, rapid sand filtration, and chemical oxidation cannot effectively eliminate MIB- and geosmin-related tastes and odors (Collins et al. 1970, Coughlin and Ezra 1968, Crittenden et al. 1986). In some cases, ozone and advanced oxidants (ozone or ultraviolet light in combination with hydrogen peroxide) can control T&O, although the effectiveness of these treatments depends largely on the background water quality (Crittenden et al. 1987, Crittenden et al. 1991, Dandekar et al. 1998, Gerber 1969, Gillogly et al. 1998a). Notably, activated carbon has repeatedly proven to be a highly effective means for controlling MIB- and geosmin-related tastes and odors in natural waters (Collins et al. 1970, Gillogly et al. 1998b, Herzing et al. 1977, Huang et al. 1996, Mallevialle and Suffet 1987, Izaguirre et al. 1982). Because MIB and geosmin are naturally occurring target organic contaminants (not from a health prospective, only from a control perspective), experience with the control of target organic contaminants by PAC and GAC (chapter 7) is relevant as the influent concentrations are also in the parts-per-trillion range. These factors include the lack of influence of the influent concentration on the fractional removal and the significant impact of the background DOM concentration.

PAC for Control of Taste and Odor

PAC Applications

PAC application is a common approach for the control of taste and odor, largely because PAC can be added on an as-needed basis when episodes occur. T&O issues are normally seasonal, and a PAC feed system can be left offline during those times of the year when tastes and odors are not present. Seasonal operation provides for lower operations and maintenance (O&M) costs than if the system were operating year-round.

Determination of Required Dose

PAC addition is perhaps the most common approach to using activated carbon for T&O control. The following sections identify and address key parameters concerning the use of PAC for removing T&O compounds. These include: (1) PAC contact time and mixing, (2) PAC product selection, (3) odorant compound influent concentration, (4) the background water quality effects, and (5) desorption. The process of planning for and designing a PAC feed system is described in chapters 4 and 12, respectively.

Contact Time and Mixing. Contact time and mixing of PAC for T&O control are similar to other systems described in chapter 4. PAC application and removal points are typically selected to optimize contact time and mixing for T&O removal. In some cases, PAC may be added directly to a raw water reservoir, although this is a relatively uncommon approach. While applying PAC to a reservoir adsorbs T&O compounds in the raw water and also prevents sunlight from penetrating the water column (which can cause the organisms responsible for producing the offending odorants to die off), a downside is that the PAC accumulates at the bottom of the reservoir. Depending on the application rate, PAC accumulation could increase the need for dredging the reservoir. Also, it is possible for activated carbon to release adsorbed odorants as described later in this chapter. Thus a risk exists that the PAC added to remove T&O compounds from within a reservoir will release those compounds back into the water after it settles to the bottom.

PAC Product Selection. Studies that have evaluated MIB adsorption by different adsorbents have emphasized the comparison of bituminous coal-based and lignite-based activated carbons (Chen et al. 1997, Chudyk et al. 1979, Gillogly et al. 1998a). Chudyk et al. (1979) demonstrated that MIB uptake by a lignite-based PAC was significantly less than the uptake by two bituminous coal-based PACs. It was shown that the surface area for the lignite PAC was 650 m^2/g compared with 1,100 m^2/g for the bituminous coal-based PACs as measured by the BET (Brunauer, Emmett, and Teller) method of measuring surface area. In addition, pore-size distributions indicated that the lignite PAC had far fewer pores in the 0–50 Å (radius) range. Chudyk et al. (1979) invoked these physical disparities as reasons for the observed differences in MIB uptake. A similar argument was made by Chen et al. (1997), who, after discovering that the isotherm for a lignite-based PAC was lower than the isotherm for a bituminous coal-based PAC, suggested that smaller surface area and higher ash content were responsible. Chen et al. (1997) listed the ash content for their experimental lignite PAC as 26.9 percent. Considering that the ash content of the experimental bituminous coal-based PAC was 5.0 percent, the argument for the influence of ash content gains importance. An important reason for inferior MIB uptake by lignite PACs may simply be that one-quarter of their mass is inert and unavailable for adsorption. Work by Gillogly et al. (1998a) that compared lignite and bituminous coal-based PACs corresponded well with the studies of Chudyk et al. (1979) and Chen et al. (1997).

In contrast to the aforementioned work, Huang et al. (1996) and MacLeod and Simpson (1993) demonstrated superior MIB removal by lignite-based PACs. A lignite-based PAC tested by Huang et al. (1996) removed more MIB after 90 min than several bituminous coal-based PACs. Similarly, MacLeod and Simpson (1993) observed superior MIB adsorption by a lignite PAC after about 33 min of contact time. Neither group included isotherms along with their kinetic data, so it is uncertain whether those trends would be apparent at equilibrium. The authors herein suggest that kinetic limitations in the more microporous bituminous coal-based PACs may account for superior MIB uptake by lignite PACs during the first 30 min of contact time. However, Cho and Summers (2007) found bituminous PAC yielded 12 to 27 percent better MIB removal compared with lignite and wood PAC after 60 min of contact and at equilibrium, where the greater surface area of the bituminous coal-based PACs is expected to result in higher ultimate uptake.

Odorant Compound Influent Concentration. Gillogly et al. (1998a) demonstrated that the MIB removal rate associated with a given PAC dose is independent of the initial MIB concentration. That is, the percent MIB removal achieved by a specific PAC dose remains relatively constant over a wide range of initial MIB concentrations. The work conducted by Gillogly et al. (1998a) used water from Lake Michigan exclusively, but this phenomenon has been observed in other waters. Graham et al. (2000b) demonstrated it for three PACs in three waters. For example, in a pilot study involving the city of Baltimore, Md., a PAC dose of 40 mg/L achieved approximately 90 percent MIB removal when the initial MIB concentration was either 200 or 800 ng/L. Similarly, a PAC dose of 60 mg/L achieved approximately 92 percent MIB removal under these same two conditions. It is expected that PAC performance would also be relatively independent of initial concentration in the case of geosmin and other odorants. Cho and Summers (2007) found similar results for three PAC types in one water.

Background Water Quality. Numerous studies involving batch adsorption tests have documented the effects of background dissolved organic matter (DOM) on MIB uptake (Chen et al. 1997, Gillogly et al. 1998a, Herzing et al. 1977, Huang et al. 1996, Lalezary et al. 1986, Lalezary-Craig et al. 1988, Newcombe et al. 1997, Pirbazari et al. 1993, Graham et al. 2000b). Collectively, they have demonstrated that MIB adsorption capacity decreases with increasing background DOM concentration. It has also been shown that the initial MIB concentration strongly influences adsorption capacity in the presence of DOM (Gillogly et al. 1998a, Huang et al. 1996, Lalezary-Craig et al. 1988). For example, Gillogly et al. (1998a) demonstrated that an MIB isotherm shifted upward when the initial MIB concentration was increased from 149 ng/L to 1,245 ng/L, and the background DOM level remained the same. Similarly, Huang et al. (1996) showed that the Freundlich K constant for MIB adsorption onto activated carbon was 14 times higher when the initial MIB concentration was 9,087 ng/L compared with 18.5 ng/L. These effects were absent from tests conducted in deionized water (Gillogly et al. 1998a, Huang et al. 1996). Notably, Lalezary-Craig et al. (1988) stated that no correlation between initial MIB concentration and MIB adsorption capacity was apparent when initial MIB levels ranged from 5 to 150 ng/L, which is the concentration range often encountered in raw water supplies used by municipal drinking water systems.

Newcombe et al. (1997) conducted MIB adsorption studies in which the effects of various molecular weight fractions of DOM were explored, namely <500, 500 to 3,000, 3,000 to 10,000, 10,000 to 30,000, and >30,000 D. They discovered that the <500 DOM fraction had the greatest effect on MIB adsorption. For example, MIB removal by a wood-based PAC was 58 percent in a solution containing 8 mg/L of the <500 DOM fraction, and MIB removal for the same PAC ranged from 90 to 95 percent in solutions with 8 mg/L of the >500 DOM fractions. Newcombe et al. (1997) also showed that MIB uptake was much lower in natural water as compared with any of the prepared DOM fractions. They suggested that some of the smallest compounds in the <500 fraction were lost during the separation process, and that these compounds might well be the most important ones with respect to competition between DOM and MIB.

Whereas others (Gillogly et al. 1998a, Herzing et al. 1977, Huang et al. 1996, Lalezary et al. 1986, Lalezary-Craig et al. 1988, Pirbazari et al. 1993) have described the competitive effects of uncharacterized DOM, Chen et al. (1997) attempted to identify some the specific organic compounds in natural water that compete with MIB for adsorption sites. They performed closed-loop stripping analysis (CLSA) on their sample water (raw water from a treatment plant in San Diego, Calif.) and identified hydrophobic organic compounds (HOCs) that were strongly adsorbed by activated carbon in the presence of MIB. Among them were several alkanes and two

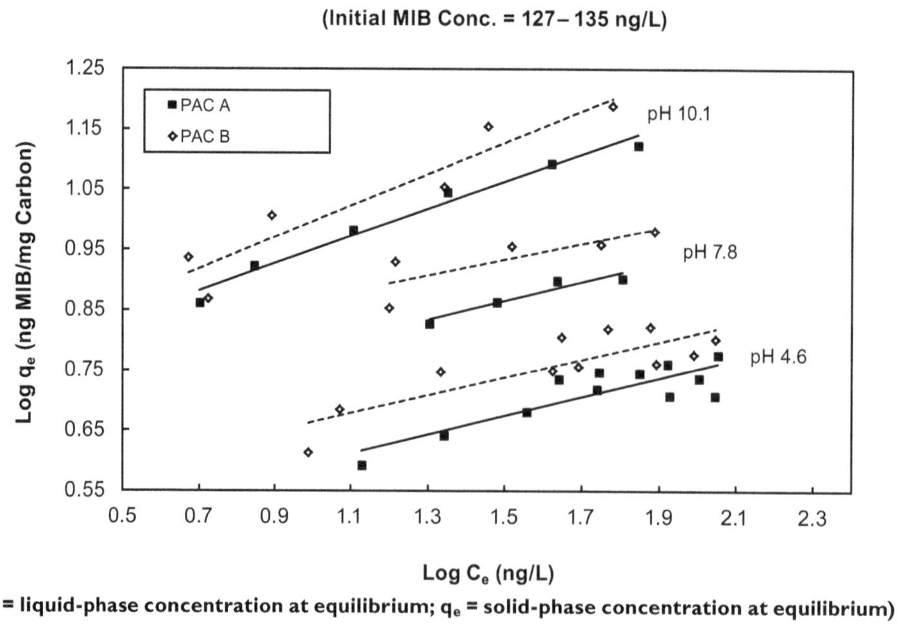

Figure 6-1. MIB adsorption isotherms prepared using the same water at three different pH values

types of organic acids. Chen et al. (1997) suggested that these strongly adsorbed HOCs competed with MIB for adsorption sites.

The pH of the water at the point of PAC addition can have a noticeable effect on odorant uptake. Although the adsorbability of odorants such as MIB and geosmin is largely unaffected by pH (Herzing et al. 1977) due to these compounds being relatively nonionizable, the adsorbability of humic materials generally increases significantly with decreasing pH (McCreary and Snoeyink 1980). Given that humic materials normally comprise a large fraction of the background organic matter in natural waters, an increase in their adsorbability can mean an increased competitive effect of the background DOM on odorant adsorption. The MIB isotherms shown in Figure 6-1 show that MIB adsorption in the presence of DOM declines with decreasing pH.

An additional consideration regarding the effect of pH on odorant removal using PAC is the potential for pH reductions within the PAC contact zone to cause odorant desorption. In many cases, the PAC feed point at a water treatment plant is upstream of the coagulant feed point. Thus the pH where the PAC enters the water is normally higher than in the rapid mix and flocculation zones (which the PAC must pass through). For a plant with a high source water pH that practices enhanced coagulation, a tremendous shift in pH can occur from the raw to the coagulated water. As a result, the adsorption conditions for odorants could be considerably more favorable prior to coagulation than after. The increased adsorbability of the background DOM following coagulation could lead to the displacement of previously adsorbed odorant molecules by DOM. There is little information in the literature concerning the displacement of odorant molecules by DOM, and thus the potential magnitude of this effect is not well understood. Site-specific testing is required to determine whether DOM-induced odorant desorption could significantly affect odorant removal rates or lead to increased odorant concentrations downstream of settling basins or clarifiers containing settled PAC.

As discussed below, odorant adsorption onto PAC can be negatively affected by the presence of other treatment chemicals, especially oxidants. Thus the odorant removal performance of a particular PAC treatment strategy will depend in part on the characteristics of the overall chemical treatment scheme, i.e., the chemicals being applied as well as the associated doses and application points.

Chlorine. Gillogly et al. (1998b) have studied the effects of chlorine on MIB adsorption extensively. They determined that aqueous chlorine significantly reduced MIB uptake by activated carbon, which confirmed earlier work by Lalezary-Craig et al. (1988). For example, Gillogly et al. (1998b) indicated that a PAC dose of 11.5 mg/L achieved 70 percent MIB removal in a sample of unchlorinated natural water (Lake Michigan). In contrast, when 3 mg/L chlorine was added simultaneously with the PAC, only about 30 percent MIB removal was achieved in the same sample water. When 3 mg/L chlorine was added 12 min after PAC addition, about 50 percent removal was achieved. Furthermore, the delayed addition of chlorine caused a release of previously adsorbed MIB back into solution. Gillogly et al. (1998b) also showed that the total chlorine dose, rather than the initial concentration, controlled the effect on MIB uptake. As an example, MIB removal by a given PAC dose was the same whether 10 mg/L chlorine was added initially or 2 mg/L was added on five separate occasions throughout the test.

Gillogly et al. (1998b) offered several possible explanations for why chlorine hindered MIB uptake and showed, using experimental evidence, that only one was plausible. To discount the possibility that MIB was being catalytically degraded by chlorine on the PAC surface, they used gas chromatography and mass spectrometry (GC-MS) to analyze water samples containing MIB that had desorbed from PAC following chlorine addition. The results indicated that no additional compounds that might have come from a chlorine-MIB reaction were present. Experiments were also conducted to establish whether the observed chlorine effect was attributable to increased competition by disinfection by-products (DBPs). For this purpose, natural water samples were prechlorinated for 24 hr prior to adsorption tests, to maximize the formation of DBPs. The remaining chlorine in these samples was eliminated with sulfite, and it was found that MIB uptake in them was identical to uptake in unchlorinated water. Thus, Gillogly et al. (1998b) suggested that the most probable reason for the observed decrease in MIB adsorption capacity was the buildup of surface oxides resulting from reactions between chlorine and the activated carbon surface. McGuire and Suffet (1984) have shown that the surface oxide content of activated carbon increases following exposure to increasing amounts of chlorine. Furthermore, more recent work has shown that an increase in surface oxides can inhibit uptake of MIB and other organic solutes (Karanfil and Kilduff 1999, Leng and Pinto 1997, Pendleton et al. 1997). Interestingly, Gillogly et al. (1998b) discovered that when an oxidized activated carbon was dried, it regained some of its lost adsorption capacity. They tentatively suggested that this increase in adsorption capacity was due to the volatilization of surface carboxyl groups, which caused a decrease in the number of surface oxides. However, unless the drying protocol involves temperatures in excess of 220 °C, it is extremely unlikely than any surface oxides were volatilized. Previous research has shown that carboxyl groups on activated carbon surfaces begin to decompose at about 227 °C (Dandekar et al. 1998; Otake and Jenkins 1993).

Other Oxidants. Activated carbon reacts with other oxidants in much the same way as it does with free chlorine. Summers et al. (2010) summarized a variety of reactions between activated carbon and several oxidants commonly used in water treatment. These included chloramines and chlorine dioxide. In each case, the oxidant–carbon reaction resulted in the formation of oxygen-containing functional groups on the activated carbon surface (i.e., surface oxides). Thus it can be surmised that the effect of these other oxidants on odorant removal by PAC would be similar to those described for free chlorine in the previous section. Summers et al. (2010) suggested that reactions between permanganate and activated carbon are similar to those for chlorine and chlorine-containing oxidants. Others have demonstrated that this also applies to reactions between ozone and activated carbon (Lee and Davidson 1999).

Coagulants. In many cases, the coagulation process in a water treatment plant involves the addition of a metal salt (coagulant) such as ferric chloride or aluminum sulfate for the purpose of destabilizing suspended particles and causing them to agglomerate. In addition, the metal ions contributed via coagulant addition normally precipitate to form iron or aluminum hydroxides. These quasi-crystalline solids combine with agglomerated particles to form floc. PAC particles are affected by the coagulation process in much the same way as other types of particles. That is, they become enmeshed in the floc. Consequently, the PAC will be less dispersed following floc formation, and individual PAC particles may become encapsulated (within the floc) to the extent that there is little or no contact between the particle and the surrounding water. The loss of contact could decrease the kinetics associated with the bulk diffusion of T&O compounds from the aqueous phase into the PAC pore structure. However, although these effects would seemingly inhibit PAC adsorption performance, researchers have found that this is not necessarily the case. Graham and Summers (1996) demonstrated that MIB uptake by PAC was seemingly unaffected by coagulant addition and subsequent floc formation. Similarly, Ho and Newcombe (2005) showed that floc formation does not automatically reduce PAC adsorption performance with respect to tastes and odors. Rather, their work indicated that at coagulant (alum) doses of less than 40 mg/L, there was little or no effect on MIB uptake. At higher alum doses, some reduction in uptake was observed.

Desorption. Both anecdotal and experimental evidence indicates that T&O compounds such as MIB and geosmin can be released from PAC following adsorption. For instance, utilities occasionally report the presence of odors downstream of clarifiers containing settled PAC when no detectable odors are present in the raw water. For PAC treatment systems, it is therefore advisable to consider such things as: (1) the need for regular sludge removal from clarifiers and settling basins that capture settled PAC and (2) the effect of PAC type on the likelihood that desorption (from settled PAC) will occur.

PAC Evaluation Methods

Jar Testing. Awareness is growing within the drinking water industry regarding the variability among commercially available activated carbons when it comes to removing T&O. Consequently, utilities have begun including performance-based comparisons as part of their PAC procurement protocols. These usually take the form of jar tests, in which a variety of PACs are exposed to the source water in question,

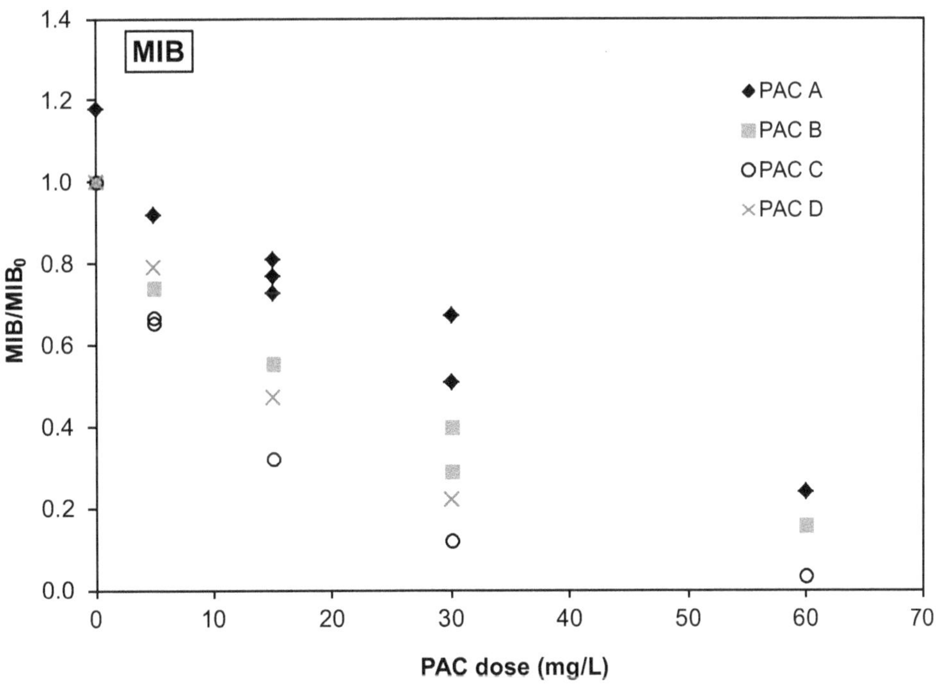

Figure 6-2. MIB remaining versus PAC dose for several commercially available PACs (background TOC concentration = 4 mg/L; 30 mg/L alum added alongside PAC)

often after it has been spiked with MIB, geosmin, or both. The jar test conditions are normally set to match the mixing intensities and hydraulic residence times of the full-scale facility. Figure 6-2 includes MIB removal data for a variety of commercially available PACs that were collected as part of a performance-based procurement process. Note the significant differences in MIB removal performance. Similar variability has been observed in numerous locations. Graham et al. (2000a) have developed a procedure for evaluating PAC for MIB and geosmin removal.

Modeling PAC Adsorption Performance. Gillogly et al. (1998a) analyzed laboratory kinetic data using the homogenous surface diffusion model (HSDM) (Qi et al. 1994, Traegner and Suidan 1989). The model can predict adsorption rates and requires the following measurable parameters: initial aqueous MIB concentration, PAC concentration (dose), media particle diameter, particle density, and Freundlich constants (1/n, K). Gillogly et al. (1998a) showed that the HDSM accurately predicted MIB uptake from natural water (Lake Michigan) for the first 4 hr of contact time, at which time a pseudo-equilibrium was achieved. Interestingly, the model fit of the kinetic data improved when Freundlich constants were determined using 4-hr equilibrium data as opposed to 4-day equilibrium data. Gillogly et al. (1998a) suggested that the slow increase in MIB adsorption capacity after 4 hr was not consistent with a mechanism characterized by a single surface diffusion coefficient (as is assumed in the HSDM model); hence the discrepancies between the predicted and experimental results when using 4-day equilibrium data. They go on to speculate that this greatly reduced adsorption rate may reflect a pore diffusion-controlled step, in which MIB must pass through pores partially blocked by DOM.

Model Predictions. Cho and Summers (2007) and Cho (2007) proposed empirical models for MIB adsorption by three PAC types: bituminous, lignite, and wood. The models were developed from the results of five waters (n = 173) with initial MIB concentrations (MIB_0) of 40 to 100 ng/L, DOC_0 of 1.3 to 5.4 mg/L, without coagulation for PAC doses between 5 and 30 mg/L and contact times between 10 and 60 min. The models are based on the following form shown in Eq 6-1:

$$DOC = \beta_1 + \beta_2 \times DOC_0 + \beta_3 \times dose + \beta_4 \times time \quad \text{(Eq 6-1)}$$

Where

- DOC = DOC concentration after PAC adsorption (mg/L)
- DOC_0 = DOC concentration of influent water (mg/L)
- dose = PAC dose (mg/L)
- time = contact time (min)
- β_1 through β_4 = fitting parameters

The values of β_1 through β_4 are presented in Table 6-1 for the three representative PACs. The model was verified with the results from three other waters (n = 80) with MIB_0 of 75 to 95 ng/L, DOC_0 of 2.4 to 3.8 mg/L, with an average variance (R^2) of 0.95.However, the models systematically underpredicted MIB removal by 2 to 10 percent. In general, the models predict that the bituminous PAC yields slightly better results compared with the other two PACs.

Table 6-1. Fitting parameters for the PAC removal of MIB (n = 173)(Cho 2007)

PAC base material	β_1	β_2	β_3	β_4	R^2
Bituminous (WPH–Calgon)	0.4947	0.0534	−0.0127	−0.0024	0.824
Lignite (HDB–Norit)	0.7000	0.0400	−0.0143	−0.002	0.851
Wood (CASP–Norit)	0.6722	0.0338	−0.0131	−0.0023	0.873

GAC for Control of Taste and Odor

GAC Applications

GAC Filter Adsorbers

Filter adsorbers, which can also be effective for removing T&O, must remain online at all times because they also serve as the primary filtration barrier. Consequently, the GAC is often processing water when no T&O are present; and during these periods, its capacity for T&O removal wanes due to the adsorption of background DOM. Thus, filter adsorbers often require frequent GAC replacement, meaning the associated O&M costs can be considerably higher than for a typical PAC system. In some cases, biologically active filter adsorbers can achieve T&O removal via biodegradation, although the short-term removal rates are generally less than what can be achieved through adsorption.

GAC Post-Filter Contactors

Post-filter GAC contactors can potentially be bypassed and thus also offer the option of conserving adsorption capacity during times when no discernable T&O are present. However, due to the high capital costs associated with post-filter contactors (discussed in chapter 13), these systems are almost never installed solely for the purpose of T&O control. As discussed in chapters 4 and 5, post-filter contactors are often implemented for DOM removal, and in those cases, the contactors provide the secondary benefit of also removing T&O. Corwin and Summers (2012) discussed the relative breakthrough of TOC and MIB and seasonal strategies to optimize GAC performance.

In situations where a utility is faced with particularly severe or persistent T&O problems, a dual-barrier approach to T&O control is sometimes implemented. A multi-barrier approach may involve a combination of PAC and GAC (either filter adsorbers or post-filter contactors) treatment. This combination approach was selected by the City of Baltimore, Md., for inclusion in its new Fullerton Water Treatment Plant (see chapter 4 for a detailed description of the pilot study and planning efforts that led to this selection).

Biologically Enhanced Activated Carbon (BAC)

GAC filters inevitably become biologically active because any disinfectant applied to the feedwater or backwash water is quickly eliminated when it contacts the GAC. In some cases, the microbial population that develops will metabolize T&O compounds, thereby allowing for odorant removal through the filter that is not linked to adsorption. Removal rates through GAC biofilters vary considerably, with some biofilters achieving no measurable removal and others achieving greater than 90 percent removal. This variation is due to a variety of factors relating to the nature of the microbial population and of the influent water. As in any organics removal application where GAC biofilters can reliably achieve the treatment targets, utilizing GAC biofilters for T&O control offers the significant benefit of not requiring regular GAC replacement, thereby reducing O&M cost.

GAC Performance Evaluation

Determination of the Use Rate

EBCT. EBCT is normally the most important criterion affecting the performance of GAC treatment systems. With respect to the EBCT requirements for achieving near-complete removal of odorants such as MIB and geosmin, Nowack and others (Nowack et al. 2004a, Nowack et al. 2004b) have repeatedly demonstrated that an EBCT of 7.5 min is sufficient to reduce the MIB in clarified water from more than 100 ng/L to below the odor threshold concentration (OTC) for extended periods. It should be noted that virgin GAC was used in these studies. Also, the influent MIB concentration in these tests consistently remained above 100 ng/L, and thus the GAC was exposed to a greater quantity of MIB than would normally be the case in a full-scale system over a similar time period. Naturally occurring T&O events normally have highly variable odorant levels and often subside quickly.

Newcombe et al. (1996) also determined that a fairly short EBCT, in this case 6.5 min, is sufficient to reduce MIB to less than the OTC. However, their work demonstrated that longer EBCTs could allow for greater GAC service life. For the water quality conditions tested, GAC filters with EBCTs of 6.5 and 13 min effectively

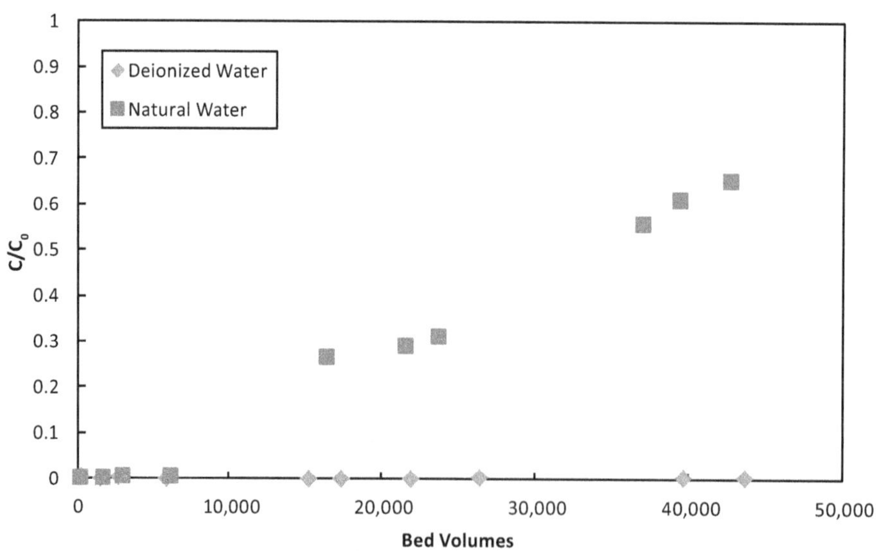

Figure 6-3. MIB breakthrough profiles (generated via RSSCTs) corresponding to deionized and natural waters (background DOC concentration for the natural water = 4 mg/L)

reduced MIB and geosmin levels to below their respective OTCs for about 10 months (of continuous operation). In contrast, a GAC filter with an EBCT of 20 min achieved that same level of performance for nearly 18 months. Thus, when designing and implementing GAC adsorbers, it is important to recognize that a long EBCT is not necessarily required to achieved near-complete removal of T&O compounds, although using a relatively short EBCT could result in a higher GAC replacement frequency than would be required for longer EBCTs. As discussed in chapter 4, the best approach to determining the most appropriate EBCT for a particular T&O control application (or for any organics removal application) is to conduct bench-, pilot- and/or full-scale testing, with realistic test conditions.

Other Factors Affecting Performance

Previous sections addressed the effects of a variety of factors on the performance of PAC for removing T&O compounds. Notably, the influence of background DOM and media type are similar for GAC. For example, the odorant removal performance of GAC adsorbers tends to decline with increasing background DOM concentration. Figure 6-3 shows MIB breakthrough profiles (generated via RSSCTs) corresponding to both deionized and natural water; the substantial difference between these two profiles highlights the effect of background DOM on MIB breakthrough performance. GAC adsorption performance can also vary considerably between GAC types. That is, GACs from different source materials, from different vendors (i.e., produced using different manufacturing protocols), and otherwise prepared differently, often do not perform similarly. Variability between products has been demonstrated in a variety of performance comparisons (Mackenzie et al. 2005), and Figure 6-4 includes the results of one such comparison. In this case, the MIB breakthrough performances of

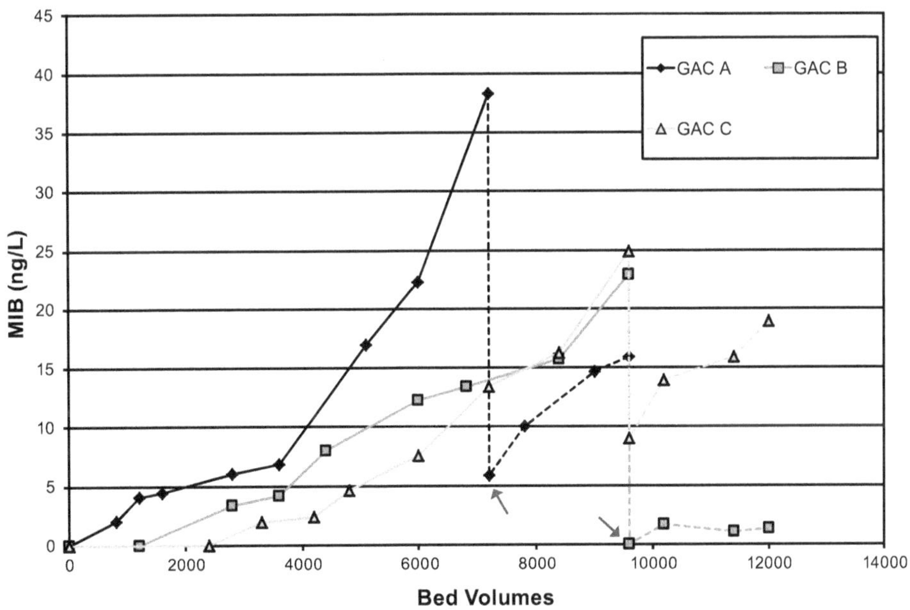

Figure 6-4. MIB breakthrough profiles for several commercially available GACs (generated via RSSCTs) (initial MIB concentration = 100 ng/L, background DOC concentration = 7–8 mg/L; arrows indicate when the influent MIB level was reduced to zero)

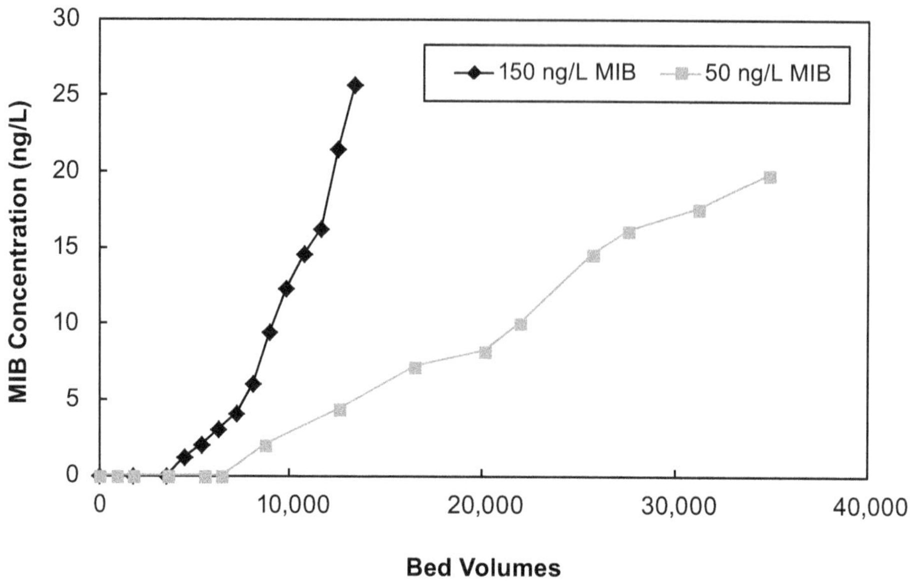

Figure 6-5. MIB breakthrough profiles (generated via RSSCTs) corresponding to influent MIB concentrations of 50 and 150 ng/L (background DOC concentration = 2–2.5 mg/L)

a bituminous coal-based, a lignite-based, and a wood-based GAC were compared by means of RSSCTs. The results clearly show significant differences in the respective performances of these GACs.

Whereas PAC performance for removing odorants is largely unaffected by the initial odorant concentration, the breakthrough profile for a GAC treatment system can be altered by changes in the influent odorant level. Figure 6-5 includes MIB breakthrough profiles corresponding to influent MIB levels of 50 and 150 ng/L. Notably, when the influent MIB level was 50 ng/L, the minicolumn used in these tests processed more than 20,000 bed volumes while maintaining an effluent MIB concentration of less than 10 ng/L (approximately the OTC for MIB). In contrast, the minicolumn processed fewer than 10,000 bed volumes (before the effluent MIB level reached 10 ng/L) when the influent MIB concentration was 150 ng/L.

However, when expressed on a normalized fraction basis (MIB/MIB_0) as shown in Figure 6-6, the influent concentration has been shown not to impact the breakthrough behavior (Kim and Summers 2006, Summers et al. 2010).

GAC Evaluation Methods

Chapter 7 explains that when evaluating GAC performance for the removal of specific organic compounds, it is far more useful to conduct flow-through adsorption tests than batch (bottle-point isotherm) experiments. This is especially true in the case of T&O compounds, due to the substantial effect of adsorption kinetics on the breakthrough profiles for these compounds. Thus, pilot-scale and bench-scale column testing are preferred; the following sections provide some guidelines concerning these methods.

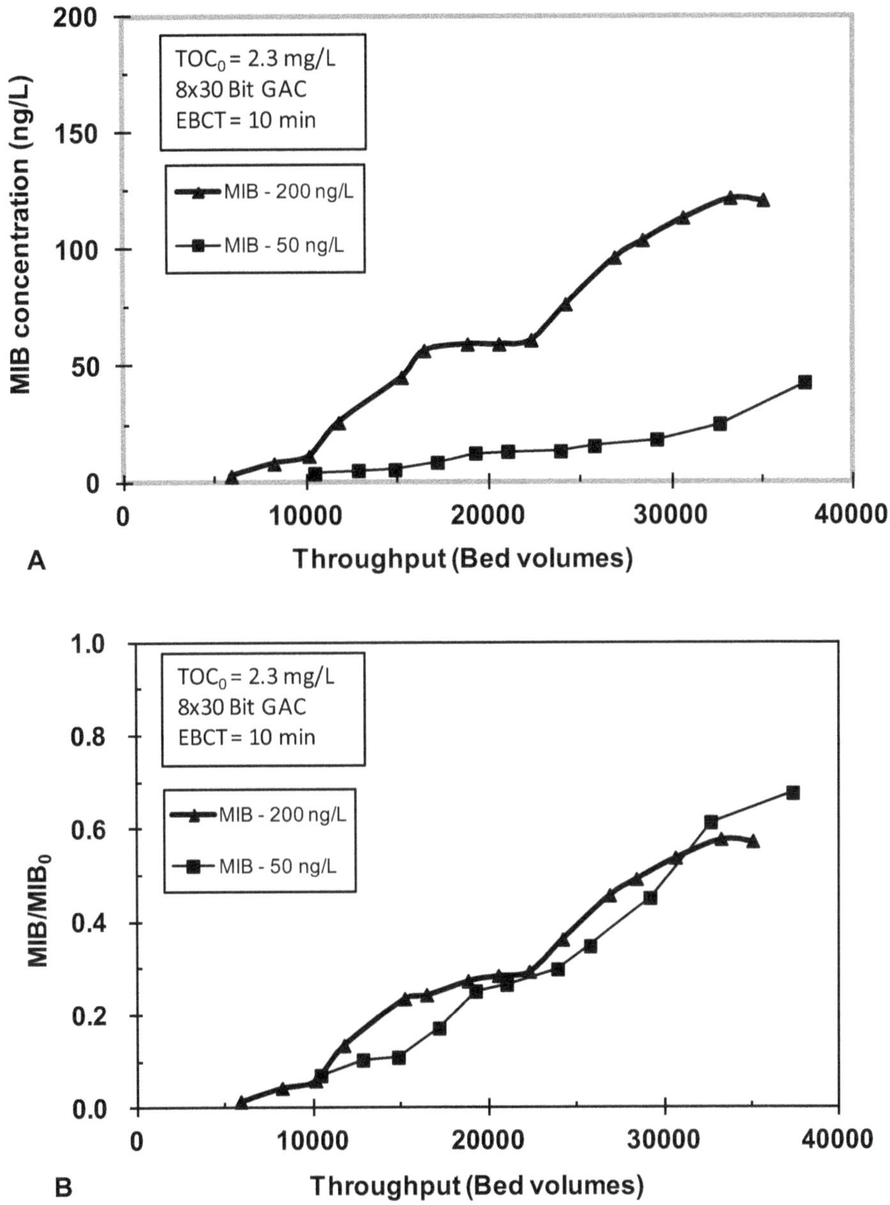

Figure 6-6. MIB breakthrough (generated via RSSCTs) on a mass concentration (a) and normalized (b) basis

Pilot-Scale Testing. As discussed in chapter 4, pilot-scale evaluations of GAC adsorption performance are perhaps the best way to predict full-scale breakthrough profiles for various organic compounds, and this also applies to odorants such as MIB and geosmin. However, in the context of evaluating odorant removal, pilot-scale testing must be carefully timed so as to occur when odorants are naturally present, or the influent to the pilot column(s) must be spiked with an odorant stock solution. The use of rapid small-scale column tests (RSSCTs) for predicting GAC breakthrough performance for MIB and geosmin has been investigated and the most appropriate mini-column design approach has not yet been identified. A more detailed discussion of RSSCT testing (in general) is provided in chapter 7. In cases where spiking the entire influent flow is prohibitively expensive (odorant stock materials tend to be costly), the stock solution can instead be applied to the influent of a few, or even a single column, provided that the necessary injection and sampling ports are available.

Bench-Scale Testing. RSSCT testing of the GAC control of MIB and geosmin has been investigated and the scale-up approach is not yet resolved. A more detailed background for RSSCT approaches is given in chapter 7. Pirbazari et al. (1993) conducted a series of adsorption rate studies that demonstrated the surface diffusivity of MIB and geosmin is independent of GAC particle size. Hence, bench-scale evaluations of MIB and geosmin removal performance should employ constant-diffusivity RSSCTs (Crittenden et al. 1986). More recently, Kim et al. (2007) and Kim (2006) found that proportional-diffusivity-designed RSSCTs work well for MIB breakthrough. They developed a model based on RSSCT results and were able to predict breakthrough of GAC columns at two different field sites. As is always the case when conducting RSSCTs, it is important to collect the water samples for these tests at a time when the background water quality conditions are most similar to the anticipated full-scale conditions during T&O events. In all likelihood, the odorant level(s) in the water collected for the RSSCTs will be lower than the preferred concentration(s) for the GAC evaluation, and thus in most cases it is necessary to spike the sample water with the target odorants.

Performance of GAC Biofilters for Taste-and-Odor Control. A variety of factors can influence odorant removal rates in GAC biofilters, including: (1) the concentration and characteristics of the biodegradable organic matter (BOM) present in the filter influent, (2) the transient nature of T&O events, (3) seasonal water temperature variations, (4) the EBCT and hydraulic loading rate (HLR), and (5) the backwashing protocol (frequency, duration, presence of chlorine, etc.).

Because T&O compounds such as MIB and geosmin normally occur at concentrations in the ng/L range, these compounds by themselves generally do not support the growth of microorganisms in biologically active filters. Hence, MIB and geosmin usually behave as secondary substrates, with the microbes in the biofilm being maintained by the BOM present in the influent (i.e., the primary substrate). Whether MIB and geosmin removal occurs in a biofilter will depend in part on whether the microbial population includes sufficient numbers of organisms capable of utilizing these compounds as secondary substrates (Westerhoff et al. 2005).

Elhadi et al. (2006) conducted a number of bench-scale tests to examine the major factors affecting MIB and geosmin removal in biofilters. The results of this study indicated that temperature, influent odorant concentration, media type, and BOM levels all had measurable effects. In particular, temperature had a significant effect on performance. A decrease in temperature from 20 to 8 °C caused a significant decline in both MIB- and geosmin-removal rates. The use of GAC rather than anthracite for biofiltration was also found to have a significant effect on MIB and geosmin removal, particularly when background BOM levels were relatively low. For example, geosmin-removal rates through a GAC biofilter were 24 and 44 percentage points higher than in a similar filter containing anthracite; and these differences corresponded to water with high and low BOM levels, respectively. Similar trends were observed for MIB-removal rates.

Westerhoff et al. (2005) also developed a model for MIB and geosmin removal in a biofilter based on pilot-plant data. The model was pseudo-first-order with respect to influent odorant concentration and depended on EBCT and biomass concentration. It took several months for the biomass to acclimate to MIB and geosmin.

Westerhoff et al. (2005) compiled a variety of pilot- and full-scale data pertaining to odorant removal in GAC biofilters, and their research produced a series of recommendations aimed at maximizing the associated odorant removal rates. These included:

- Backwashing with nonchlorinated water.
- Minimizing changes to influent water quality, especially with respect to pH.
- Maximizing EBCT, as it was shown that removal rates increased with increasing EBCT (up to an EBCT of 15 min).

The work of Westerhoff et al. (2005) strongly emphasizes the importance of pilot testing in developing a strategy for utilizing GAC biofiltration for T&O control.

Desorption

The potential effects and likely causes of odorant desorption on the performance of GAC treatment systems was addressed previously in this chapter. Both anecdotal and experimental evidence indicates that T&O compounds such as MIB and geosmin can be released from GAC following adsorption. For instance, odors have been detected downstream of GAC filter adsorbers when no odors were detected in the raw water. The occurrence of odorant desorption is a reasonable explanation for the apparent increase in odorant levels following an activated carbon treatment step. More specific evidence of odorant desorption appears in Figure 6-4, which shows MIB breakthrough profiles for several minicolumns containing various types of activated carbon. In these tests, the minicolumns were first exposed to clarified water containing approximately 100 ng/L MIB and 7 to 8 mg/L DOC. After the columns had processed 7,000 to 10,000 bed volumes of water, the influent MIB concentration was reduced to zero while the background DOC level remained the same. The effluent MIB level for two of the three columns exceeded the nominal odor threshold concentration for MIB during this phase. Desorption occurs in the absence of MIB molecules in the water entering the column, and the resultant shift in the equilibrium conditions for MIB in the aqueous and adsorbed phases creates a driving force for desorption that leads to the release of MIB molecules from the activated carbon (shown previously in Figure 3-2b).

References

Ashitani, K., Y. Hishida, and K. Fujiwara. 1988. Behavior of musty odorous compounds during the process of water treatment. *Water Sci. Technol.* 20(8–9):261–267.

Chen, G., B.W. Dussert, and I.H. Suffet. 1997. Evaluation of granular activated carbons for removal of methylisoborneol to below odor threshold concentration in drinking water. *Water Res.* 31(5):1155–1163.

Cho, H. 2007. 2-methylisoborneol and natural organic matter adsorption by powdered activated carbon. Masters Thesis, University of Colorado, Boulder, Colo.

Cho H., and R.S. Summers 2007. MIB removal by powdered activated carbon: impact of NOM, kinetics and pretreatment. In *Proc. Amer. Water Works Assoc. Conference, Toronto, Ontario, Canada.* Denver, Colo: AWWA.

Chudyk, W.A., V.L. Snoeyink, D. Beckmann, and T.J. Temperly. 1979. Activated carbon versus resin adsorption of 2-methylisoborneol and chloroform. *Journal AWWA* 71(9):529–538.

Collins, R.P., L.E. Knaak, and J.W. Soboslai. 1970. Production of geosmin and 2-exo-hydroxy-2-methylbornane by *Streptomyces odorifer. Lloydia* 33:199–200.

Corwin, C.J., and R.S., Summers 2012. Controlling trace organic contaminants with granular activated carbon adsorption. *Journal AWWA* 104(1):43–44.

Coughlin, R.W., and F.S. Ezra. 1968. Role of surface acidity in the adsorption of organic pollutants on the surface of carbon. *Env. Sci. Tech.* 2(4):291–297.

Crittenden, J.C., J.K. Berrigan, and D.W. Hand. 1986. Design of rapid small scale adsorption tests for a constant diffusivity. *Jour. WPCF* 58(4):312–319.

Crittenden, J.C., J.K. Berrigan, D.W. Hand, and B.W. Lykins Jr. 1987. Design of rapid fixed bed adsorption tests for non constant diffusivities. *Jour. Envir. Engrg.* 113(2):243–259.

Crittenden, J.C., P.S. Reddy, H. Arora, J. Trynoski, D.W. Hand, D.L. Perram, and R.S. Summers. 1991. Prediction of GAC performance using rapid small scale column tests. *Journal AWWA* 83(1):77–87.

Dandekar, A., R.T.K. Baker, and M.A. Vannice. 1998. Characterization of activated carbon, graphitized carbon fibers and synthetic diamond powder using TPD and DRIFTS. *Carbon* 36(12):1821–1831.

Elhadi, S.L.N., P.M. Huck, and R.M. Slawson. 2006. Factors affecting the removal of geosmin and MIB in drinking water biofilters. *Journal AWWA* 98(8):109–119.

Gerber, N.N. 1969. A volatile metabolite of actinomycetes, 2-methylisoborneol. *Jour. Antibiot* 22:508–509.

Gillogly, T.E.T., V.L. Snoeyink, J.R. Elarde, C.M. Wilson, and E.P. Royal. 1998a. 14C-MIB adsorption on PAC in natural water. *Journal AWWA* 90(1):98–108.

Gillogly, T.E.T., V.L. Snoeyink, A. Holthouse, C.M. Wilson, and E.P. Royal. 1998b. Effect of chlorine on PAC's ability to adsorb MIB. *Journal AWWA* 90(2):107–114.

Graham M., and R.S. Summers. 1996. The role of floc formation and presence on adsorption by powdered activated carbon. In *Proc. of the AWWA Annual Conference and Exposition, Toronto, Ontario, Canada.* Denver, Colo.: AWWA.

Graham, M.R., I. Najam, M.R. Simpson, B. MacLeod, R.S. Summers, and L. Cummings 2000a. *Optimization of powdered activated carbon application for geosmin and MIB removal.* Report 909. Denver, Colo.: AWWA [now Water] Research Foundation.

Graham, M.R., R.S. Summers, M.R. Simpson, and B.W. Macleod, 2000b. Modeling equilibrium adsorption of 2-methylisoborneol and geosmin in natural waters. *Water Res.* 34:2291–2300.

Herzing, D.R., V.L. Snoeyink, and N.F. Wood. 1977. Activated carbon adsorption of the odorous compounds 2-methylisoborneol and geosmin. *Journal AWWA* 69(4):223–228.

Ho, L., and G. Newcombe. 2005. Effect of NOM, turbidity and floc size on the PAC adsorption of MIB during alum coagulation. *Water Res.* 39(15):3668–3674.

Huang, C., J.E. Van Benschoten, and J.N. Jensen. 1996. Adsorption kinetics of MIB and geosmin. *Journal AWWA* 88(4):116.

Izaguirre, G., C.J. Hwang, S.W. Krasner, and M.J. McGuire. 1982. Geosmin and 2-methylisoborneol from cyanobacteria in three water supply systems. *Appl. Envir. Microbiol.* 43:708–714.

Karanfil, T., and J.E. Kilduff. 1999. Role of granular activated carbon surface chemistry on the adsorption of organic compounds. 1. Priority pollutants. *Environ. Sci. Technol.* 33:3217–3224.

Kim, S. 2006. Understanding and predicting 2-methylisoborneol (MIB) adsorption by granular activated carbon and process selection approaches for controlling taste and order. Doctoral Thesis, University of Colorado, Boulder, Colo.

Kim, S.M., and R.S. Summers. 2006. Control of MIB and geosmin by granular activated carbon. In *Proc. AWWA Conference, San Antonio, Texas.* Denver, Colo.: AWWA.

Kim, S.M., R.S. Summers, and H. Ahn, H. 2007. Predicting MIB breakthrough in granular activated carbon adsorbers. In *Proc. AWWA Conference, Toronto, Ontario, Canada.* Denver, Colo.: AWWA.

Lalezary, S., M. Pirbazari, and M.J. McGuire. 1986. Evaluating activated carbons for removing low concentrations of taste- and odor-producing organics. *Journal AWWA* 78(11):76–82.

Lalezary-Craig, S., M. Pirbazari, M.S. Dale, T.S. Tanaka, and M.J. McGuire. 1988. Optimizing the removal of geosmin and 2-methylisoborneol by powdered activated carbon. *Jour. AWWA*, 80(3):73–80.

Lee, P., and J. Davidson. 1999. Evaluation of activated carbon filters for removal of ozone at the PPB level. *Amer. Ind. Hyg. Assoc. Jour.* 60(5):589–600.

Leng, C.-C., and N.G. Pinto. 1997. Effects of surface properties of activated carbons on adsorption behavior of selected aromatics. *Carbon* 35(9):1375–1385.

Mackenzie, J.A., M.F. Tennant, and D.W. Mazyck. 2005. Tailored GAC for the effective control of 2-methylisoborneol. *Journal AWWA* 97(6):76–88.

MacLeod, B.W., and M.R. Simpson. 1993. Enhanced coagulation: a case study of full-scale operation at Manatee Water Treatment Plant. In *Proc. AWWA Water Quality Technology Conference, Miami, Fla.* Denver, Colo.: AWWA.

Mallevialle, J., and I.H. Suffet, eds. 1987. *Identification and treatment of tastes and odors in drinking water.* Denver, Colo.: AWWA [now Water] Research Foundation.

McCreary, J.J., and V.L. Snoeyink. 1980. Characterization and activated carbon adsorption of several humic substances. *Water Res.* 14:151–160.

McGuire, M.J., and I.H. Suffet. 1984. Aqueous chlorine/activated carbon interactions. *Jour. Envir. Eng.* 110(3):629–645.

Newcombe, G., A. Collett, M. Drikas, and B. Roberts. 1996. Granular activated carbon pilot plant studies: an Australian perspective. *Water Supply* 14(2):281–284.

Newcombe, G., M. Drikas, and R. Hayes. 1997. Influence of characterised natural organic material on activated carbon adsorption: II. Effect of pore volume distribution and adsorption of 2-methylisoborneol. *Water Res.* 31(5):1065–1073.

Nowack, K.O., F.S. Cannon, and D.W. Mazyck. 2004a. Enhancing activated carbon adsorption of 2-methylisoborneol: methane and steam treatments. *Envir. Sci. & Tech.* 38(1):276–284.

Nowack, K.O., F. Cannon, A. Pinto, and D. Mazyck. 2004b. Enhancing activated carbon adsorption of MIB and geosmin from drinking water. In *Proc. of the AWWA Annual Conference and Exposition. Orlando, Fla.* Denver Colo.: AWWA.

Otake, Y., and R.G. Jenkins. 1993. Characterization of oxygen-containing surface complexes created on a microporous carbon by air and nitric acid treatment. *Carbon* 31(1):109–121.

Pendleton, P., S.H. Wong, R. Schumann, G. Levay, R. Denoyel, and J. Rouquerol. 1997. Properties of activated carbon controlling 2-methylisoborneol adsorption. *Carbon* 35(8):1141–1149.

Pirbazari, M., V. Ravindran, B.N. Badriyha, S. Craig, and M.J. McGuire. 1993. GAC adsorber design protocol for the removal of off-flavors. *Water Res.* 27(7):1153–1166.

Qi, S., S.S. Adham, V.L. Snoeyink, and B.W. Lykins Jr. 1994. Prediction and verification of atrazine adsorption by PAC. *Jour. Envir. Engrg.* 120(1):202–218.

Snoeyink, V.L., and R.S. Summers. 1999. Adsorption of organic compounds. In *Water quality and treatment*, 5th ed. R.D. Letterman, ed. New York: McGraw-Hill and AWWA.

Suffet, I.H., D. Khiari, and A. Bruchet. 1999. The drinking water taste and odor wheel for the millennium: beyond geosmin and 2-methylisoborneol. *Water Sci. & Tech.* 40(6):1–13.

Summers, R.S., D.R.U. Knappe, and V.L. Snoeyink. 2010. Adsorption of organic compounds by activated carbon. In *Water quality and treatment,* 6th ed. J.K. Edzwald, ed. New York: McGraw-Hill and AWWA.

Traegner, U.K., and M.T. Suidan. 1989. Parameter evaluation for carbon adsorption. *Jour. Envir. Engrg.* 115(1):109–128.

Westerhoff, P., R.S. Summers, Z. Chowdhury, and S. Kommineni. 2005. *Ozone-enhanced biofiltration for geosmin and MIB removal.* Denver, Colo.: AWWA [now Water] Research Foundation.

Young, W.F., H. Horth, R. Crane, T. Ogden, and M. Arnott. 1996. Taste and odour threshold concentrations of potential potable water contaminants. *Water Res.* 30(2):331–340.

Chapter 7

Activated Carbon for Organic Contaminant Control

Activated carbon may be used to control dissolved organic contaminants other than DBP precursors and T&O compounds. Target contaminants for removal may include (1) VOCs and SOCs for regulatory purposes, and (2) unregulated trace contaminants such as endocrine-disrupting compounds, pharmaceuticals, and personal care products for broad protection of public health and maintenance of consumer confidence.

This chapter addresses the use of activated carbon to remove target organic contaminants during the process planning stage. It first discusses the application of PAC, how PAC performance is evaluated, and how the PAC dose is determined. Next, it turns to GAC applications, evaluating GAC performance and predicting the GAC use rate. The factors affecting the control of MIB and geosmin by PAC and GAC discussed in chapter 6 extend to the control of target organic contaminants when the influent concentrations are in the parts-per-trillion range. These factors include having a fractional removal of contaminants that is independent of influent concentration and the significant effect of the background DOM concentration on contaminant removal.

PAC for Removal of Organic Contaminants

PAC Applications

PAC systems employed for control of organic contaminants are generally limited to surface water utilities concerned with intermittent occurrences. Intermittent occurrences typically include seasonal spikes of pesticides or herbicides above the MCL during spring runoff or the control of spills. PAC is generally not used in groundwater applications because an additional process is typically needed to remove the PAC.

PAC Performance Evaluation

The goal of the PAC performance evaluation is to determine whether PAC will be an economical solution to control a target organic contaminant. As such, the PAC performance evaluation ultimately consists of determining the PAC dose that should be applied to meet the treatment objective. In this section, first the factors that determine the PAC dose are discussed; then, methods used to determine the PAC dose are presented.

Determination of Required Dose

Several parameters should be investigated in order to determine the PAC dose needed to meet the treatment objective: contact time and mixing, PAC product, influent concentration of the target compound, background water quality, and treatment objective.

Contact Time and Mixing. Contact time and mixing are important variables in the performance of PAC; however, they are often constrained to the point that the engineer has limited ability to control them. In a typical PAC application, the contact time is fixed by the hydraulic residence times between the points of PAC addition and removal, with the engineer only able to specify the PAC addition and removal points. Similarly, mixing conditions are typically optimized for coagulation, flocculation, and settling performance, leaving the engineer with little opportunity to optimize the process for PAC. A PAC contact basin (discussed in chapter 4) allows the engineer to control the contact time and mixing conditions but requires a new capital facility and footprint area on the plant site. Generally, contact times of at least 15 min are needed for efficient PAC use, with times up to 90 min or more reducing the required doses.

PAC Product Selection. PAC products are manufactured from several base materials and are available with different activation processes. The result is multiple products from each manufacturer with widely variable cost and adsorption performance. The engineer, with the help of the PAC suppliers, may test several products to determine which will be most economical. Figure 7-1 shows the dose-response curves for an organic contaminant for three PAC products. If the treatment objective was a C/C_0 of 0.2 for this compound, PAC C requires a dose of 20 mg/L; whereas PAC B requires a dose of 60 mg/L, and the PAC A dose will be greater than 60 mg/L. However, the cost of each PAC product should also be taken into account to determine which PAC will be most economical.

Target Compound Influent Concentration. When a compound must be reduced to a specific concentration, such as an MCL, the PAC dose increases as the influent concentration of the target compound increases. If the concentration of the target compound is small compared with the background organic matter, the percent removal of a compound stays constant with PAC dose (Gillogly et al. 1998, Knappe et al. 1998, Graham et al. 2000, Matsui et al. 2003). In this case, the results of a single dose-response can be used to predict performance for a range of influent concentrations. For example, the data shown in Figure 7-1 were collected with an influent concentration of about 100 ng/L. Therefore, if the treatment objective is 80 percent removal, it will be met with PAC C at a

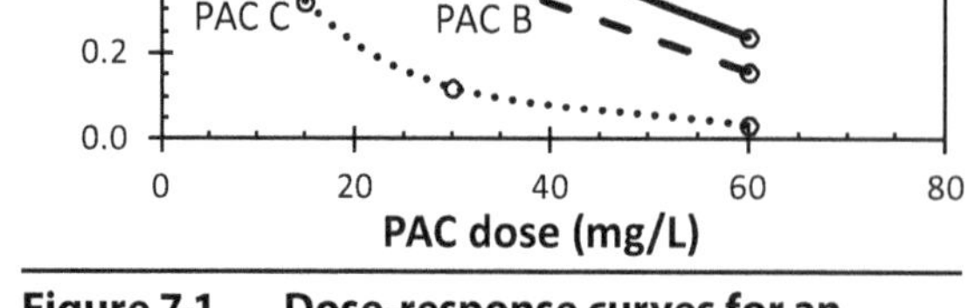

Figure 7-1. Dose-response curves for an organic contaminant for three PAC products (C_0 < 1 µg/L)

dose of 20 mg/L for any influent concentration below 1 μg/L. However, if higher influent concentrations occur, then a higher PAC dose will be needed to meet the treatment objective. This was discussed for PAC control of MIB in chapter 6. More detail is provided on the role of influent concentration in the GAC section below.

Background Water Quality. The concentration of DOC in the water has the largest influence on PAC adsorption of the target organic contaminant. The background DOC competes with the target adsorbate for adsorption sites available in the PAC. Most target organic adsorbates have a relatively low molecular weight compared with the bulk of the background DOM, resulting in faster diffusion of the target adsorbate than the DOM. The faster diffusion results in the target adsorbate only having to compete with a small portion of the background DOM with similar diffusion rates for adsorption sites. Thus, the influence of competition from background DOM may be less pronounced in PAC adsorption of target organic contaminants compared with GAC adsorption. However, the concentration of the background DOC is generally orders of magnitude higher than the target compound, resulting in a major reduction in adsorption capacity due to competition.

Treatment Objective. The treatment objective and the influent concentration of the target organic contaminant determine the level of removal required by PAC. In some cases, the treatment objective is based on a regulation, and in other cases, the utility should determine an appropriate treatment objective based on its overall water quality goals. When little regulatory guidance is available, often the treatment objective becomes a dependent variable based on the amount of removal that can be economically achieved.

PAC Evaluation Methods

Typically, PAC performance is evaluated with a jar test. In some instances, model predictions may also be useful.

Jar Testing. Jar testing is a simple, flexible, and effective method of generating the dose-response curve for removing an organic contaminant with PAC. A standard jar test apparatus is used to simulate the contact time and mixing conditions the PAC will experience. Source water should be used and may need to be spiked with the target compound(s) to ensure the influent concentration is high enough to provide meaningful results considering the detection limits of the analytical method used. The jar test should simulate the process train to the maximum extent possible and include chemical additions such as coagulants, pH control, and preoxidants. Variable doses of PAC are applied to each jar to generate the dose-response curve.

The jar test can be repeated to investigate other PAC products, varying target initial concentrations, varying background DOC concentrations, as well as varying other contact times and mixing conditions that model alternative PAC addition and removal points, as appropriate. Alternatively, the jars in the jar test apparatus can have a constant PAC dose applied and be sampled after varying times and mixing conditions to simulate all the PAC addition and removal points under consideration. Graham et al. (2000) developed a procedure for evaluating PAC for MIB and geosmin removal, and it can also be directly implemented for removal of other target organic contaminants.

Model Predictions. Because jar testing is both inexpensive and reliable, modeling is only of limited value in PAC performance predictions. No widely accepted models exist of PAC performance for target organic contaminant removal. Traditional dose calculations based on published isotherms should be avoided due to the error associated with them. To be of any value, the isotherm data should be collected in the actual water to be treated and with the specific PAC product to be used.

The homogeneous surface diffusion model (HSDM) is a mechanistic model of the batch adsorption process that may be useful. The HSDM requires input variables for adsorption capacity and kinetics (Freundlich isotherm parameters and target adsorbate surface diffusion coefficient) that are typically not readily known because they depend highly on the background water quality and the specific PAC product. However, the HSDM may be calibrated to experimental jar test kinetic data to fit the capacity and kinetic parameters. These variables are then assumed constant (for the same water, PAC product, and mixing conditions), and the HSDM can be used to investigate other conditions, such as PAC dose, initial concentration of the target compound, and contact time. The model results should always be verified with a jar test. The competition from background organic matter may be incorporated into multicomponent HSDM models that include the ideal adsorbed solution theory (IAST) calculations. However, the complexity of the models, requirements, and accuracy of the results generally make such an undertaking unproductive.

GAC for Removal of Organic Contaminants

GAC Applications

GAC can be effectively used for the control of a wide array of organic contaminants whether applied as full-depth or a GAC cap. However, organic contaminants have a wide range of adsorbabilities so economics of the adsorption process depend largely on the target contaminant and the treatment objective. GAC may be particularly effective when removal of multiple organic compounds is desired. Adsorption performance is also specific to the background water quality and the GAC product characteristics. Application of GAC caps may be limited by the generally short EBCTs, which require frequent media replacements. BAC filter performance depends highly on the compound biodegradability and the frequency of occurrence, and thus is limited where high levels of removal of a wide range of organic contaminants are required. However, GAC contactors that are allowed to operate biologically may degrade some organic contaminants, providing more complete treatment and better protection of public health at little or no extra cost.

GAC Performance Evaluation

A GAC performance evaluation is undertaken in order to determine whether the treatment objective(s) can be economically achieved with GAC. Ultimately, the GAC use rate determines the economics of adsorption and is considered the figure of merit for GAC performance evaluation. Evaluating GAC is typically more complex than evaluating PAC because PAC operates at a steady state whereas GAC is an unsteady-state process. The factors that affect GAC use rate and the methods to determine the economics of adsorption are discussed in this section.

Determination of the Use Rate

Like most GAC processes, determining the GAC use rate for target organics depends on EBCT, the GAC product, the influent concentration of the target organic, the background water quality, and the treatment objective. The effects of these variables are explored in this section along with the role of desorption on adsorber performance.

EBCT. As shown in chapter 1, the EBCT does not affect breakthrough on a throughput basis in a single-adsorbate system. However, a dependence on EBCT was shown for trace organic contaminants breakthrough in the presence of DOM by Corwin and Summers (2012). Figure 7-2 shows this behavior for the target compound bisphenol A. Figure 7-2 shows that when breakthrough of bisphenol A is less than 10 percent, longer EBCTs perform better. At breakthroughs above 10 percent, shorter EBCTs perform better. Thus, the optimum EBCT will depend on the treatment objective. The breakthrough behavior of bisphenol A shown in Figure 7-2 is typical of other trace organic contaminants, but the crossover point where EBCTs result in equal performance may vary.

Performance dependence on EBCT occurs due to the effects of competitive adsorption. When the fresh adsorber is first put in operation, target organics and the background DOM matrix compete for adsorption sites. Generally, the target organic will have molecular weight in the range of 80 to 700 D, with most around 200 to 300 D, which results in a constant mass transfer zone for each compound. The background DOM matrix may have a molecular weight distribution including molecules up to 10,000 D, although coagulation is effective at removing the largest size fraction of the DOM shifting the distribution to the <4,000-D range (Matilainen et al. 2006). The molecular-weight distribution of the DOM results in a distributed mass transfer zone, with the bulk of the mass transfer zone being longer than that of the target adsorbate. Thus, in the top of the fresh adsorber, the target adsorbate only competes for adsorption sites with the fraction of DOM in the same molecular weight range, which is termed *co-loading*. However, as adsorption capacity for the target adsorbate is exhausted in the top of the column, the mass transfer zone of the target adsorbate starts to migrate deeper into the column where adsorption of the DOM has been occurring since the start of operation; in other words, the GAC has been preloaded by DOM. Preloading by the DOM results in lower adsorption capacity for the target adsorbate by two competitive adsorption mechanisms: direct competition and pore blocking.

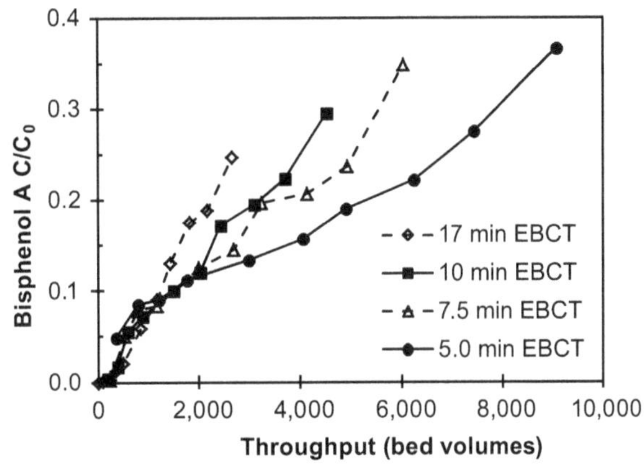

Figure 7-2. Bisphenol A breakthrough exhibits dependence on EBCT

The crossover point in Figure 7-2 is the point at which adsorption transitions from co-loading to preloading. Once the preloading conditions dominate, the shorter EBCTs outperform longer EBCTs (on a use-rate basis) because the longer columns result in more preloading and, consequently, less adsorption capacity for the target adsorbate. The crossover point is typically located between 10 and 30 percent breakthrough, often in the same range as that of the treatment objective, which may simplify selection of the optimum EBCT.

GAC Product Selection. GAC products are available in a variety of sizes from several available base materials. Bituminous, lignite, coconut, and wood GACs are the most common base materials for the removal of organic contaminants in water treatment applications. Bituminous products are most common for removal of organic contaminants because they are widely available and provide a good balance between cost and performance.

The particle-size distribution of the GAC product affects adsorber performance. Smaller GAC particles have better adsorption kinetic properties, but they cause higher head losses. Larger GAC particles may have a lower adsorption capacity on a per-mass-of-GAC basis, which likely results from the activation process not being able to penetrate the larger GAC particles as effectively.

Chemically activated and surface-treated GAC products are usually avoided for the control of target organics in municipal water treatment because these products are more costly and typically are difficult to reactivate to virgin capacity economically.

Influent Concentration of the Target Compound. The influent concentration of a target compound has a great effect on its adsorption behavior in GAC systems. Under typical conditions where the Freundlich equation describes the isotherm, higher influent concentrations result in a higher percentage of the target organic adsorbed because of the greater driving force to the GAC surface. However, because the mass loading rate is also higher, breakthrough occurs faster for higher influent concentrations.

When the influent concentration of the target adsorbate is several orders of magnitude lower than the competing background organic matter, the isotherm tends to linearize (Knappe et al. 1998, Kilduff et al. 1998), which can fundamentally affect adsorption phenomena. Linearization results in two significant adsorption behaviors: a lack of influence of influent concentration on breakthrough and a lack of competitive effects from other adsorbates.

Linearization of the isotherm at low concentrations can be explained in two ways. First, the adsorption theory behind the Langmuir isotherm results in a linear isotherm at low concentrations. However, clean-water isotherms at low concentrations are typically better described by the Freundlich isotherm (Yu et al. 2008). The second, and better, explanation is made from examination of the IAST due to competitive effects. Knappe et al. (1998), Kilduff et al. (1998), and Graham et al. (2000) showed theoretically that the IAST predicts a linearization of a Freundlich isotherm for an adsorbate when competing against another adsorbate at much higher concentrations. The threshold concentration where linearization occurs is not well defined, but it is typically in the low-μg/L range. The threshold becomes higher as adsorbability decreases, because weakly adsorbing compounds cannot compete against the background matrix as effectively.

Several researchers have previously reported the lack of influence of initial concentration on the fraction remaining in batch isotherm studies in the presence of DOM (Gillogly et al. 1998, Knappe et al. 1998, Matsui et al. 2003). Matsui et al. (2003) and Corwin and Summers (2012) observed that very low initial concentrations did not affect normalized breakthrough for continuous flow adsorbers. Figure 7-3 shows

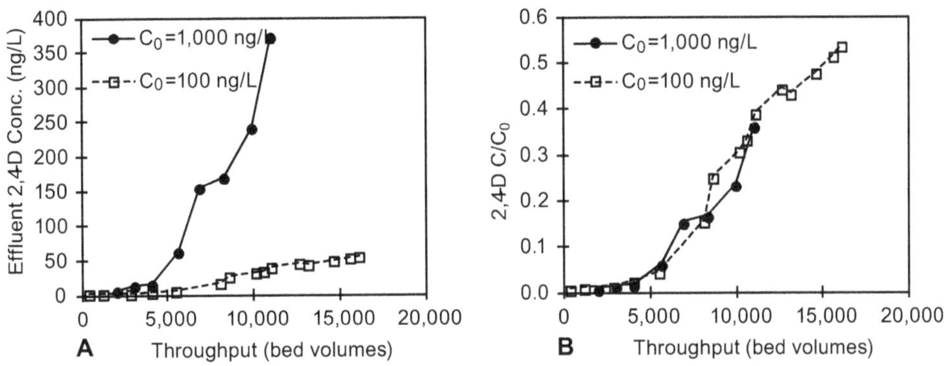

Figure 7-3. 2,4-dichlorophenoxyacetic acid breakthrough at two concentrations shown on concentration and normalized concentration basis

2,4-dichlorophenoxyacetic acid breakthrough data at 100 and 1,000 ng/L under otherwise identical conditions. Figure 7-3a presents breakthrough on a concentration basis and indicates that the higher influent concentration breaks through faster. However, if the data are plotted on a concentration-normalized basis as in Figure 7-3b, the breakthroughs are the same. Similar results were shown for GAC control of MIB in Figure 6-6. From a practical standpoint, if the treatment objective is a set concentration, then the GAC media will have a shorter bed life as concentration increases. Importantly, a single breakthrough curve can be used to predict performance of all concentrations low enough to linearize the isotherm. For example, if the treatment objective of 2,4-D is 50 ng/L, Figure 7-3a shows the bed life will be about 5,000 BV for an influent of 1,000 ng/L and 15,000 BV for the 100 ng/L influent. The bed life at an influent concentration of 500 ng/L can be determined from 7-3b. The 2,4-D C/C_0 is the 50 ng/L treatment objective divided by 500 ng/L influent, which results in a C/C_0 of 0.1. The bed life from Figure 7-3b to a C/C_0 of 0.1 is about 6,500 BV.

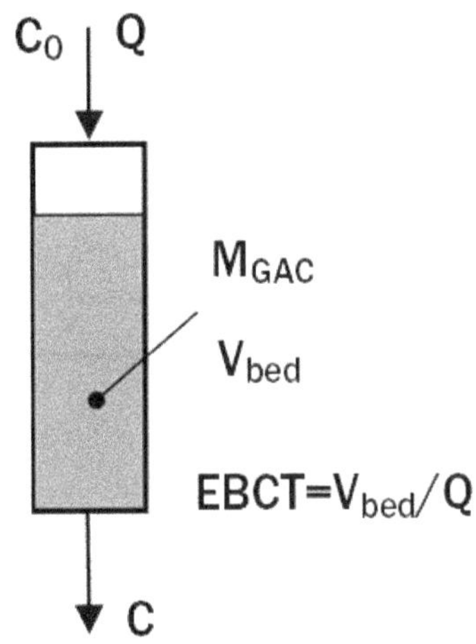

Figure 7-4. Mass balance on GAC adsorber

A simplified IAST approach can be used to explain the lack of dependence on initial concentration with a linear isotherm for continuous-flow adsorbers. A mass balance is written around the adsorber (shown in Figure 7-4) for the mass adsorbate removed per mass of the GAC in the adsorber and equated to the solid-phase concentration in the Freundlich equilibrium, assuming local equilibrium in Eq 7-1.

$$\frac{(C_0 - C) \times Q \times t}{M_{GAC}} = K \times C_0^{1/n} \quad \text{(Eq 7-1)}$$

Where

C_0 = influent concentration
C = effluent concentration
Q = flow rate
t = operation time
M_{GAC} = mass of GAC
K = Freundlich adsorption capacity parameter
$1/n$ = Freundlich adsorption intensity parameter

Eq 7-1 is rearranged to solve for the normalized breakthrough, C/C_0, in Eq 7-2:

$$\frac{C}{C_0} = 1 - \frac{K \times M_{GAC}}{Q \times t} \times \frac{C_0^{1/n}}{C_0} \qquad \text{(Eq 7-2)}$$

Eq 7-2 includes C_0 terms on the right-hand side of the equation, which shows that normalized breakthrough depends on initial concentration when $1/n \neq 1$.

Graham et al. (2000) demonstrated the lack of competition between two trace organics in the presence of background organic matter at much higher concentrations using the IAST. Thus, when many trace organic contaminants are present in water at very low concentrations compared with the background organic matter, the adsorption of any one of the trace organic contaminants is determined by its competition with the background organic matter and not the presence of other trace organic contaminants. When these conditions occur, breakthrough curves may be considered valid for all conditions as they do not depend on the concentration of other micropollutants in the water matrix.

Background Water Quality. The background water quality affects adsorption behavior of target organics primarily by providing DOM that competes with the target organics for adsorption sites. Thus, the most important measure of background water quality is DOC. However, DOM is highly variable, and a single surrogate measurement is not always adequate to predict the effect of DOM.

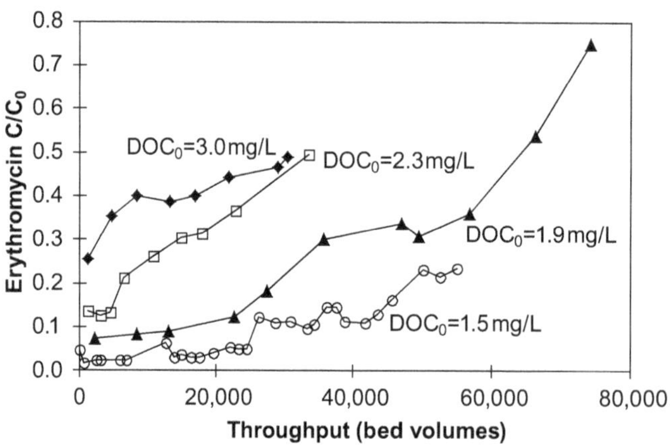

Figure 7-5. Effect of influent DOC on erythromycin breakthrough

Figure 7-5 shows the breakthrough of erythromycin in four coagulated waters with DOC concentrations ranging from 1.5 to 3 mg/L and otherwise similar conditions (EBCT = 8 min, Calgon F300 GAC, erythromycin C_0 = 100 ng/L). The breakthrough of erythromycin becomes progressively earlier as the DOC concentration increases. The behavior can be rationalized by noting increases in the DOC concentration provides more adsorbates to compete with erythromycin for a fixed number of adsorption sites. Also of note in Figure 7-5 is that the doubling of the DOC concentration from 1.5 to 3 mg/L decreases bed life (measured at a treatment objective of C/C_0 = 0.2 for convenience) from 50,000 BV to about 5,000 BV, or 90 percent, which is much more than the corresponding 50 percent decrease in DOC.

Pretreatment can affect both the concentration and character of the DOM, which affects the way DOM competes with the target organic for adsorption sites. Coagulation is the most common pretreatment prior to GAC. Coagulation results in the removal of a substantial portion of the total DOC and nearly all the larger material above 4,000 D (Matilainen et al. 2006), which tends to homogenize the DOM. Most coagulated waters can be well characterized by DOC in reference to GAC adsorption performance.

Oxidation is another common pretreatment that tends to affect the adsorption characteristics of the background water matrix. Oxidation, particularly with chlorine-based oxidants and ozone, tends to shift the molecular-weight distribution of the DOM toward lower molecular weights, which may result in less pore blockage and more direct competition for adsorption sites. Oxidation of the DOM also tends to make the DOM more hydrophilic and less strongly adsorbed. A decrease in UVA with no change in DOC is typically observed after oxidation, resulting in a lower specific ultraviolet absorbance (SUVA). The effect of oxidation on adsorption should be assessed on a case-by-case basis.

Ion exchange also affects the adsorption characteristics of DOM. Ion exchange may significantly lower the TOC concentration, which results in less competition for adsorption sites, much like coagulation. However, ion exchange tends to favor removal of the low-molecular-weight fraction of the DOM that competes directly for adsorption sites with target organics. Therefore, ion exchange may be particularly effective for improving GAC performance for target organics, but it must be shown on a case-by-case basis. Any improvement gained in the GAC process from ion exchange (or any pretreatment process) should be weighed against the cost of pretreatment.

In some cases, membranes may be used upstream of a GAC process. Low-pressure membranes are sometimes used upstream of GAC, but they only remove particulate organic matter, which is typically less than 10 percent of the TOC. Thus, low-pressure membranes have little effect on GAC performance. In cases where coagulant is applied upstream of the membrane, significant removal of DOC may occur with the resulting DOM being similar to other coagulated waters. In rare cases, GAC may be used downstream of high-pressure membranes to provide a barrier against contaminants that are not well rejected by the membrane. In this case, pretreatment by the high-pressure membrane results in removal of most of the competing dissolved organics, yielding very long GAC life.

Treatment Objective. As with any process, treatment objective along with the influent concentration of the target compound determine the level of removal needed by the GAC process. Typically, treatment objectives for target organics result in high levels of removals required in the GAC process (>80 percent). As the level of removal required increases, the importance of kinetics increases (chapter 1), and the more likely series operation will be beneficial (chapter 4).

Desorption. Desorption of an organic contaminant is possible when the influent concentration drops below the equilibrium point of the surface concentration of the

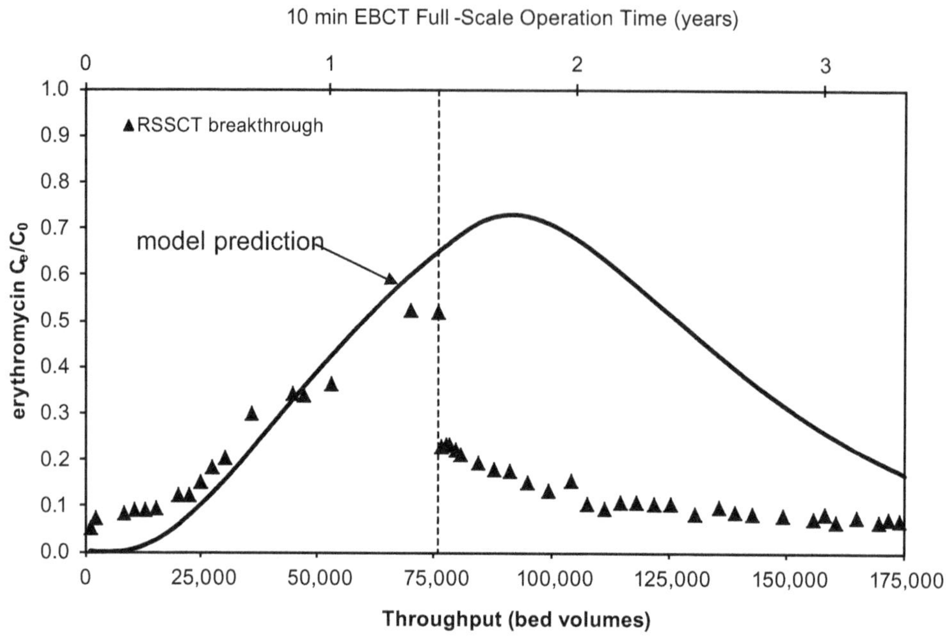

Figure 7-6. Desorption behavior compared to a model prediction calibrated to the adsorption performance

adsorbate. Desorption conditions may occur in two common scenarios: (1) when a pulse of contaminant passes through the adsorber and (2) after backwashing. Corwin and Summers (2011) studied both of these conditions and found that the kinetics of desorption of organic contaminants was 20 to 50 times slower than the kinetics of adsorption. Kinetics of back diffusion was hypothesized to be hindered by accumulated organic matter in the pore. The very slow desorption kinetics lead to very low levels of target compound desorbing after a pulse loading for a very long time. Figure 7-6 shows desorption of erythromycin after 75,000 bed volumes of adsorption. Desorption occurs at about 10 percent of the influent loading concentration and is still occurring after 100,000 bed volumes of desorption. At the end of the test, only 18 percent of the mass of the adsorbed compound had desorbed. The behavior shown in Figure 7-6 has two implications for adsorbers used for the control of organic contaminants. First, where the treatment objective is based on an MCL, desorption is unlikely to result in out-of-compliance operation. Second, where removal of mass is the treatment objective, desorption is a concern but occurs so slowly that some removal will likely occur when the GAC media are replaced.

Backwashing of GAC adsorbers results in a mixing of the GAC media, which disrupts the mass transfer zone. Disruption of the mass transfer zone may lead to desorption if GAC particles saturated with adsorbate are transferred to the bottom of the column where there is no adsorbate in the liquid phase because it has been adsorbed by the remaining capacity in the above GAC media. In this case, the concentration gradient reverses and desorption occurs. However, Corwin and Summers (2011) found the slow desorption results in a negligible effect on breakthrough due to mixing of the GAC media. Figure 7-7 shows data from columns that were unmixed, mixed in three segments, and completely mixed. Mixing of the GAC media does not affect breakthrough for the five relatively weakly adsorbed compounds over a range

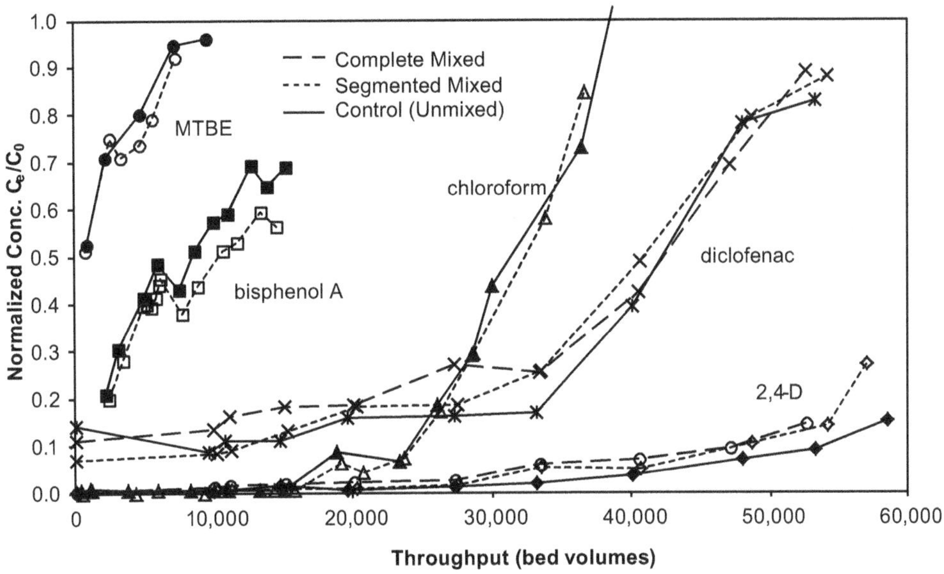

Figure 7-7. Backwashing shown to have a negligible effect on breakthrough of target organic contaminants (Courtesy of IWA)

of influent concentrations. Mixing of GAC media does not have an effect because the slow desorption occurs on a much longer time scale than breakthrough.

GAC Evaluation Methods

Several methods for GAC performance evaluation are available to the engineer. These include pilot-scale testing, bench-scale testing, and model predictions. Much of the cost of evaluating GAC for organic contaminants is in the analysis of samples for the target compound or, in many cases, a suite of compounds. Thus, minimizing the cost of GAC evaluation is achieved by reducing the number of samples and the number of analyses performed on each sample by selecting compounds that can be analyzed with a single method.

Pilot-Scale Testing. Pilot-scale testing is currently the most commonly used and accepted method to predict GAC performance. However, because pilot testing is time and resource consuming, it is inefficient for preliminary assessment of GAC economic feasibility. Pilot testing generally takes place over several months at the water source. Depending on the conditions at the pilot plant, the target compounds may or may not be able to be spiked into the influent water. Spiking the influent can be informative in situations where target compounds are not present in the influent water at all times of the year or to raise the concentration of the target compound to allow for easier analysis (where raising the influent concentration is unlikely to affect adsorption performance). In surface water plants, the background water quality will also vary throughout the pilot test. When a pilot test is used to confirm performance of a design, these conditions are preferred as being representative of actual service conditions. However, in pilot studies where design conditions are to be investigated, the variable influent water quality makes it more difficult to interpret the results and determine the optimal adsorber configuration.

Bench-Scale Testing. Two bench-scale tests that may provide useful data are the rapid small-scale column test (RSSCT) and isotherm testing.

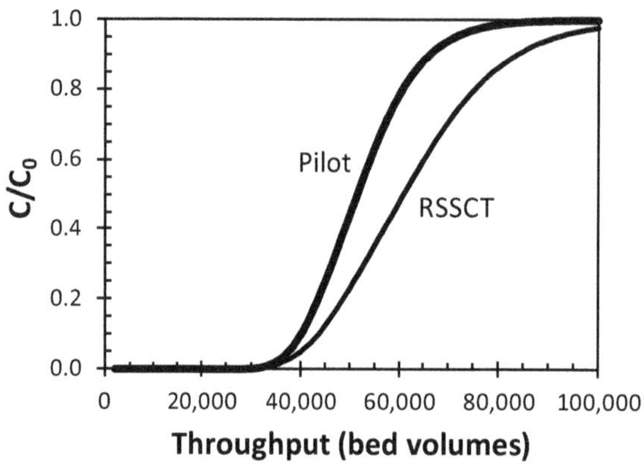

Figure 7-8. Typical results of RSSCTs versus pilot results

The RSSCT is a bench-scale GAC test that uses the principles of similitude to scale down the adsorption process based on the dispersed flow pore and surface diffusion model. GAC is crushed to a smaller particle size, and the scaling equations ensure the spread of the mass transfer zone due to the mass transfer mechanisms will be the same in the large and small columns. However, two notable problems have been reported in comparing RSSCT data to full-scale data for organic contaminants. First, the dependence of adsorbate diffusion on GAC particle size must be known. The breakthrough data in Figure 7-8 exhibit different slopes in the pilot versus the RSSCT, which is the manifestation of not matching adsorption kinetics. Second, RSSCTs exhibit more adsorption capacity than full-scale data. Capacity can be determined from a breakthrough curve by integrating the area above the breakthrough curve to the influent concentration. Because breakthrough curves are approximately symmetrical, capacity can be estimated by the throughput to 50 percent breakthrough. The RSSCT shown in Figure 7-8 exhibits more adsorption capacity than in the pilot column.

For most target organic contaminants, intraparticle diffusivity is not constant with GAC particle size. Crittenden et al. (1987) showed that the RSSCT can still be used in cases where diffusivity depends on GAC particle size, but the intraparticle diffusivity must control adsorption kinetics because film mass transfer will not be perfectly matched. The dependence of diffusivity on particle size appears to be compound specific. Therefore, a separate RSSCT would have to be run for each compound in order to match breakthrough curves perfectly.

Corwin and Summers (2010) found that grinding of the GAC particles affected fouling in a way that was not scalable. Grinding of the GAC opens more of the internal surface area available to the bulk flow, making the small GAC less susceptible to pore blockage. Corwin and Summers (2010) presented a method to correct for the capacity differences, but data at two particle sizes are required. Performing RSSCTs at two particle sizes is probably impractical for most situations, but in some cases, the RSSCT data may be combined with pilot data to yield useful results. Often, spiking contaminants is not an option for pilot studies for a number of reasons. In this case, if just one or two compounds can be analyzed in the pilot data, they can then be matched to the RSSCT data for the same compounds. Then the rest of the RSSCT data can be adjusted because the capacity difference due the scalability of fouling does not appear to be compound specific. The RSSCT data can now be used for performance predictions for all the compounds included in the RSSCT.

Even without the perfect prediction of adsorption capacity or kinetics, the RSSCT provides useful information. RSSCTs can be used to investigate different GAC products, the effect of EBCT, and the effect of influent concentrations. Obtaining this information on the bench-scale early in the design process can be used to design more efficient and less costly pilot studies. Bench-scale studies also have the advantage of being able to be performed with a small amount of water in a central laboratory, making spiking of target organics more practical. The stable influent water quality also makes analysis of the data straightforward.

Most waters contain significant enough background DOM to make surface diffusion negligible compared with pore diffusion. When pore diffusion controls, Biot numbers typically favor intraparticle mass transfer control. Thus, a nonconstant RSSCT design is appropriate. The proportional diffusivity RSSCT (PD-RSSCT) will be the best choice and will also yield an accurate DOC breakthrough curve, but the breakthrough curves will not perfectly match the full-scale adsorption kinetics. Often, not matching kinetics leads to much less error than not matching capacity and is thus acceptable.

The value of isotherm testing is generally limited for evaluating GAC because only adsorption capacity information is obtained. The capacity data obtained are generally not useful in predicting GAC performance for two reasons. First, the competitive mechanisms of adsorption are not well simulated in the batch test because preloading effects are not representative of a column system. Second, because high levels of removal are usually required by the treatment objective, the kinetics of adsorption control the use rate as shown in chapter 1.

Isotherm testing may be beneficial in the preliminary screening stage. If GAC is to be used to control a suite of target organic contaminants, isotherm testing can be used to determine which contaminants are the mostly weakly adsorbing of the targets. Thus, further testing can be performed on just a subset of all the target compounds reducing the analytical load and cost. Isotherm testing can also be used for preliminary screening of GAC products, but it must be recognized that because isotherm testing does not accurately simulate fouling conditions, only large differences in capacity should be considered meaningful. Small differences in adsorption capacities should be investigated further by side-by-side continuous flow testing.

Model Predictions. Model predictions are of limited value in evaluation of GAC performance. To date, no widely accepted empirical models are available to predict GAC performance for the adsorption of organic compounds. The future of such models may be limited because of the huge number of adsorbates of concern and the fact that performance depends highly on the (1) GAC product, (2) adsorbate, and (3) background DOM. Furthermore, adsorbate behavior is not easily predicted by physical or chemical properties such as molecular weight, ionic state, and hydrophobicity.

Theoretical models may be of some value in much the same way the HSDM may be used with PAC jar testing. The pore and surface diffusion model (PSDM) is a mechanistic model of continuous flow column adsorption (Crittenden et al. 1986, Hutzler et al. 1986) with a computer package available in the AdDesignS package available from Michigan Technological University.

Jarvie et al. (2005) proposed a hybrid model that uses empirical fouling correlations used to adjust the Freundlich K value in the PSDM. The method basically consists of selecting a surrogate water, calculating the Freundlich K value for trichloroethylene (TCE) to account for fouling after a certain time, then calculating the Freundlich K for the specific organic compound of interest using another correlation.

Surrogate water coefficients and properties are reported in Table 7-1. Each of the fouling correlations yields a very different result and does not appear to trend with

Table 7-1. Surrogate water coefficients and properties (Jarvie et al. 2005)

Background Water Matrix	DOC (mg/L)	A_1	A_2 day^{-1}	A_3	A_4 day^{-1}
Surface water with significant anthropogenic input (Rhine River, Germany)	2.25	35.0	8.86×10^{-4}	65.0	1.29×10^{-1}
Surface water with small anthropogenic input (Portage Lake, Mich.)	1.6	51.0	1.33×10^{-1}	49.0	4.03×10^{-2}
Groundwater in Germany found similar to six other German groundwaters (Karlsruhe, Germany)	0.7	65.0	9.66×10^{-2}	35.0	1.44×10^{-1}
Rural Midwestern groundwater (Wausau, Wis.)	8.1	83.0	1.31×10^{-1}	17.0	3.82×10^{-1}
Rural Northern groundwater (Houghton, Mich.)	1.5	66.0	2.23×10^{-2}	34.0	1.05×10^{-1}

Table 7-2. Correction factors for different classes of compounds (Jarvie et al. 2005)

Group	Surrogate Compound	B_1	B_2
Halogenated alkanes	1,1,1-Trichloroethane	1.2	−0.2
Halogenated alkenes	Trichloroethylene	1.0	0
Trihalomethanes	Chloroform	1.0	0
Aromatics	Toluene	0.9	0.1
Nitro compounds	3,4-Dinitrotoluene	0.75	0.25
Chlorinated hydrocarbons	1,4-Dichlorobenzene	0.59	0.41
Phenols	2,4-Dichlorophenol	0.65	0.35
Polynuclear-aromatics	Methylene blue	0.32	0.68
Pesticides	Atrazine	0.0	0.05

DOC, so care must be used in selecting the surrogate water. The percent reduction in the Freundlich K for TCE is calculated by Eq 7-3:

$$\left[\frac{K(t)}{K}\right]_{TCE} = \frac{A_1 - A_2 \times t + A_3 e^{-A_4 \times t}}{100} \qquad \text{(Eq 7-3)}$$

The Freundlich K of the target compound can now be calculated with the correlation in Eq 7-4:

$$\frac{K(t)}{K} = B_1 \times \left[\frac{K(t)}{K}\right]_{TCE} + B_2 \qquad \text{(Eq 7-4)}$$

Coefficients for Eq 7-4 can be found in Table 7-2, separated by class.

Surface diffusion of trace organic contaminants has been shown to be negligible for systems containing DOM (Hand et al. 1984, Carter and Weber 1994, Matsui et al. 2001). The pore diffusion coefficient (D_p) is calculated by dividing the liquid diffusion coefficient in water (D_L) of the target compound by the tortuosity. The liquid diffusion coefficient in water of each target compound may be estimated via the Hayduk and Laudie (1974) correlation shown in Eq 7-5.

$$D_L = \frac{13.26 \times 10^{-5}}{\mu^{1.14} V_b^{0.589}} \quad \text{(Eq 7-5)}$$

Where

D_L = Liquid diffusivity in water, cm^2/sec
μ = viscosity, centipoise
V_b = Molal volume at boiling point, cm^3/mol

The Gnielinski correlation, shown in Eq 7-6, may be used to estimate the film mass transfer coefficient for spherical particles under laminar flow conditions (Gnielinski 1978).

$$k_f = \frac{(1 + 1.5(1 - \varepsilon))D_L}{d_p} \left[2 + 0.644 Re^{1/2} Sc^{1/3}\right] \quad \text{(Eq 7-6)}$$

Where

k_f = film mass transfer coefficient, cm/sec
ε = bed porosity, unitless
d_p = GAC particle diameter, cm
D_L = liquid diffusivity in water, cm^2/sec

$$Re = \frac{d_p V}{\varepsilon \upsilon}$$

$$Sc = \frac{\upsilon}{D_L}$$

Where

V = filter velocity
υ = kinematic viscosity

A tortuosity (τ) value of three was found to result in a good match between the PSDM to the experimental data and is in the range considered typical for GAC in water treatment (Sontheimer et al. 1988, Cussler 2009).

With all the equilibrium and mass transfer parameters defined, the system can be investigated for any design using the PSDM. While this model procedure may be worthwhile as a desktop investigation in the absence of other data, it has a lot of implicit uncertainty. When some breakthrough data are available, Corwin and Summers (2011) presented a methodology for using the PSDM to calibrate the model to observed performance for trace organic contaminants. The calibrated model can then be used to investigate other conditions such as EBCT, loading rate, and temperature for the same water and GAC.

The Freundlich equilibrium parameters can be calculated by assuming that the apparent capacity (i.e., the volume of water treated at 50 percent breakthrough) of a nonideal adsorber is near the quantity of water treated at exhaustion of an ideal plug flow adsorber. The assumption is good if the mass transfer zone is symmetrical.

The starting point is the mass balance shown in Eq 7-1. For the assumed ideal plug flow adsorber, C is equal to zero. The product of Q and t is the volume of water treated at breakthrough and is equal to the number of bed volumes treated multiplied by the volume of the bed (V_{bed}). These substitutions are made and the equation solved for K, which is now replaced with K* to denote that this is the apparent capacity of a specific GAC for the target compound in a specific water in Eq 7-7:

$$K^* = \frac{BV_{50} \times V_{bed}}{M_{GAC}} \times \frac{C_0}{C_0^{1/n}} \qquad \text{(Eq 7-7)}$$

Where

K* = apparent capacity parameter
BV_{50} = bed volumes to 50 percent breakthrough
V_{bed} = volume of the empty GAC bed

Eq 7-7 may be used to calculate K* for higher concentration ranges by assuming the 1/n value stays constant as shown in Sontheimer et al. (1988). In this case, the 1/n value can be estimated by the values found in the literature for clean-water isotherms. For systems with concentrations low enough to linearize the isotherm, the substitutions $1/n = 1$ and $\rho_{bed} = M_{GAC}/V_{bed}$ are made in Eq 7-7 to the form shown in Eq 7-8, which is identical to the specific throughput at 50 percent breakthrough, which equals the inverse of the GAC use rate.

$$K^* = \frac{BV_{50}}{\rho_{bed}} \qquad \text{(Eq 7-8)}$$

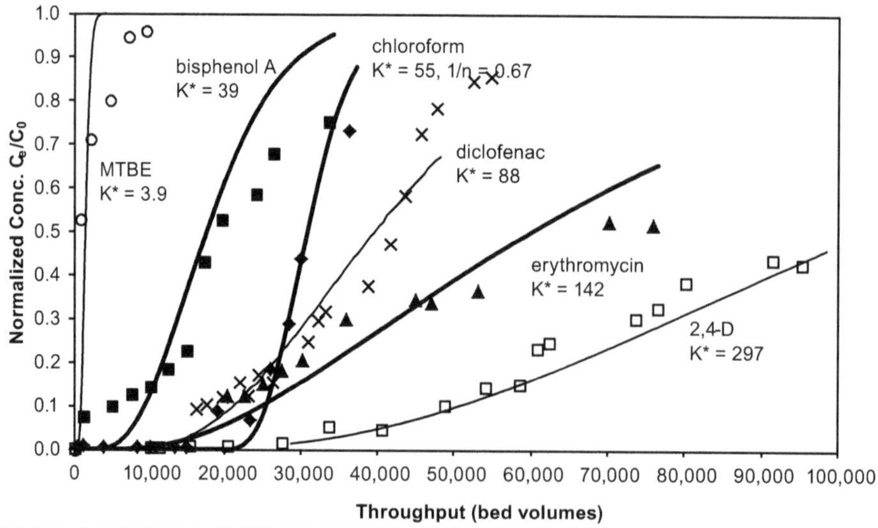

Figure 7-9. Apparent capacity model simulations (Courtesy of IWA)

Results of the apparent capacity simulation approach using the PSDM with no surface diffusivity and τ =3 are shown in Figure 7-9 for six different probe compounds with fresh and DOC preloaded GAC. The simulation provides a good match of the breakthrough curves for the seven probe compounds over a wide range of adsorbabilities and diffusivities. Also, data in Figure 7-9 show the widely varying slopes of the breakthrough curves are well described by using a pore diffusion model with D_p based on the calculated liquid diffusivity and τ = 3.

Common Organic Contaminants Controlled Using Activated Carbon

All surface waters and groundwaters contain a wide range of organic compounds, many that are naturally occurring. However, anthropogenic contaminants are also prevalent. These include a number of regulated compounds that have been categorized as volatile or synthetic organic compounds (VOCs and SOCs), as well as numerous unregulated organic contaminants of emerging concern (often deemed "emerging contaminants" or "micropollutants").

Volatile Organic Compounds

Volatile organic compounds (VOCs) are a class of man-made compounds that are used in a variety of industrial processes. These are relatively low-molecular-weight compounds with a high vapor pressure, which allows them to readily volatilize and evaporate from the water into the air. Because of their volatility, they are primarily a groundwater concern and rarely found in surface waters. VOCs leach into groundwater from many sources, such as chemical spills that migrate through the subsurface. This section highlights the VOCs that are commonly detected in drinking water and outlines activated carbon treatment considerations.

TCE and PCE

Trichloroethylene (TCE) and tetrachlorethylene (PCE) are solvents used in metal degreasing, dry cleaning, and textile processing applications for their solvent properties. TCE and PCE often co-occur and are among the most frequently detected VOCs in groundwater in the United States (Fischer et al. 1987, Zogorski et al. 2006). Groundwater generally becomes contaminated with TCE and PCE due to spills and discharges from industrial activities. Urban land-use practices and hazardous-waste facilities can also be sources of TCE and PCE contamination (Zogorski et al. 2006). These chemicals are considered toxic at low concentrations, meaning relatively small amounts of either chemical can contaminate large volumes of groundwater. USEPA has established an MCL of 0.005 mg/L or 5 µg/L for TCE and PCE each, but both are currently under review by the USEPA with MCLs as low as 0.5 µg/L being considered.

BTEX Compounds

The BTEX group of contaminants consists of benzene, toluene, ethyl benzene, and xylene. These organic chemicals make up a significant percentage of petroleum products such as crude oil, diesel, and gasoline. For example, standard gasoline is about 18 percent BTEX by weight. BTEX chemicals are highly soluble in water and thus can easily move through soil into the groundwater. A major source of BTEX contamination of groundwater is the release of petroleum products from leaking underground storage tanks (Pillai 2009).

BTEX contaminants are known to have some acute and long-term toxic effects. Benzene is considered a human carcinogen (NTP 2001, IRIS 2001, IARC 1987), and the associated MCL is 0.005 mg/L. Ethyl benzene is considered a possible human carcinogen (IARC 2000) and has an MCL of 0.7 mg/L. Although not currently classified as human carcinogens, toluene and xylene are toxic by other modalities (IRIS 2001 and IARC 1999a, 1999b), and the MCLs for these compounds are 1 and 10 mg/L, respectively.

MTBE

Methyl-*tert*-butyl-ether (MTBE) is a fuel oxygenate. MTBE may be added to gasoline to increase its oxygen content for better combustion and reduced exhaust emissions (e.g., carbon monoxide and ozone). In the United States, MTBE began being added to gasoline in 1979, and its use significantly increased in 1992 as a result of the oxygenate requirements (for gasoline) set by Congress in the 1990 Clean Air Act Amendments (USEPA 2010). By the early 2000s, nearly 30 percent of all gasoline sold in the United States contained MTBE in the amount of 11 percent by volume (Baus et al. 2005). MTBE has been detected in groundwater and surface water throughout the United States as a result of leaking underground storage tanks and pipelines, spills, emissions from marine engines into lakes and reservoirs, and, to some extent, air deposition (Carter et al. 2006, USEPA 2010).

Although USEPA has not yet established a national drinking water standard for MTBE, its offensive T&O at low concentrations has caused public concern. Stocking et al. (2001) conducted a consumer study to evaluate the odor threshold of MTBE and determined that it was approximately 15 µg/L. In 2000, the USEPA issued a

Table 7-3. State guidelines and enforceable standards for MTBE in drinking water

State	Guidelines (µg/L)	Basis
Arizona	35	Health-based guideline
California	5	Secondary MCL
	13	Proposed primary MCL
Connecticut	100	Guideline
Florida	50	Drinking water standard
Massachusetts	70	Proposed guideline
Michigan	50	Guideline
New Hampshire	100	Guideline
New Jersey	70	Health-based MCL
New York	50	MCL
Rhode Island	50	Guideline
Vermont	40	Drinking water standard

drinking water advisory recommending that MTBE levels in drinking water be kept below 40 μg/L to minimize T&O. The advisory also indicated that MTBE levels in this range would likely not cause negative health effects (USEPA 1997). USEPA is continuing to study both the potential health effects and the occurrence of MTBE with its inclusion on the Contaminant Candidate List.

A number of states have banned the use of MTBE, and several have set guidelines or enforceable standards for MTBE in drinking water supplies (Table 7-3). Most notably, California has established an enforceable secondary MCL of 5 μg/L and set a tentative primary MCL of 13 μg/L. Some states have set groundwater cleanup goals as low as 5 μg/L.

Activated Carbon Treatment of VOCs

Activated carbon has long been recognized as an effective means for removing VOCs from water. In contrast, conventional treatment systems (i.e., coagulation, sedimentation, filtration, and chlorination) are generally ineffective for reducing VOC concentrations in drinking water (Mowawada 1994, Robeck and Love 1983, Najm et al. 1991). Surface water treatment plants may use PAC to remove VOCs, as VOC occurrence in surface waters is usually due to spills or other intermittent occurrences. However, the largest post-filter GAC system in the world, found at the Greater Cincinnati Water Works (GCWW) water treatment plant (WTP), was installed partly to protect against the presence of VOCs in the Ohio River (chapter 4; Case Study No. 1). The Ohio River is subject to contamination from accidental chemical spills, which can introduce VOCs into the river that may not volatilize before the water enters the GCWW plant intake.

The most common application of activated carbon treatment for removing VOCs is the use of GAC contactors in treating contaminated groundwater. GAC is widely used for treating groundwater contaminated with VOCs, both as the primary treatment process and as a polishing step following air stripping. GAC systems for groundwater treatment often include multiple GAC contactors, typically in series configuration. Coconut-based GACs have been shown to be particularly effective for MTBE, which may make investigation of coconut GAC worthwhile for other target organic contaminants. The improved performance has been attributed to coconut GACs having more of their surface area contained in micropores, which competing adsorbates such as DOM cannot access.

Synthetic Organic Chemicals

All man-made organic compounds can be categorized as synthetic organic chemicals (SOCs). However, here the definition of SOCs is limited to the subset of SOCs regulated in the national primary drinking water standards. The SOCs of primary concern include pesticides, solvents, and metal degreasers (e.g., trichlorobenzene), polycyclic aromatic hydrocarbons (PAHs), and polychlorinated biphenyls (PCBs). Notably, the VOCs described in the previous section are also SOCs, yet they are usually considered separately as per the categorizations within the federal drinking water standards. This section will primarily focus on pesticides because these are among the most commonly encountered SOCs in the drinking water industry.

Pesticides

Pesticides are substances or a mixture of substances used for agricultural or public health purposes with the intention of preventing, destroying, repelling, or mitigating a pest (i.e., insect, weed, rodent, etc.). Pesticides can be classified according to their chemical class (e.g., organochlorine, carbamate, organophosphorus, chlorophenoxy compounds) or according to the type of pest they are intended to control (e.g., fungicide, herbicide, algicide). Pesticides are transported from land into streams primarily by runoff or drainage resulting from rainfall or irrigation. Pesticides can affect both groundwater and surface water. Pesticide concentrations in stream water vary through the year, usually characterized by long periods with low or undetectable concentrations of most pesticides, punctuated by seasonal pulses of much higher concentrations during spring runoff.

Once in a stream, a pesticide may be chemically or biologically transformed (i.e., degraded into other compounds), be taken up by aquatic organisms, attach to suspended particles and be deposited in bed sediment, or volatilize into the atmosphere. Degradates are particularly important in groundwater, which moves relatively slowly through soils and aquifers, providing the extended time and conditions favorable for transformation of pesticides into their degradates.

The human health effects of a pesticide depend on the type of pesticide, its concentration, and the duration of exposure. Pesticides can affect the nervous or endocrine systems and could even be carcinogens. The health effects of long-term exposure to low-level mixtures of pesticide compounds in drinking water are still not well understood and need to be studied further (Gilliom et al. 2006). Currently, 24 pesticides are regulated under the SDWA, of which 14 are registered for use. In addition, USEPA has issued health advisories for many other pesticides.

Atrazine is the most commonly detected pesticide in US waters (Gilliom et al. 2006), is water soluble, and is relatively conservative in the environment due to little loss by volatilization, low sediment partitioning, and relatively slow rates of degradation. Atrazine is an endocrine disruptor that can interfere with the body's hormonal activity and the development of reproductive organs. Although the European Union banned it in 2004, atrazine is still widely used in the United States for the control of annual broadleaf and grassy weeds. In an extensive study conducted by the US Geological Survey (USGS) between 1992 and 2001, atrazine was detected in approximately 75 percent of stream water samples and in about 40 percent of groundwater samples in agricultural areas (Gilliom et al. 2006). USEPA has established a primary MCL of 3 µg/L for atrazine, but like many pesticides, it may be an endocrine disruptor at much lower concentrations.

Activated Carbon Treatment of SOCs

Both PAC and GAC have been demonstrated to be highly effective processes for removing organic chemicals, including certain pesticides (Miltner et al. 1989, USEPA 2001, Knappe et al. 1998). PAC is the more common method for seasonal occurrences because it can be used in concert with conventional water treatment systems with minimal additional capital investment. GAC, under the SDWA, is the best available technology for removing synthetic organic chemicals (SOCs).

PAC can effectively remove SOCs in drinking water, and it has the advantage of being brought online as needed to treat seasonal runoff episodes. PAC effectiveness is also a function of the carbon dose. Miltner et al. (1989) found that PAC applied at doses used to control T&O compounds can be effective for pesticide control if moderate percentage removal is required to meet an MCL.

GAC has been used to remove pesticides as a primary treatment or a secondary barrier following PAC addition or advanced oxidation. GAC could be used at a surface water treatment plant either in conventional gravity filters (replacing anthracite) or in a post-filter application. In the filter application, GAC serves as a filtering medium and provides limited adsorption capacity but will attenuate the pulse loading.

Micropollutants

Advancements in analytical methods have allowed for the detection of an increasing number of micropollutants at trace concentrations (i.e., approximately 0.1 ng/L) in waters around the world. More research is needed about their human health effects, occurrence, fate and transport, and treatability before regulation can be considered. However, regardless of regulatory status, public concern has grown over the presence of trace levels of pesticides, endocrine-disrupting compounds, pharmaceuticals, and personal care products in drinking water sources since first reported by Kolpin et al. (2002). These micropollutants are not well removed during conventional drinking water treatment (Westerhoff et al. 2005). The public concern over these compounds stems from potentially hazardous health effects and is causing water purveyors to consider additional treatment to remove them and maintain consumer confidence. Although it is probable that the vast majority of the trace organic contaminants pose little health risk at the concentrations typically encountered, Pomati et al. (2006) found cytotoxicity to human embryo cells with a mixture of 15 trace organic contaminants at environmentally relevant concentrations. Wastewater treatment plant (WWTP) effluents constitute major sources and routes for micropollutants to enter surface water and groundwater (Oppenheimer 2008, Snyder et al. 2008). Other possible sources include combined sewer overflows (CSOs), septic systems, untreated stormwater flows, and urban runoffs. Many of these compounds are incompletely degraded or removed during wastewater treatment and are persistent in the aquatic environment.

Endocrine-disrupting compounds (EDCs) and pharmaceuticals and personal care products (PPCPs) are two groups of micropollutants that are often discussed in conjunction because some pharmaceutical compounds can interfere with endocrine systems.

Endocrine Disruptors

EDCs are naturally occurring and synthetic chemicals that interfere with the normal functioning of the human endocrine system. There are hundreds of purported EDCs, and opinions as to what defines an EDC vary greatly. A definitive list of EDCs does not exist, as most chemicals in commerce have not been tested for endocrine toxicity. Substances known or suspected of being EDCs include natural and synthetic hormones excreted by humans and animals, phytoestrogens and mycoestrogens, certain natural and anthropogenic anions such as perchlorate and nitrate, metals such as cadmium (Stoica et al. 2000), select pesticides (e.g., atrazine, dichlorodiphenyl trichloroethane [DDT]), and different types of industrial chemicals (e.g., PCBs, PAHs, furans, phenols, and phthalates).

The three endocrine-disruption pathways currently under discussion are estrogenic, androgenic, and thyroidal (Snyder et al. 2003). The majority of drinking water research thus far has focused on estrogenic compounds. More recent attention has been drawn to thyroidal (also called goitrogenic) pathways, with discussions about perchlorate. The risk of EDCs to humans and aquatic species is strongly influenced by their fate and behavior in the environment. Among their health effects, EDCs can

- Mimic or partly mimic naturally occurring hormones in the body like estrogen, androgen, and thyroid hormones, potentially producing overstimulation.
- Bind to a receptor within a cell and block the body's hormones from their normal binding functions. Examples of chemicals that block or antagonize hormones are anti-estrogens or anti-androgens.
- Interfere or block the way natural hormones or their receptors are produced or controlled, for example by blocking their metabolism in the liver.

Adverse health effects in wildlife species exposed to EDCs in ambient surface water supplies are well documented and mainly relate to developmental abnormalities. The issue of combined or synergistic effects of multiple contaminants is poorly understood and the subject of much debate and discussion. A recent study demonstrated that the combination effects of EDCs can be estimated by "dose addition" (Kortenkamp 2007) and concluded that joint effects can occur even when all mixture components are present at levels below doses that cause observable effects. Based on this study, grouping of EDCs according to their ability to induce similar effects may be an appropriate approach to consider for mixture exposure assessment. However, competitive effects are also possible; additional research is needed to clarify the role of mixtures of EDCs on health effects.

Pharmaceuticals and Personal Care Products

Pharmaceuticals are chemicals that are used for diagnosis, treatment, alteration, or prevention of diseases and affect the function of the human body. Pharmaceutical products include contraceptives, antibiotics, pain medication, antidepressants, and antiepileptics. Personal care products include compounds found in a wide array of substances such as fragrances, skin creams, sunscreens, and disinfectants. Collectively, pharmaceuticals and personal care products (PPCPs) have been receiving increased attention within the drinking water field (i.e., utilities, regulatory agencies) and the media, due primarily to advancements in analytical methods that allow detection of these compounds at ng/L concentrations in source waters.

PPCPs enter WWTPs when medicated people excrete pharmaceuticals or their metabolites, rinse them from their bodies during bathing, or carelessly dispose of unused medications down the sink or toilet. WWTPs also receive effluents containing pharmaceuticals from hospitals, veterinary medical facilities, and pharmacies. Although PPCPs are likely to be nontoxic at the low levels encountered, many utilities are addressing them as a consumer confidence issue in the same way they treat T&O compounds.

Activated Carbon Treatment for Micropollutants

Both PAC and GAC can be effective for removing micropollutants (Snyder et al. 2007, Westerhoff et al. 2005, Ternes et al. 2002, Stackelberg et al. 2007). Snyder et al. (2007) demonstrated greater than 90 percent removal by PAC and GAC of the compounds evaluated. Powdered activated carbon is generally used in treating surface water, with the advantage of being applied on an as-needed basis to remove contaminants. Westerhoff et al. (2005) found that increasing PAC dose improved the percent removal of

EDCs and PPCPs, when considering the range of PAC typically applied at drinking water treatment plants. Doses of 20 mg/L effectively removed greater than 80 percent of the compounds included in the study, and trends were comparable between two different PAC brands.

GAC may be particularly effective for controlling micropollutants because removal of a broad array of contaminants is required, all with varying influent concentrations throughout the year. Table 7-4 lists several micropollutants in their relative order of adsorbability from weakly adsorbing to strongly adsorbing. The table was compiled primarily from Corwin and Summers (2012), Snyder et al. (2007), and Cardenas (2011). Each study had more than 20 compounds and at least 6 overlapping compounds all with influent concentrations below 1 μg/L. The overlapping compounds were used to normalize the data using breakthrough to 10 percent, and the data confirm that adsorption of the compounds remained relative in each study. The adsorbability of the compounds was then normalized to iopromide, the most weakly adsorbed compound in the studies. Though not included in the referenced studies; MTBE, bisphenol A, and MIB were also included because they are known to be weakly adsorbed and are common treatment objectives. Relative adsorption factors were estimated for MTBE, bisphenol A, and MIB by comparing breakthrough curves relative to other compounds on the list in a manner similar to how the table was generated. The data in Table 7-4 can be used to assist in choosing target adsorbates for analysis. For example, if sulfamethoxazole, atrazine, and caffeine have been found at relatively high concentrations in a source water, sulfamethoxazole will be the most weakly adsorbed and will control adsorber operation. Additionally, even if hormones are of great concern to the stakeholders, GAC adsorption will be very effective at removing them.

Table 7-4. Relative adsorbability of selected compounds

	Adsorption	Compound	Use
weakest	weakly adsorbing	MTBE	fuel oxygenate
		bisphenol A	plasticizer
		MIB	T&O compound
		iopromide	x-ray contrast media
		clofibric acid	pesticide
		sulfamethoxazole	pharmaceutical
		ibuprofen	pharmaceutical
		cotinine	pharmaceutical
		meprobamate	pharmaceutical
		warfarin	pharmaceutical
	moderately adsorbing	dimethoate	pesticide
		metolachlor	pesticide
		dilantin	pharmaceutical
		prometon	pesticide
		diclofenac	pharmaceutical
		aldicarb	pesticide
		acetochlor	pesticide
		TCEP	fire retardant
		2,4-dichlorophenoxyacetic acid	pesticide
		gemfibrozil	pharmaceutical
		malaoxon	pesticide degradate
		DEET	personal care product
		naproxen	pharmaceutical
		daidzein	plant estrogen
		deethylatrazine	pesticide degradate
		atrazine	pesticide
		caffeine	pharmaceutical
		carbamazepine	pharmaceutical
		diazepam	pharmaceutical
		pentoxifylline	pharmaceutical
		hydrocodone	pharmaceutical
		acetaminophen	pharmaceutical
	strongly adsorbing	trimethoprim	antimicrobial
		genistein	plant estrogen
		deisopropylatrazine	pesticide degradate
		diphenhydramine	pharmaceutical
		flutolanil	pesticide
		simazine	pesticide
		estriol	steroid
		ethinylestradiol	steroid
		estradiol	steroid
		testosterone	steroid
		androstenedione	steroid
		estrone	steroid
		fluoxetine	pharmaceutical
		oxybenzone	personal care product
		progesterone	steroid
strongest		triclosan	antimicrobial

References

Baus, C., H.W. Hung, F. Sacher, M. Fleig, and H. Brauch. 2005. MTBE in drinking water production —occurrence and efficiency of treatment technologies. *Acta Hydrochim. Hydrobiol.* 33(2):118–132.

Cardenas J.C. 2011. Granular activated carbon adsorption of dissolved organic matter and trace organic contaminants: effects of influent dissolved organic matter concentration and pretreatment. Masters Thesis, University of Colorado, Boulder, Colo.

Carter, J.M., S.J. Grady, G.C. Delzer, B. Koch, and J.S. Zogorski. 2006. Occurrence of MTBE and other gasoline oxygenates in CWS source waters. *Journal AWWA* 98(4):91–104.

Carter, M.C., and W.J. Weber. 1994. Modeling adsorption of TCE by activated carbon preloaded by background organic-matter. *Environ. Sci. Technol.* 28(4):614–623.

Corwin, C.J., and R.S. Summers. 2010. Scaling trace organic contaminant adsorption capacity by granular activated carbon. *Environ. Sci. and Technol.* 44(14):5403–5408.

Corwin, C.J., and R.S. Summers. 2011. Adsorption and desorption of trace organic contaminants from granular activated carbon adsorbers after intermittent loading and throughout backwash cycles. *Water Res.* 45(2):417–426.

Corwin, C.J., and R.S. Summers. 2012. Controlling trace organic contaminants with granular activated carbon adsorption. *Journal AWWA* 104(1):43–44.

Crittenden, J.C., N.J. Hutzler, D.G. Geyer, J.L. Oravitz, and G. Friedman. 1986. Transport of organic compounds with saturated groundwater flow: model development and parameter sensitivity. *Water Resources Res.* 22(3):271–284.

Crittenden, J.C., J.K. Berrigan, D.W. Hand, and B. Lykins. 1987. Design of fixed-bed adsorption tests for nonconstant diffusivities. *Jour. Envir. Engrg.* 133(2):243–259.

Cussler, E.L. 2009. *Diffusion: Mass transfer in fluid systems.* 3rd ed. Cambridge, U.K.: Cambridge University Press.

Fischer, A.J., E.A. Rowan, and R.F. Spalding. 1987. VOCs in ground water influenced by large scale withdrawals. *Ground Water* 25(4):407–414.

Gilliom, R.J., J.E. Barbash, C.G. Crawford, P.A. Hamilton, J.D. Martin, E. Callender, N. Nakagaki, L.H. Nowell, J.C. Scott, P.E. Stackelberg, G.P. Thelin, and D.M. Wolock. 2006. *Pesticides in the nation's streams and ground water, 1992–2001.* US Geological Survey Circular 1291.

Gillogly, T.E.T., V.L. Snoeyink, J.R. Elarde, C.M. Wilson, and E.P. Royal. 1998. 14C-MIB adsorption on PAC in natural water. *Journal AWWA* 90(1):98–108.

Gnielinski, V. 1978. Gleichungen zur berechnung des warmeund stoffaustausches in durchstromten ruhenden kugelschuttungen bei mitteren und grossen pecletzahlen. *Verf. Tech.* 12(6):363–366.

Graham, M.R., R.S. Summers, M.R. Simpson, and B.W. MacLeod. 2000. Modeling equilibrium adsorption of 2-methylisoborneol and geosmin in natural waters. *Water Research* 34(8):2291–2300.

Hand, D.W., J.C. Crittenden, and W.E. Thacker. 1984. Simplified models for design of fixed-bed adsorption system. *Jour. Environ. Engrg.* 110(2):441–456.

Hayduk, W., and H. Laudie. 1974. Prediction of diffusion coefficients for nonelecrolytes in dilute aqueous solutions. *J. Amer. Inst. Chem. Engrg.* 20(3):611–615.

Hutzler, N.J., J.C. Crittenden, J.S. Gierke, and A.S. Johnson. 1986. Transport of organic compounds with saturated groundwater flow: experimental results. *Water Resources Res.* 22(3):285–295.

IARC (International Agency for Research on Cancer). 1987. Benzene group 1: A. Evidence for carcinogenicity to humans (sufficient). in: *IARC monographs on the evaluation of carcinogenic risks to humans. Overall evaluations of carcinogenicity: an updating of IARC Monographs Volumes 1 to 42.* Lyon, France: World Health Organization, International Agency for Research on Cancer.

IARC. 1999a. Monographs on the evaluation of carcinogenic risks to humans. Toluene. In Volume 71 Part Two. *Re-evaluation of some organic chemicals, hydrazine, and hydrogen peroxide.* Lyon, France: World Health Organization.

IARC. 1999b. Monographs on the evaluation of carcinogenic risks to humans. Xylenes. In Volume 71 Part Three. *Re-evaluation of some organic chemicals, hydrazine, and hydrogen peroxide.* Lyon, France: World Health Organization.

IARC. 2000. Some industrial chemicals. Ethylbenzene. Volume 77. In *IARC Monographs on the evaluation of carcinogenic risks to humans.* Lyon, France: World Health Organization, International Agency for Research on Cancer.

IRIS (Integrated Risk Information System). 2001. Xylenes. Integrated Risk Information System, U.S. Environmental Protection Agency. Available online: www.epa.gov/iris/subst/0270.htm.

Jarvie, M.E., D.W. Hand, S. Bhuvendralingam, J.C. Crittenden, and D.R. Hokanson. 2005. Simulating the performance of fixed-bed granular activated carbon adsorbers: removal of synthetic organic chemicals in the presence of background organic matter. *Water Research* 39(11):2407–2421.

Kilduff, J.E., T. Karanfil, and W.J. Weber Jr. 1998. Competitive effects of nondisplaceable organic compounds on trichloroethene uptake by activated carbon 1. Thermodynamic predictions and model sensitivity analyses. *J. Colloid and Interface Sci.* 205(2):271–279.

Knappe, D.R.U., Y. Matsui, V.L. Snoeyink, P. Roche, M.J. Prados, and M.M. Bourbigot. 1998. Predicting the capacity of powdered activated carbon for trace organic compounds in natural waters. *Environ. Sci. Technol.* 32(11):1694–1698.

Kolpin, D.W., E.T. Furlong, M.T. Meyer, E.M. Thurman, S.D. Zaugg, L.B. Barber, and H.T. Buxton. 2002. Pharmaceuticals, hormones, and other organic wastewater contaminants in US Streams, 1999–2000: a national reconnaissance. *Environ. Sci. Technol.* 36(6):1202–1211.

Kortenkamp, A. 2007. Ten years of mixing cocktails: a review of combination effects of endocrine-disrupting chemicals. *Environ Health Perspect* 115:98–105.

Matilainen, A., N. Vieno, and T. Tuhkanen. 2006. Efficiency of the activated carbon filtration in the natural organic matter removal. *Environ. Int.* 32:324–331.

Matsui, Y., A. Yuasa, and K. Ariga. 2001. Removal of a synthetic organic chemical by PAC-UF systems—I: Theory and modeling. *Water Research* 35(2):455–463.

Matsui, Y., Y. Fukuda, T. Inoue, and T. Matsushita. 2003. Effect of natural organic matter on powdered activated carbon adsorption of trace contaminants: characteristics and mechanism of competitive adsorption. *Water Research* 37(18):4413–4424.

Miltner, R.J., D.B. Baker, T.F. Speth, and C.A. Fronk. 1989. Treatment of seasonal pesticides in surface waters. *Journal AWWA* 81:43–52.

Najm, I.N., V.L. Snoeyink, B.W. Lykins Jr., and J.Q. Adam. 1991. Using powdered activated carbon: a critical review. *Journal AWWA* 83(1):65–76.

NTP (National Toxicology Program). 2001. Ninth report on carcinogens. U.S. Department of Health and Human Services, Public Health Service, National Toxicology Program. Available online: ehis.niehs.nih.gov/roc/toc9.html. June 2001.

Oppenheimer, J., R. Stephenson, and J. Decarolis. 2008. Emerging contaminants: Insights to the most effective EDC and PPCP treatment strategies. *Opflow* 34(5):12–16.

Pillai, M.P. 2009. BTEX-contamination and remediation. Musalliar College of Engineering and Technology. Pathanamthitta, India. Available online: www.scribd.com/doc/25860772/BTEX-Contamination-and-Remediation

Pomati, F., S. Castiglioni, E. Zuccato, R. Fanelli, C. Vigetti, C. Rossetti, and D. Calamari. 2006. Effects of a complex mixture of therapeutic drugs at environmental levels on human embryonic cells. *Environ. Sci. Technol.* 40(7):2442–2447.

Robeck, G.G., and O.T. Love. 1983. Removal of volatile organic contaminants from groundwater. *Environ. Micro.* 53:949–954.

Snyder, S.A., P. Westerhoff, Y. Yoon, and D.L. Sedlak. 2003. Pharmaceuticals, personal care products, and endocrine disruptors in water: implications for the water industry. *Environmental Engineering Science* 20(5).

Snyder, S.A., S. Adham, A.M. Redding, F.C. Cannon, J. DeCarolis, J. Oppenheimer, E.C. Wert, and Y. Yoon. 2007. Role of membranes and activated carbon in the removal of endocrine disruptors and pharmaceuticals. *Desalination* 202(1–3):156–181.

Snyder, S.A., B.J. Vanderford, J. Drewes, E. Dickenson, E.M. Snyder, G.M. Bruce, and R.C. Pleus. 2008. State of knowledge of endocrine disruptors and pharmaceuticals in drinking water. AwwaRF report no. 91228. Denver, Colo.: Water Research Foundation.

Sontheimer, H., J.C. Crittenden, and R.S. Summers. 1988. *Activated carbon for water treatment.* Karlsruhe, Germany: DVGW-Forschungsstelle am Engler-Bunte Institut der Universitat Karlsruhe.

Stackelberg, P.E., J. Gibs, E.T. Furlong, M.T. Meyer, S.D. Zaugg, and R.L. Lippincott. 2007. Efficiency of conventional drinking-water-treatment processes in removal of pharmaceuticals and other organic compounds. *Science of the Total Envir.* 377:255–272.

Stocking, A.J., I.H. Suffet, M.J. McGuire, and M.C. Kavanaugh. 2001. Implications of an MTBE odor study for setting drinking water standards. *Journal AWWA* 93(3):95–105.

Stoica, A., B.S. Katzenellenbogen, and M.B. Martin. 2000. Activation of estrogen receptor-alpha by the heavy metal cadmium. *Mol Endocrinol.* 14:545–553.

Ternes, T.A., M. Meisenheimer, D. McDowell, F. Sacher, H.-J. Brauch, B. Haist-Gulde, G. Preuss, U. Wilme, and N. Zulei-Seibert. 2002. Removal of pharmaceuticals during drinking water treatment. *Environ. Sci. Technol.* 36(17):3855–3863.

USEPA (US Environmental Protection Agency). 1997. Drinking water advisory: Consumer acceptability advice and health effects analysis on methyl tertiary-butyl ether (MtBE). EPA-822-F-97-009.

USEPA. 2001. The Incorporation of water treatment effects on pesticide removal and transformations in Food Quality Protection Act (FQPA) drinking water assessments. Available online: www.epa.gov/pesticides/trac/science/water_treatment.pdf

USEPA. 2010. MTBE (methyl-*tert*-butyl ether) in drinking water. Available online: water.epa.gov/drink/contaminants/unregulated/mtbe.cfm. Accessed on November 16, 2010.

Westerhoff, P., Y. Yoon, S. Snyder, and E. Wert. 2005. Fate of endocrine-disruptor, pharmaceutical, and personal care product chemicals during simulated drinking water treatment processes. *Environ. Sci. Technol.* 39(17):6649–6663.

Yu, Z., S. Peldszus, and P.M. Huck. 2008. Adsorption characteristics of selected pharmaceuticals and an endocrine disrupting compound—naproxen, carbamazepine and nonylphenol—on activated carbon. *Water Res.* 42(12):2873–2882.

Zogorski, J.S., J.M. Carter, T. Ivahnenko, W.W. Lapham, M.J. Moran, B.L. Rowe, P.J. Squillace, and P.L. Toccalino. 2006. Volatile organic compounds in the nation's ground water and drinking-water supply wells. U.S. Geological Survey Circular 1292.

PART 3

DESIGN AND PROCUREMENT OF ACTIVATED CARBON SYSTEMS

Once a decision has been made to proceed with the use of activated carbon in a treatment works, using the guidelines provided in chapter 4, the next major step is designing the facilities. The process planning phase is expected to establish the most appropriate application method of activated carbon for the desired water quality goals. The final design, construction, startup, and operation are aimed at getting a working activated carbon system in place.

The objective of this section is to provide widely accepted, time-proven approaches; however, even these may not be appropriate for all individual circumstances. Part 3 addresses:

- *Gravity GAC Treatment System Design.* The use of and design of GAC in gravity treatment systems for liquid-phase water treatment (chapter 8).
- *Pressure GAC Treatment System Design.* The use of and design of GAC using prefabricated pressure vessels for liquid-phase water treatment (chapter 9).
- *Vapor-Phase GAC Treatment System Design.* The use of and design of GAC in vapor-phase treatment systems associated with potable-water treatment (chapter 10).
- *On-Site GAC Reactivation Design.* The alternatives available and design of on-site GAC reactivation facilities (chapter 11).
- *PAC Feed System Design.* The application of powdered activated carbon and the design of associated facilities (chapter 12).
- *Activated Carbon System Implementation and Economic Considerations.* The available capital project delivery methods, activated carbon procurement approaches, and cost estimation procedures (chapter 13).

Chapter 8

Gravity GAC Treatment System Design

This chapter addresses the use of GAC in gravity treatment systems for liquid-phase treatment; pressure systems are discussed in chapter 9. Gravity GAC treatment systems are usually selected over pressurized systems in large-scale (>10 mgd) water treatment applications for control of T&O episodes, reduction of TOC to reduce DBP formation, and removal of other organic contaminants. The following benefits apply when comparing gravity (concrete basin) treatment units to pressure vessel treatment units:

- Lower construction cost per unit production volume
- More compact footprint
- Optimized basin sizing (geometry, empty bed contact time, etc.)
- Lower operation and maintenance requirements
- Potential avoidance of pumping to GAC treatment units if they are placed within the constraints of the plant's hydraulic profile

The following GAC gravity treatment systems are discussed in this chapter:

- *Post-filter contactors.* Treatment units similar to traditional filters located after the particle filters in the treatment process. GAC is used for adsorption of dissolved organics.
- *Filter adsorbers.* Replacement of the granular filter media with GAC in gravity filters. GAC is used for adsorption and particle filtration.
- *GAC caps.* Addition of GAC to a granular filter as a supplementary filtration layer. GAC is added for its adsorption characteristics, but an additional filtration benefit is also realized.
- *Biologically enhanced activated carbon (BAC).* Gravity filters equipped with GAC, which serves as media for particle removal as well as support media for biological growth, which encourages biodegradation of organic carbon. While some adsorption occurs when new media are present, adsorption is typically not important to long-term BAC performance.

Design and Construction

Design begins with reviewing the results of the Water Quality Master Plan, which includes treatability and process planning data. Development of the master plan was presented in chapter 4.

Review Key Design Parameters

After a gravity adsorption GAC treatment technology is selected for implementation, the first step is a review the following key design parameters from the master plan to determine how each affects the final design:

- GAC application type
- Design flow rate
- Empty bed contact time (EBCT)
- Influent concentrations of target and background dissolved organic carbon (DOC)
- Treatment objectives
- GAC product selection
- Operational mode

GAC Application Type

The first consideration in final design is whether the GAC will be used as a filter adsorber (including GAC caps), a post-filter contactor, or a BAC filter.

Filter Adsorbers. Filter adsorbers may be full-depth GAC but often include a layer of sand below the GAC to aid in particle removal and are common in modern dual-media filters. Filter adsorbers are designed for regular backwashing. Backwashing is affected by the presence of a sand layer because expansion of both layers is needed, but intermixing of layers should be minimized for optimal GAC media replacement. A sand layer also makes GAC media replacement more difficult because GAC must be separated from the sand if it has to be reactivated.

Post-Filter Contactors. Post-filter contactors are full-depth GAC designed for frequent periodic replacements. The primary concern in final design is the effect of adding additional head loss to the plant's hydraulic profile. The adsorption process is not as affected by fluctuations in flow rate, because dislodging of filtered particles is not a concern. In fact, increasing flow rate reduces film mass transfer resistance, which may improve adsorber performance.

Biologically Enhanced Activated Carbon Filters. The EBCT is typically not part of a BAC facility conceptual design. It is more common to take into account the surface loading rate and media depth for BAC. Like the EBCT for adsorption processes, these parameters set the total volume of GAC required for the treatment process, but in the case of BAC, the media depth is set as well, leaving only the number of treatment units and the footprint per treatment unit to be determined. BAC may be added with a new facility or retrofitted in an existing facility. Retrofitting a granular media filter treatment unit to BAC will set the remaining parameters.

Design Flow Rate

The minimum and maximum design flow rates for the entire GAC process affect the size and number of treatment units required. The maximum design flow rate is typically equal to the plant design capacity, but it may be determined based on future design capacity as well. The minimum design flow rate is typically equal to the minimum plant flow rate. Both may differ if the facility plans to partially bypass the GAC process. Partial or full bypassing of the GAC process should be determined during the conceptual design phase and coordinated with the local regulatory agency as some locations, including California, do not allow bypassing of treatment units. At final design, additional considerations such as production of backwash water and performance with filters out of service are taken into account. For example, maximum flow through an individual filter may be higher than the maximum plant flow divided by the number of filters because (1) additional flow must be produced to backwash the filters and (2) individual filters may not be available for production due to backwash or media replacement.

Empty Bed Contact Time (EBCT)

The EBCT is the most critical design parameter for sizing an adsorption GAC treatment process because it affects both the size of the process and efficiency of its operation. The optimum EBCT is usually determined in the preliminary design phase, and facilities are sized to achieve the desired EBCT at the design flow rate. In final design, the effect of variable plant flow is considered and partially determines how adsorbers are to be operated. Because design flow rates are generally used to size systems and achieve EBCT, the actual operating EBCTs of adsorbers are often longer than the specified EBCT due to operating flows being lower than the design flow. If EBCT is allowed to increase during lower flows, then the surface loading rate of the filter decreases. Lower filtration rates increase the film mass transfer resistance and may decrease adsorption of some adsorbates; however, film mass transfer control is not typical. Lower loading rates also decrease head loss through the bed.

GAC use rate may depend on EBCT for some target compounds. For instance, the use rate for DOC typically decreases with an increasing EBCT; however, Corwin and Summers (2011) showed that the use rate of some target organics increases with an increasing EBCT. In the latter case, it would be economical to take adsorbers out of service to achieve an EBCT closest to the design EBCT as plant production decreases. For DOC removal, all adsorbers are kept in-service to take advantage of the longer EBCTs.

Influent Concentrations of Target and Background DOC

Influent concentrations of target compounds and the background organic matter often fluctuate throughout the year. Higher concentrations result in a higher mass loading rate of the available adsorption sites, which results in higher use rates. Concentration effects should be investigated during process planning. At final design, concentration effects generally manifest themselves in performance monitoring of the treatment objective. However, in some instances they can drive operation. For example, a utility that uses GAC to control T&O episodes that occur seasonally may decide to replace GAC media just before the anticipated episode each year. After the episode has passed, the residual capacity of the GAC adsorbers is used as a polishing step to filtration.

Treatment Objectives

The target effluent concentration corresponds to the concentration desired in the GAC effluent for the process to meet water quality treatment goals. The treatment objective ultimately determines the size of facility required.

GAC Product Selection

Different types of GAC are available. GAC can be classified by its source (examples include bituminous, lignite, coconut shell) and by its size (examples include 8×20 mesh and 12×40 mesh). The type of GAC and GAC treatment technology recommended by the master plan or additional studies will guide GAC selection. The GAC type should be determined early as it affects the EBCT and hydraulics of the GAC process.

The size of the GAC media is largely determined by the GAC treatment technology selected. A post-filter contactor, which benefits by upstream removal of turbidity, can use a smaller GAC media size than a filter adsorber that must also achieve turbidity removal. Smaller GAC sizes (for example, 12×40 mesh) allow faster adsorption kinetics. However, clean-bed head loss is higher than for larger media sizes of the same EBCT. GAC manufacturers can be consulted for clean-bed head loss information and backwash expansion data to aid design of the GAC treatment units. An example of GAC head loss information is shown in Figure 8-1. In Figure 8-1, the pressure drop through the clean GAC bed is shown for various surface loading rates (also called the superficial velocity) for two different types of GAC.

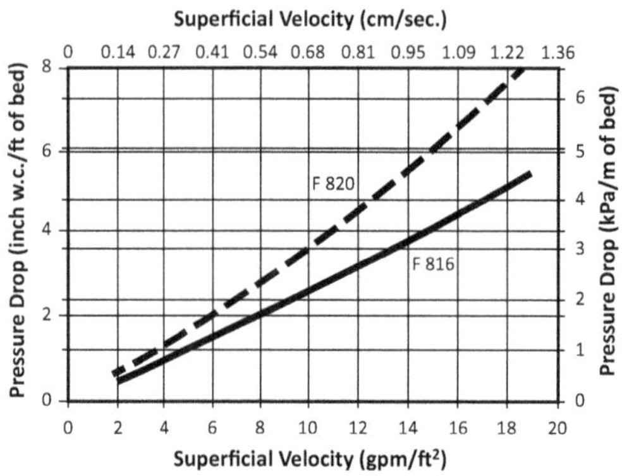

Figure 8-1. Example head loss curves for GAC (F816 and F820, provided by Calgon Carbon Corporation)

The uniformity coefficient (UC) also varies between GAC products. Typical UCs vary from 1.3 to 2.4, which is typically higher than for traditional filtration media. The higher UCs are considered desirable because the distribution of particle sizes causes the bed to restratify after backwashing, which limits disruption of the mass transfer zone. However, the higher UCs also cause the media to expand at different flow rates during backwash, which may require more distance to the washwater trough to prevent blowout of the media. Media base material affects its hardness, which can lead to attrition of the GAC media and production of "fines" that affect process operation. Initial procurement of the GAC media is significant to the total construction cost, so it is advantageous to the owner not to limit the bid to a single product, which is complicated by the above issues. GAC product procurement is explored in detail in chapter 13.

Operation Mode

The master plan will address how the beds are to be operated, in either series or parallel mode. Adsorbers in series have to include piping and valving to allow flow to be fed to either stage and collected from either stage as described in chapter 4. Adsorbers in parallel may be operated in single replacement or staged replacement mode. In single replacement mode, the media in all adsorbers are replaced at the same time to potentially lower operating costs. In staged parallel mode, the media in the beds are replaced one at a time to optimize use rate.

Filter Layout

The EBCT sets the total volume of GAC necessary to achieve the treatment goals for an adsorption process. However, the EBCT does not set the distribution of the GAC media volume. The number of treatment units, surface area, and depth of each unit need to be determined. If an existing facility is being modified, such as converting granular media filters into filter adsorbers, these parameters may not be able to be manipulated. For a new adsorption GAC facility, optimizing the dimensions of the adsorbers may be an iterative process, particularly if the GAC process must fit within the footprint of an existing treatment facility.

The EBCT is related to the number of units, surface area, and depth by Eq 8-1:

$$\text{EBCT} = \frac{A_f \times n \times d}{Q_t} = \frac{d}{v_f} \qquad \text{(Eq 8-1)}$$

Where

A_f = filter area of single unit
n = number of filters
d = filter depth
Q_t = total flow
v_f = filter velocity (surface loading rate)

Eq 8-1 demonstrates that to achieve the design EBCT, as the flow increases, the filter area, number of filters, or filter depth must be increased.

Figure 8-2 shows the effect of surface loading rate on total filter area and filter depth requirements based on a 10-min EBCT and 12×40 GAC. As the surface loading rate increases, the total filter area asymptotically decreases while the media depth linearly increases, as expected from Eq 8-1. The head loss increases more rapidly at surface loading rates greater than 7 gpm/ft^2. For this GAC product, the total depth of the filter increases to undesirable levels above 8 gpm/ft^2, while the decrease in filter area starts to level out.

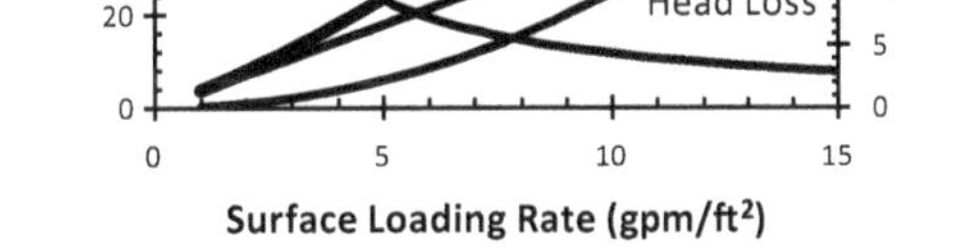

Figure 8-2. Relationship between filter area and filter depth; total filter depth does not include freeboard (EBCT = 10 min, 12×40 GAC)

Filter Velocity

Filter velocity, or surface loading rate, is equal to the total flow divided by the total filter area. Increasing the filter velocity can lead to a smaller total filter area but requires deeper filter beds. Increasing filter velocity and filter depth both results in increased head loss through the bed, which may limit minimization of footprint area (Figure 8-2). Thus, filter velocity is typically the first variable to be constrained. For filter adsorbers, filter velocities are in the traditional media filtration range of 2 to 6 gpm/ft^2. Post-filter contactors may have higher rates, typically in the

4 to 8 gpm/ft^2 range. The surface loading rate of 6 gpm/ft^2 often provides a good balance between required filter area and filter depth.

Filter Surface Area

The total filter surface area can be calculated directly by dividing the design flow by the filter velocity. Area of each filter is then determined by the number of filters, considering operation with one or more filters out of service. The maximum size of an adsorber is limited by providing even distribution of backwash water, backwash flow rate, structural considerations, and meeting plant flow requirements with adsorbers out of service. Filters are typically rectangular with aspect ratios between 2:1 and 4:1.

Number of Filters

Generally, the operation and construction costs are minimized by the lowest number of filters that can provide satisfactory performance. A minimum of two adsorbers is often recommended so that one is left in service while the other is in backwash or having its media replaced. However, an advantage of using smaller adsorbers is that the larger number of adsorbers results in more adsorbers to operate in staged parallel mode, optimizing use rate and having less effect with a filter out of service. The smaller adsorbers also contain less media, which requires less effort to change out.

Though it is typical to only consider performance with one adsorber out of service at a time, filter adsorbers may provide an exception. Media replacements may take up to several days to complete, which may require another adsorber to go into backwash, which results in having two adsorbers out of service at the same time. Problems may be avoided if the media replacements can be timed during periods where maximum plant flow is not required.

Filter Depth

Total depth of GAC is typically 10 to 12 ft for post-filter contactors and 4 to 6 ft for filter adsorbers. The depth of media directly affects the depth of the gravity GAC treatment unit. Shallower-bed adsorbers take up more space on a treatment plant site and require more piping, underdrains, and backwash troughs; in some cases, they also require larger buildings to house/cover the treatment units.

Flow Control

Flow control for adsorbers is similar to flow control for traditional filters. The most common control system is downstream control with a modulating flow control valve. A rate of flow control system is the most flexible in operation and allows full plant automation, but it is also the most complex. Upstream control with a split flow weir may also be used for satisfactory equal rate adsorber performance. Upstream control with constant-head, declining rate filters is typically avoided because it creates too much variability in EBCT, which requires deep beds to ensure adequate EBCT is maintained.

Design Details of Treatment Units

This section addresses the issues pertinent to design of gravity GAC treatment systems. Some issues regarding GAC treatment that need to be considered in the design of GAC treatment systems include the following items:

- GAC is corrosive to some of the common materials of construction for the structures and systems associated with GAC treatment. For example, carbon steel is corroded by GAC.
- GAC is abrasive and is typically transferred in slurry form to prevent abrasion damage to piping.
- GAC is lower in density than silica sand, which needs to be taken into account in design of backwash systems.

Design of GAC gravity treatment systems assumes common-wall concrete structures (e.g., filter boxes) as the treatment units. The detailed design of gravity GAC treatment systems involves selecting and sizing the components associated with individual treatment units, including the structure, GAC media, underdrains, piping and valves, backwash system, auxiliary facilities required, and site design considerations.

Structure

GAC is installed in reinforced concrete structures (e.g., filter boxes). Square- to rectangular-shaped units with common wall construction are a practical option for a large facility. Whether the GAC treatment units should be enclosed in a building or be open may be decided based on multiple factors:

1. GAC post-filter contactors need to be covered to protect the filtered water.
2. A building might be required to provide a reasonable working environment for operations and maintenance staff in harsh environments.
3. The methods available for removing and replacing GAC need to be considered.

Whether the facility is constructed within a building or as an open structure, the facility should be functional and easy to access for maintenance and cleaning.

GAC Media

As previously discussed, GAC media selection, including type and size, is typically performed in the conceptual design stage. Media selection, including evaluation of equivalent products, should be performed early, because the performance of the media can affect the detailed design including selecting the optimum EBCT and fitting the process to the plant hydraulic profile.

Underdrains

A feature that has a large effect on the overall success or failure of a GAC treatment system is the design of the underdrains in the treatment units. From an operational standpoint, the underdrains perform several important functions:

- During normal downflow operation, underdrains collect water at a uniformly low rate, which ensures retention of the GAC in the treatment unit and prevents binding or plugging of the collection system.
- During backwash, underdrains uniformly distribute backwash water over the entire area of the treatment unit.
- During transfer of GAC into or out of the GAC unit, GAC can be fluidized at a low backwash rate to redistribute GAC across the treatment unit.

Except for their role in fluidizing GAC to redistribute it across the treatment units, the underdrains perform many of the same functions as in a conventional granular media filter. Several major differences between GAC and sand or anthracite should be considered when designing GAC underdrains:

- GAC contains numerous, very fine particles (termed "fines") that are less than original size due to abrasion and attrition during handling and backwashing. Without adequate backwashing, fines can penetrate through smaller openings than sand or anthracite and accumulate in areas of reduced velocity or with restricted openings, causing serious plugging problems and, in some cases, structural failure. Optimizing fines removal is discussed further in the section "Fines Removal During Filling of GAC Treatment Units" later in this chapter.
- GAC is corrosive, while sand and anthracite are not.

Many different types of underdrains exist; however, not all work well with full-depth GAC. Some considerations in underdrain selection include:

1. *Material compatibility:* Plastic block underdrains (examples include standard product offerings of ITT/F.B. Leopold Inc. and Roberts Filter Group) are compatible with GAC. For metal underdrain options, stainless steel is typically required to prevent corrosion of the underdrains.
2. *Media retention:* A media retention layer is normally required between the underdrain and the media for a number of different underdrain types. On many older filter designs, the media retention layer was gravel of particular depth and gradation. With GAC, the gravel makes it more difficult to remove GAC from the treatment units. More recently, plastic block underdrain manufacturers have used porous media caps manufactured of sintered, high-density polyethylene beads attached directly to their underdrains. These porous caps have lower profiles (1–2 in.) than the gravel media retention designs (10–12 in.), which allows a greater depth of GAC to be installed. Stainless-steel underdrain alternatives do not require a media retention layer: the openings in the underdrains are small enough to preclude media from penetrating the underdrains.
3. *Use of Nozzles:* Nozzles can be subject to plugging.
4. *False Bottom or Plenum:* GAC "fines" can accumulate in underdrain systems that have a false bottom or plenum.
5. *Additional Uplift Pressure:* Where nozzles and false bottoms are used in combination, plugging of the nozzles may create uplift pressures during backwash that can cause structural failure of the false bottoms.

Three underdrain types are discussed in detail in the following sections: plastic block, stainless-steel wedgewire, and stainless-steel folded-plate underdrains. Examples of each type are shown.

Plastic Block Underdrains. Plastic block underdrains are grouted in place immediately adjacent to each other on the bottom of the filter, forming a tile pattern. The blocks are hollow to convey water out of the filters and into a common treated water pipe or conduit (Figure 8-3). As mentioned above, these underdrains require either a media support layer of gravel or a porous media cap to retain the GAC in the filters.

Advantages of plastic block underdrains include:

- Inert construction material.
- A lower profile version (4 in. less in height) can be used in lateral lengths less than 16 ft from the flume/gullet entrance to increase GAC bed depth (ITT/F.B. Leopold Inc. 1999).

Figure 8-3. Example of a plastic block underdrain with media retention layer (Leopold® Universal® Underdrain with IMS® cap)

Figure 8-4. Example of a semi-circular wedgewire screen underdrain (Johnson Screens® Triton® Underdrain)

- Provides for low-entrance velocities during downflow operation and proper distribution of upflow during backwashing and fluidization for GAC transfer.
- Lower capital cost.
- Lightweight material makes material handling easier.

Stainless-Steel Wedgewire Underdrains. Wedgewire underdrains consist of metal channels wrapped with wedgewire. In downflow operation, water passes between the wedgewire strands into the interior channels and then into the main portion of the underdrain lateral. The wedge-shaped wire prevents the media from entering the underdrain during downflow operation and provides high exit velocities during backwashing to achieve uniform distribution of flow over the entire area of the bed. These underdrains are customized for each application to optimize flow distribution and media retention. Stainless steel is used due to the corrosive nature of GAC. Two versions are readily available: a semicircular version (Figure 8-4) that is laid directly on the floor of the treatment unit and a cylindrical version (Figure 8-5) that is supported just above the floor. The semicircular version is better suited to retrofits and treatment units with flat-bottom floors.

Figure 8-5. Example cylindrical wedgewire screen underdrain (Johnson Screens® Underdrain)

Advantages of the wedgewire underdrains include that they

- Are accessible for inspection or cleaning.
- Can be cleaned in place or easily disassembled.
- Do not require a gravel support layer or the use of other materials that can become mixed with the GAC in the filter.
- Have a lower profile than plastic block underdrains, allowing a greater depth of GAC media for the same adsorber depth.
- Provide for low entrance velocities during downflow operation so as to prevent screen blinding and for high exit velocities during fluidization or backwashing to ensure the desired uniform distribution of flow over the entire area of the bed.
- Are easily installed.

Figure 8-6. Example folded plate underdrain (AWI Phoenix Underdrain)

Stainless-Steel Folded-Plate Underdrains. Folded-plate underdrains (Figure 8-6) are similar to wedgewire screens in that they have small slots in the outer layer of the underdrain that lead to an intermediate channel, from which the water enters the main portion of the underdrain lateral. However, the outer orifices in the underdrain are slots instead of spaces between wire strands. Orifices in the walls separating the intermediate channels from the main portion of the lateral are custom-sized to optimize flow distribution for each specific application. Stainless steel is used due to the corrosive nature of GAC.

Advantages of the folded-plate underdrains include that they

- Are accessible for inspection or cleaning.
- Do not require a gravel support layer or the use of other materials that can become mixed with the GAC in the filter.
- Have a lower profile than plastic block underdrains, allowing a greater depth of GAC media for the same adsorber depth.
- Provide for low entrance velocities during downflow operation so as to prevent blinding of the openings and for high exit velocities during fluidization or backwashing to ensure the desired uniform distribution of flow over the entire area of the bed.
- Are easily installed.

It should be noted that the underdrain alternatives discussed here are not comprehensive. A summary of information about them is provided in Table 8-1.

Underdrain System Design. Design of the underdrain and water distribution/collection system can vary but will generally consist of several laterals that connect to a distribution flume, pipe, or gullet. The underdrain manufacturers being

Table 8-1. Comparison of GAC underdrain alternatives

Parameter	Underdrain Alternatives		
	Wedgewire	Plastic Block	Folded Plate
Example manufacturer	Johnson Screens®	Leopold	AWI
Model	Triton® SS* Vee-Wire Screen Underdrain	Universal Block† with IMS® Cap	Phoenix SS Underdrain
Material(s) of construction	Stainless steel	High density polyethylene	Stainless steel
Air scour capability	Yes; integral in the underdrain	Yes; integral in the underdrain	Yes; integral in the underdrain
Underdrain height (inches)	4 to 8	14	6.5
Head loss during filtration (surface loading rate = 5.1 gpm/ft^2)	2.4 in. H_2O	5.0 in. H_2O	3.0 in. H_2O
Head loss during backwash (backwash rate = 15 gpm/ft^2)	22 in. H_2O	23 in. H_2O	24 in. H_2O

*SS = stainless-steel; Johnson also has a version of the Triton® manufactured using PVC instead of stainless steel.

†Leopold also has the Type SL underdrain that is 4 in. less in height and can be used in laterals less than 16 ft in length from the flume/gullet entrance.

considered should be consulted in the development of the configuration to discuss the following design considerations:

- Configuration of existing structure, in the case of a retrofit, or a new structure to optimize water distribution.
- Incorporation of piping for air scour.
- Means of water distribution to/from the laterals.
- Maximum recommended lengths of underdrain laterals.
- GAC replacement frequency/ease of media replacement.
- Overall life-cycle cost, including energy costs related to head loss.

Several example underdrain and distribution/collection system designs are shown in Figure 8-7 through Figure 8-9. While these are not exhaustive, they provide some examples of how the underdrain types presented previously may be incorporated into a GAC treatment unit.

The goal of underdrain design is to provide uniform flow through the filter in production and backwash modes and to facilitate efficient changeout of the GAC media. Each system has its advantages and disadvantages; final design and layout should be developed with the owner and manufacturer(s) being specified.

GAC replacement should be considered during underdrain configuration. The wedgewire and folded-plate underdrains have space between laterals where GAC may be difficult to remove. Wash-down systems installed around the perimeter of a treatment unit can assist GAC removal from each treatment unit, which is covered later in this chapter. A hose or hose connections that allow plant staff to manually assist media removal should also be considered. For all underdrain types, where possible, sloping the bottoms of gravity GAC treatment units toward the spent GAC piping

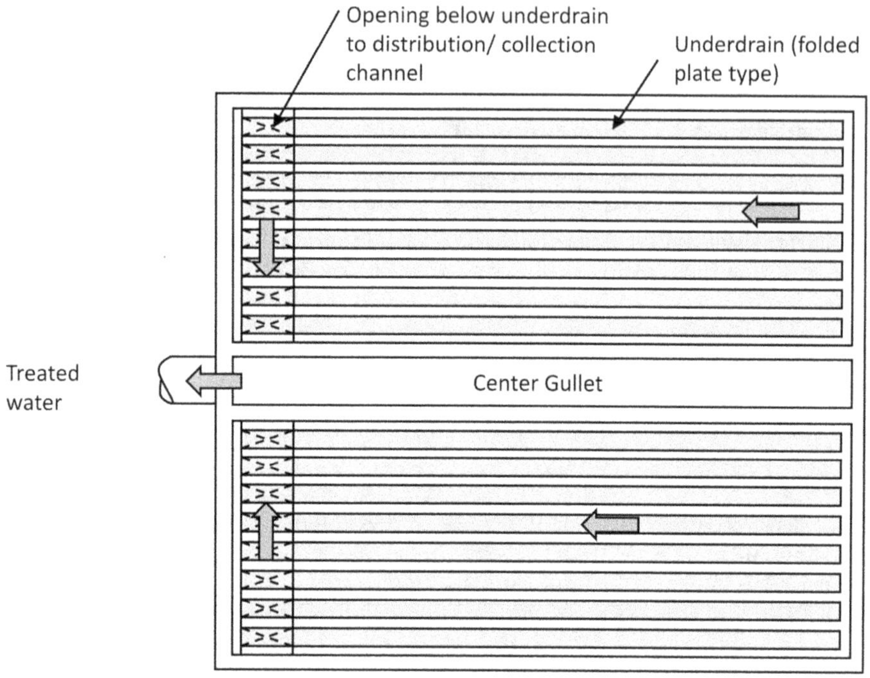

Figure 8-7. Example longitudinal underdrain layout

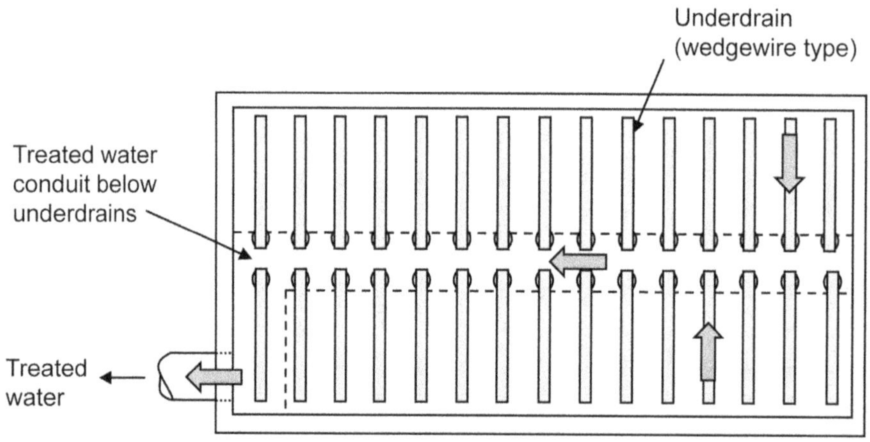

Figure 8-8. Example radial underdrain layout

should be considered to facilitate GAC removal. Adsorbers may employ a floor that is sloped from side walls to the centerline of the adsorber to facilitate GAC removal. When sloped floors are used, underdrains that are suspended above the floor of the adsorber can be used, or laterals can be placed along the slope to facilitate GAC removal. Removal of GAC from treatment units with floor-mounted underdrains will not be as efficient; it may require maintenance staff to enter the filter to remove the media or an owner to consider leaving the last few inches of media in place during a change out (which will have a minimal effect on adsorber performance).

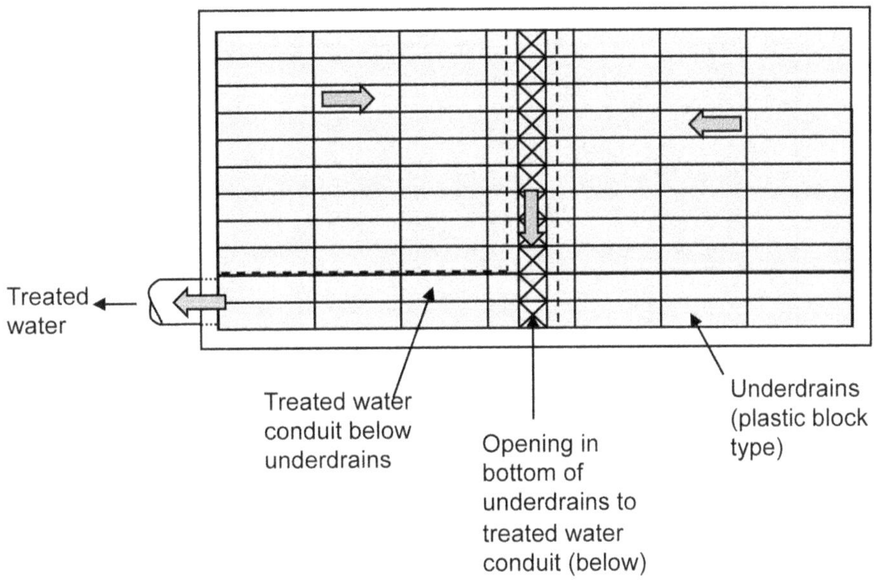

Figure 8-9. Example tile/block underdrain layout

Backwash Systems

Backwashing typically occurs when one of the following conditions is observed:

- High treated-water turbidity
- High head loss through the treatment unit
- Low flow through the treatment unit when rate of flow control valve is at the maximum open position
- Media compaction
- Filter run time

Extended backwashing is used after GAC replacement to remove fines. Post-filter contactor backwashes typically use water only, not air scour, to remove particles from the media. The water used for backwashing a post-filter contactor does not have to be GAC-contacted water; filtered water from the upstream filtration process may be acceptable. Unchlorinated water should be used if available, because chlorine degrades the GAC, resulting in production of more fines. Chlorine may reduce the adsorptive capacity of the GAC and result in formation of chlorinated organic by-products at the GAC surface. For filter adsorbers, air scour is sometimes used in addition to backwashing using water to reduce formation of mudballs within the media (Amirtharajah et al. 1991). Surface washing can also be used to help break up particles on the top of the GAC bed and accumulated biomass near the surface of a BAC filter.

Backwash Requirements. For retrofits, the engineer should determine if the existing backwash and equalization facilities are sufficient for the new GAC treatment units or if a new or rehabilitated pump station is required. The requirements for backwashing GAC may be very different from the previous media used. Likewise, a new supplemental facility such as post-filter contactors often requires a new pump station if the backwash requirements differ significantly from those of the existing filters. GAC filter retrofits may need a new or refurbished pump station to provide

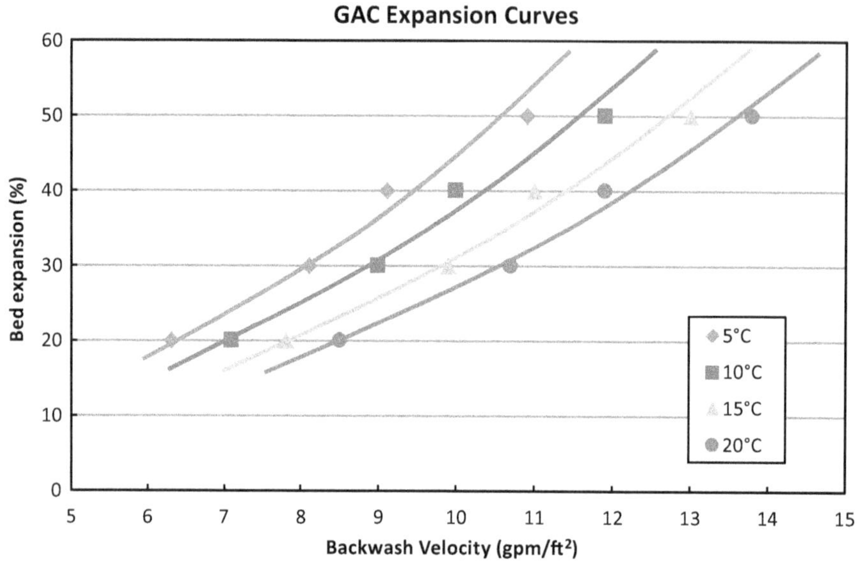

Figure 8-10. Variation of GAC bed expansion during backwash at multiple temperatures

better process control; variable-frequency drives (VFDs) on the backwash pumps or a programmable, modulating backwash flow control valve can provide better control of backwash rates than constant speed pumps alone, reducing the risk of washing out the GAC during the backwash process.

GAC manufacturers provide information regarding proper bed expansion of the media during backwashing and the backwash rates needed to accomplish that expansion. Figure 8-10 is an example showing the variation in bed expansion that can occur at a given backwash rate with varying water temperature. As shown in the figure, to achieve a particular bed expansion, the highest backwash pumping rates are required at the highest water temperatures, when the water viscosity is lowest. Backwash pumps should be sized for the backwash flow required at the facility's highest-expected water temperature. One backwash pump should be able to backwash one GAC treatment unit.

For GAC gravity treatment systems employing more than one type of media (usually GAC and sand), the method proposed by Kawamura (2000) of determining the proper combination of media is recommended. Kawamura presents an approach that normalizes the effective size, uniformity coefficient, and specific gravity of each filter media to a backwash rate so that effective cleaning of all filter media is achieved.

The backwash rate needs to be evaluated experimentally once GAC is placed in full-scale treatment units. Multiple facilities report using lower backwash rates than indicated by manufacturer's data because media loss has occurred when recommended backwash rates were used (Kearns 2004, Scott 2006).

Ramping the backwash rate can also help reduce media loss. Figure 8-11 provides an example of how the backwash flow can be ramped up to the target backwash rate and back down at the conclusion of a backwash. Flow ramping helps prevent air entrainment in the backwash water or the media bed, reduces the possibility of the entire bed lifting as a monolithic unit, and helps bed expansion be evenly distributed over the filter area. VFDs or a backwash flow control valve can be programmed to accomplish the ramping process.

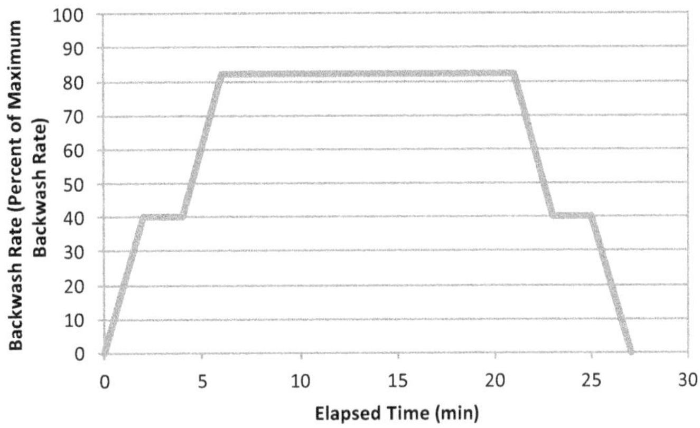

Figure 8-11. Example of stepped backwashing sequence

One way to monitor bed expansion is to install bed depth markers around the perimeter of the GAC treatment unit at the maximum expanded bed height during backwash. The backwash rate can be adjusted based on visual observation during backwashing. Examples of bed-depth markers can be a stainless-steel strip or stripe of white paint certified by NSF International (NSF).

Air Scour. Granular media filters have historically been provided with water-only backwash capabilities or water-only backwash supplemented with surface wash. However, older granular media filter designs were typically shallow (2–3 ft in depth). Air scour is a relatively new approach to filter backwashing that has improved filter cleaning on deep bed (~5–6 ft) granular media filters, including GAC filters. Air is introduced into the filter through the underdrain system and bubbles up through the filter media as part of the backwash process. An air scour system requires air-ready underdrains, a blower facility, and air piping from the blower facility to the underdrains.

The effect of air on adsorption performance of GAC is largely uncertain. Dry GAC requires a wetting period to displace air from the pore spaces in GAC media, and guidance for adsorption processes has historically discouraged the introduction of air to GAC media. Pilot testing at the Glendale Cholla WTP (Case Study No. 9) was conducted to determine the effect(s) of air scour on performance of filter adsorbers. No benefit in turbidity reduction resulted from using air scour. Additionally, pilot testing indicated some negative effects of air scour, including:

- The turbidity peak after backwash (filter ripening period) occurred a few minutes later in filter adsorbers cleaned with air scour compared with those backwashed with water only.
- The turbidity peak during filter ripening was broader.
- The rate of head loss increased during filtration after backwash in filter adsorbers backwashed with air scour.

Air scour has frequently been used for backwashing BAC to improve filter media cleaning, including mitigating mudball formation. One such example is the Greenway WTP in Peoria, Ariz. (Case Study No. 10), which is an ozone-BAC facility. Operations staff at that facility have experimented with air scour over the course of operations and have determined that it is not desirable for every backwash because it increases fines production; typically air scour is employed with every other backwash on each BAC filter (Kearns 2004).

Surface Wash. Surface wash is used during backwashing to help break up solids that have accumulated on the surface of the GAC. Surface wash systems are installed slightly above the media bed and typically consist of a header pipe with several rotating arms, or "agitators," at the end of which are small spray nozzles directed down toward the media. The header pipe and agitators should be constructed of a material that is suitable for use with GAC, such as stainless steel.

It is important to install the surface wash arms at the proper height above the GAC bed, accounting for expansion of the bed due to backwashing. Surface wash arms should be installed close enough to the media bed that effective washing can occur, but the arms should never be allowed to be buried by the settled media.

Backwash Troughs. Backwash troughs are designed to distribute water evenly into the filters and collect backwash wastewater. Typically, troughs for filters are spaced so that each serves an equal area of the filter. Troughs are located so that the maximum horizontal travel for particles to reach the trough does not exceed 3 ft.

Because particle filtration is not the focus of post-filter contactor treatment, these trough design requirements are not mandatory. Post-filter contactors may be equipped with concrete, U-shaped wall-hugging troughs installed around the perimeter of the contactor. Such an arrangement precludes the need for features such as a center gullet to distribute the water to multiple troughs, simplifying the treatment unit design. Figure 8-12 shows a U-shaped trough design. Influent and backwash waste pipes enter directly into the trough, and water is distributed to/from the post-filter contactor by overflowing the weir wall of the trough.

Filter-to-Waste. Following backwashing, the GAC filtrate is sometimes discharged to waste for a period of time. For filter adsorbers and BAC, the practice of "filter-to-waste" recycles the water from the filter while the post-backwash turbidity spike/filter ripening period occurs. For post-filter contactors, providing facilities to accommodate a short period of "contactor-to-waste" reduces GAC fines entering treated water piping and reservoirs downstream of the contactors. While the amount of fines expected to pass through the underdrains after backwashing is minimal, facilities online to date have provided filter-to-waste capability and have confirmed its utility. After a set period of time and/or when any elevated turbidity in the GAC-treated

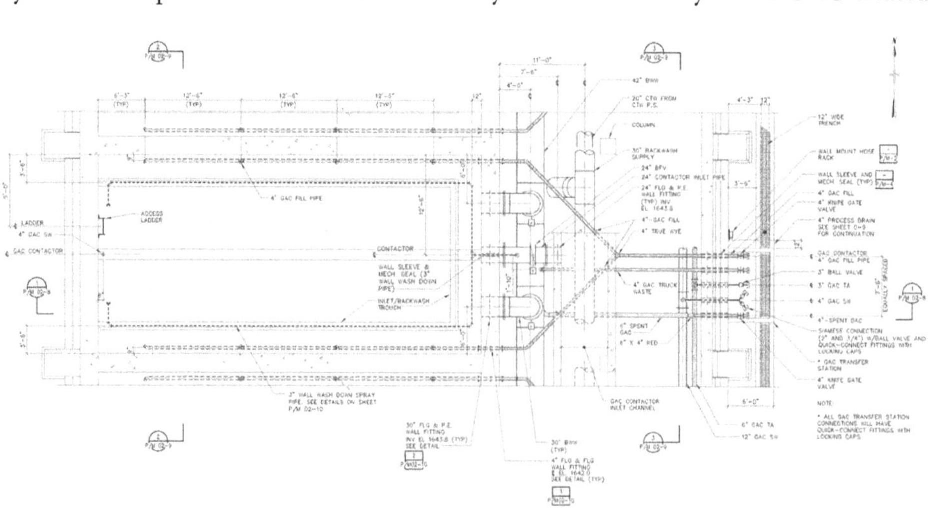

Figure 8-12. Example post-filter contactor with U-shaped trough (courtesy Malcolm Pirnie)

water dissipates, wasting is discontinued, and the GAC treatment unit is returned to service. Filter-to-waste flows are recycled within the treatment facility per the guidance provided in the Filter Backwash Recycling Rule (USEPA 2001).

Water Service Piping and Valves

The following piping systems are associated with GAC treatment units:

- Large-diameter piping for influent, GAC-treated water, backwash supply water, backwash wastewater, and filter-to-waste water.
- Smaller-diameter piping for service water for GAC transfer, GAC flushing/wash systems, and sample piping. If air scour or surface wash is used, space should be provided in the filter gallery.

Some guidelines for design of pipe systems for GAC treatment units follow:

1. Pipe in contact with GAC, such as surface wash piping systems within a GAC treatment unit, should be made of a corrosion-resistant material (e.g., stainless steel).
2. Piping systems for process water entering or leaving the GAC treatment unit (e.g., process feedwater, backwash water, flushing/wash water) can be mortar-lined/cement-lined ductile-iron or carbon steel pipe.
3. Mortar-lined/cement-lined ductile iron pipe is suitable for all buried and below-slab treated water piping.

Many valve types can be used for these services.

- Slide gates or butterfly valves for isolation of a filter.
- Butterfly valves for rate of flow control through a filter.
- Gate valves for service, plant, and utility water.
- Butterfly valves for air systems.
- Knife-gate valves for GAC slurry service.

Auxiliary Systems

In addition to the GAC treatment units and backwash systems mentioned above, additional auxiliary systems may be required. Table-8-2 presents a summary of auxiliary systems potentially required for each GAC treatment technology.

Table 8-2. Potential auxiliary systems required for GAC treatment

Auxiliary System	Post-Filter Contactors	Filter Adsorbers	BAC Filters	GAC Caps
Seal well	X			
Feed pump station*	X			
Recycle stream equalization	X	X	X	X
Surface wash pump station		X	X	X
Air scour blower facility		X	X	X
GAC transfer systems	X	X		
On-site GAC storage	X	X		X

* Filter GAC technologies are assumed to be incorporated in the hydraulic profile of the water treatment facility whether new or retrofits. Post-filter contactors may also be incorporated into the hydraulic profile of a new water treatment facility, or may require a new feed pump station if retrofitting an existing facility.

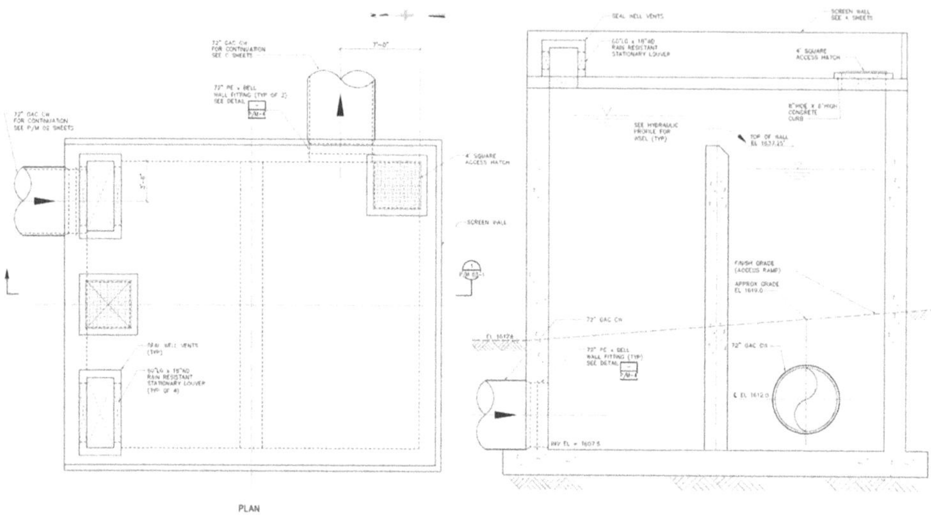

Figure 8-13. Seal well (courtesy Malcolm Pirnie)

Design of feed and surface wash pump stations and air scour blower facilities is not covered in this volume. The reader is referred to other sources for further information on these topics (Sanks 1989, Kawamura 2000).

Seal Well. Some facilities feature a hydraulic structure referred to as a *seal well* downstream of the GAC treatment units. The sole purpose of the seal well is to prevent accidental drainage of the GAC media and consequential air binding. Allowing air into a GAC treatment unit is not desirable because the air may occupy pore volume, reducing the amount of surface area available for adsorption. The air can be displaced by water, but the treatment unit has to be taken out of service and soaked for a period of time (operating facilities have reported allowing GAC to soak from 24 hr to 1 week for air displacement). The primary feature of a seal well is a weir over which the water must flow after the GAC treatment facility. A drawing of a seal well is shown in Figure 8-13.

Seal wells have been used as a fail-safe approach. The worst-case scenario is when flow is at or near zero through the gravity treatment units, because little or no head loss occurs between the top of the media and the seal well. Setting the weir height to the same elevation as the top of the media prevents dewatering of the media under any conditions. Some facilities have accounted for head loss at a design minimum flow rate and lowered the weir height accordingly.

Seal wells should be equipped with a means of collecting GAC fines that pass through the filters, such as a sump or sump/drain valve installed in the seal well upstream of the weir. Provisions should also be made to remove the collected fines from the seal well using a drain or pump. A drain valve also allows the media to be stored dry if the adsorber is taken out of service for long periods.

Recycle Stream Equalization. Several recycle streams are associated with GAC treatment (e.g., backwash waste, filter-to-waste). Where the recycle streams are to be conveyed should be determined during design. Backwash wastewater from granular media filters is typically sent to a residuals treatment facility. Residuals treatment can be another entire treatment process in a water treatment facility to process the solids for disposal and recycle water back to the head of the plant, or it can be a solids collection basin such as a sedimentation lagoon. The process receiving the recycle streams should be evaluated to make sure it can handle the anticipated additional flows and solids.

One approach to reduce the effect of recycle streams on residuals handling processes is to provide equalization for those flows. Intermittent high-flow recycle streams are sent to the equalization basin and then pumped to the receiving processes at a lower rate. Considerations for design of an equalization process include:

1. Determine the maximum frequency, flow rate, total volume, and solids loading of each recycle stream expected to be sent to the equalization basin:
   a) Backwash waste.
   b) Filter-to-waste.
   c) Contactor-to-waste.
   d) Other streams as applicable.
2. Determine the receiving process(es) for the recycle stream flows. Determine the available capacity, if any, in terms of volumetric flow rate and mass flow rate of solids for the subject process(es).
3. Size the equalization basin to provide enough storage to accept the recycle streams for a period of time (for example: 24 hr, backwash of one GAC treatment unit, etc.). One method is to size the basin to handle the volume of backwash water generated during media loading and fines backwashing, which takes longer than a normal backwash. A large equalization basin is better from an operational perspective, but the benefits should be weighed against the higher cost and footprint required.
4. Size the equalization pump station to deliver the recycle streams to the receiving process(es).
   a) Determine the minimum flow rate out of the equalization pump station required to maintain it at or below normal operating levels within the time period of interest (determined under (3), above).
   b) Determine the maximum flow rate the receiving process(es) can accept.
5. If (4b) is greater than (4a), the equalization system is adequate. If (4b) is less than (4a), additional evaluations need to be conducted:
   a) Can the capacity of the receiving process(es) be expanded?
   b) Can the GAC treatment process be modified to reduce the recycle streams? For instance, can the footprint of the gravity treatment units be reduced to reduce the flow per backwash and/or flow per wasting event going to the equalization basin?

Recycle Streams With Low Solids. The recycle streams from a post-filter contactor are not expected to contain a significant mass of solids because turbidity removal occurs in the filtration process. The primary source of particle loading in post-filter contactors is expected to come from the GAC fines during the adsorber filling process. According to Norit Americas Inc., a manufacturer of GAC, the amount of fines generated during installation of GAC in an adsorber can vary widely and depend on how the GAC is handled. Fines are always less than 5 percent of the total weight of the shipment and are usually less than 1 percent (Naylor 2003).

Recycle streams from a post-filter contactor process can be sent to the inlet or head of the water treatment facility instead of being sent to the equalization process. Recycle to the headworks avoids the need for substantial additional equalization volume and pumping capacity and prevents dilution of thickened solids at the residuals treatment facility. Depending on the effect GAC fines will have on the water treatment processes, backwash waste generated during filling of the GAC treatment units may be sent to the inlet/head of the water treatment facility as well; the fines will settle out in the sedimentation basins. If fines cause an issue for the water treatment processes (for example, clogging of a membrane filtration process, or direct filtration with no sedimentation basins), backwash waste containing fines should be sent to the residuals treatment facility.

If the filter-to-waste stream from a filter adsorber, BAC, or GAC-cap GAC treatment process is determined to be sufficiently low in solids, that stream may also be recycled to the head of the water treatment facility rather than the residuals treatment process. To provide maximum flexibility for the water treatment facility, all recycle streams can be directed to the equalization facility for subsequent pumping to either the head of the water treatment facility or the residuals treatment facility, depending on the solids loading.

For flows sent to the inlet/head of the water treatment facility, if the flexibility exists in the water delivery system, raw water flow to the facility can be reduced by the increment of the recycle flow, thus reducing the effect to operations.

Site Design

The GAC treatment facility layout incorporates the footprint of the GAC treatment units and the backwash, auxiliary, and storage facilities that are required. The following items also need to be considered in the facility layout:

1. Access for vehicles and equipment transferring GAC to or from the facility.
   a) If storage silos are not included in the facility, then access should be provided to each GAC treatment unit. Fill stations for GAC transfer can be located at each individual treatment unit or can be centralized. In general, GAC transfer piping runs are minimized to reduce the numbers of bends and long runs of pipe, thus minimizing plugging potential and fines generation. As a result, a number of facilities have been designed with individual transfer stations located near each treatment unit or with the capability to run temporary lines short distances to each treatment unit.
   b) Vehicle access to the GAC facility should account for the sizes and turning radii of vehicles expected to service the site. Flatbed trucks are usually used to deliver and remove 1,000- or 2,000-lb bags and pallets of 50-lb bags of GAC, while tanker-style trucks are used for bulk delivery. For larger facilities where loading and/or unloading of multiple GAC treatment units would occur simultaneously, with several trucks on-site at the same time, additional driveway width and length can be provided adjacent to the GAC treatment facility. Providing a loop drive around the water treatment facility allows delivery vehicles to navigate the site without the difficulty of backing up. Providing flexibility at the site to accommodate different delivery vehicles should be weighed against the additional costs of design and construction.
   c) The site paving should be designed to handle the load of a full GAC delivery truck. The design of paving in the area adjacent to the GAC fill stations is especially important, because tanker-style trucks loaded with

20,000–40,000 lb of GAC will have approximately 4,000–6,000 gal, or an additional 30,000–50,000 lb, of water added to them while trucks are parked to slurry the GAC into the treatment units. High-strength asphalt paving or reinforced concrete paving is recommended for the loading/unloading area to prevent excessive wear and tear on the pavement.

2. Proximity of backwash and auxiliary facilities to the GAC treatment units.
3. Proximity of GAC storage and reactivation facilities to the GAC treatment units. Different facilities have taken different approaches to GAC storage proximity:
   a) The Richard G. Miller WTP in Cincinnati, Ohio (Case Study No. 1), located the GAC storage and reactivation facilities centrally and accomplished GAC transfer among the facilities via hard-piped eductors and pumps.
   b) The Lake Pleasant WTP in Phoenix, Ariz. (Case Study No. 8), has the GAC storage and reactivation facilities located remotely from the GAC treatment units. GAC is transferred between the facilities by intermediate transfer to/from trucks.
4. A vehicle scale for trucks to check the weight of GAC being delivered to the site or road weight prior to leaving the facility.

GAC Handling

GAC replacement is a routine operation for post-filter contactors and filter adsorbers. Spent GAC should be removed from the adsorbers and replaced with virgin or reactivated GAC to maintain adsorption of contaminants.

GAC Delivery. Depending on the manufacturer supplying GAC for a water treatment facility and its location relative to the facility, virgin or reactivated GAC can be delivered in bulk tanker trucks or delivered in large bags (1,000 or 2,000 lb each) by flatbed truck. Manufacturer's bulk delivery trucks carry 20,000 to 40,000 lb of dry GAC. Figure 8-14 shows examples of both types of deliveries.

Bulk delivery tanker trucks have hard piping installed on the trucks that can be directly connected via flexible hose to piping at a transfer station installed at the GAC treatment facility. The bags can be emptied in one of two ways:

1. Emptied into a skid-mounted hopper-eductor unit that is connected via flexible hose to the transfer station (Figure 8-15).

Figure 8-14. Bulk delivery of GAC by tanker truck (a); 1,000-lb virgin GAC bags staged for loading a GAC treatment unit (b)

2. Lifted by crane over an open treatment unit and cut open. When bags are emptied directly into the adsorber, the GAC treatment unit should be filled with enough water to at least cover the underdrain system to mitigate GAC abrasion.

Figure 8-15. 1,000-lb virgin GAC sack and skid-mounted hopper-eductor unit

Figure 8-16. Spent GAC dewatering equipment used with 1,000-lb bags (provided by Calgon Carbon Corp.)

Spent GAC Removal. For removing GAC from a treatment unit, GAC is transferred as a slurry via eductor within the GAC treatment facility to a bulk tanker truck or skid-mounted dewatering hopper. As of 2012, it appears that the only dewatering hopper has been developed by Calgon Carbon Corporation to facilitate removal of spent GAC from water treatment facilities without the need for bulk delivery trucks. Water enters a separation tank on the upper level of the skid, and dewatered GAC is dropped into one of two 1,000-lb bags located on the lower level (Figure 8-16). The bags are removed by forklift and either placed directly on a flatbed truck or placed in temporary storage at an on-site staging facility for further draining and drying. Further draining is recommended to decrease the weight of each bag and allow more bags to be loaded per truck, minimizing shipping costs and truck traffic on-site. An on-site staging facility with drainage collection capability is required.

GAC transfer operations can require substantial time to complete:

- One tanker truck containing approximately 1,400 ft^3 of GAC can unload in approximately 15–20 min, not including time for hookups, truck draining, and so on.
- During initial filling of the post-filter contactors at the Scottsdale CAP WTP (Case Study No. 7), one contactor was filled with approximately 15,000 ft^3 of GAC in three 8-hr days via 1,000-lb bags.

GAC Transfer Facilities. A transfer station adjacent to the GAC treatment facility should be provided to facilitate loading/unloading the GAC treatment units. Figure 8-17 shows a post-filter contactor facility from the access drive: Fill stations for the first five contactors in the facility can be seen along the building wall.

Figure 8-17. Post-filter contactor facility access drive with GAC transfer station connections at each contactor location

The fill station connections and systems required to accommodate GAC loading/ unloading operations are summarized below:

1. *Service water line with quick-connect fitting to provide water to fluidize the fresh GAC in the tanker truck and/or for eduction of GAC into the treatment units.* Typically 4,000 to 6,000 gal of water are added to tanker trucks containing 20,000-lb shipments of GAC to initially wet the GAC. During the transfer process, water is added at a rate of 100 to 200 gpm to educt GAC into a treatment unit.
2. *GAC fill line with a quick-connect fitting and valve.* Tanker trucks transfer the fluidized GAC to the fill line by applying pressure (by water and compressed air) to the tanker. Portable eductor units transfer GAC with water only.
3. *Compressed air supply and connections if not provided by the GAC delivery truck.* Some GAC suppliers provide truck-mounted air compressors for air padding the delivery trucks. The air padding aids in conveying GAC slurry out of the truck storage bins and into the GAC transfer piping. The air padding system should be capable of supplying 200 scfm at 15 psi to the truck.
4. *Spent GAC line with quick-connect fitting and valve to deliver spent GAC to empty tanker trucks.* Spent GAC can be transferred from the adsorber into the spent GAC line by an eductor within the GAC facility that uses water as the motive force. The eductor should be sized to fill a tanker truck with 40,000 lb of wet GAC (drained weight) in a reasonably short amount of time (i.e., 1–2 hr). More information on sizing eductors is provided in the next section.
5. *GAC truck waste line with valve and quick-connect fitting.* After delivery of spent GAC to the delivery truck, the transfer water can be drained out of the truck through the waste line and into the recycle flows pipeline.

Communicating with potential GAC suppliers during design will identify the needs of the delivery trucks so the necessary connections and services are provided at the site.

Moving the GAC. GAC can be conveyed either by water or by air. However, air transport results in high GAC attrition losses and rapid erosion of pipe and fittings,

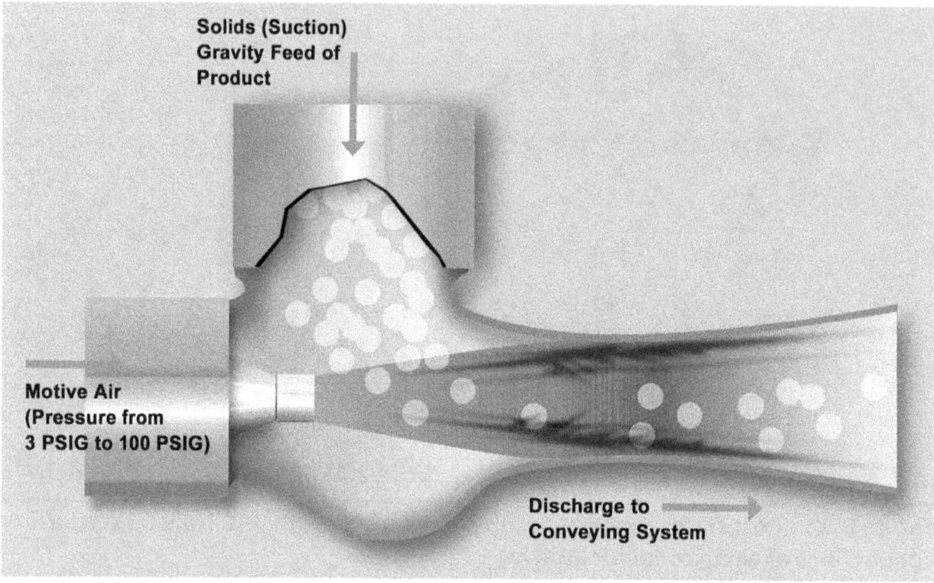

Figure 8-18. Schematic of an eductor (courtesy Schutte and Koerting)

particularly from sandblasting effects at pipe bends. GAC transport using water is recommended because it avoids the difficulties mentioned above and is a cleaner, dust-free operation. Some general guidelines for water transport of GAC:

1. About 1 gal of water per pound of GAC is required to form a suitable slurry.
2. The minimum diameter for pipelines carrying GAC slurry should be 3 in. to minimize the risk of plugging.
3. Maintain slurry velocities of at least 3 ft/sec to prevent settling out of GAC but not more than 5 ft/sec to minimize GAC abrasion and pipe wear.

Two examples of GAC slurry transfer equipment are explained further below: eductors and recessed-impeller centrifugal pumps.

Eductors. Spent GAC can be removed from the adsorbers using an eductor with plant water as the motive force. A schematic of an eductor is shown in Figure 8-18. Water running longitudinally through the eductor induces a vacuum, which pulls slurry in through the side opening.

Eductor manufacturers can provide performance curves relating the amount of solids/slurry (i.e., GAC) moved per flow of motive water for the given system pressure (Figure 8-19). The GAC removal rate of the eductor directly affects the amount of time it takes to complete a GAC replacement. Therefore, the size of the eductor is selected based on the desired duration of the spent GAC removal process and the plant service water pressure and flow available. Using the example performance curve in Figure 8-19, for a plant service water pressure of 60 psi and a discharge pressure of 5 psig (at the discharge of the eductor), a 1-in. eductor can remove approximately 17 ft^3/hr. With this eductor, a GAC treatment unit containing 1,000 ft^3 of GAC will be emptied in approximately 60 hr. By comparison, a 3-in. eductor can remove 9×17 ft^3/hr, or 153 ft^3/hr, requiring only 6-½ hr to empty a GAC treatment unit (not including time for hookups, truck draining, etc.). Note that the service water flow requirement for the larger eductor is also increased by a factor of 9. Service water pressure remains the same.

Recessed-Impeller Centrifugal Pumps. GAC can be transferred out of a GAC treatment unit by a recessed-impeller centrifugal pump. These pumps are installed low in the filter gallery so GAC slurry flows to the inlet by gravity can then be pumped elsewhere. Pumps should be horizontal centrifugal, recessed-impeller, rubber-lined pumps, with a maximum speed of 900 rpm. Note that the impellers will need to be replaced regularly.

GAC Transfer Service Piping and Valves. GAC transfer piping conveys GAC slurry into and out of the GAC treatment units. Minimizing the length of GAC transfer lines and turns results in smoother operation and less production of fines. Long-radius bends help to keep the GAC slurry moving uniformly. GAC transfer piping should be easily disassembled to facilitate cleaning of pipeline plugs and replacement of pipe and fittings. Stainless-steel piping is recommended to withstand the abrasiveness of GAC slurry flow.

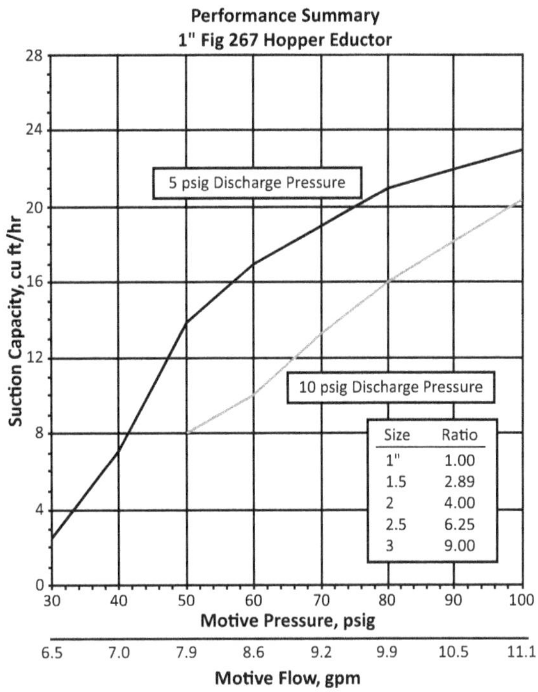

Figure 8-19. Example eductor performance curves (Figure 267 Eductor, courtesy Schutte and Koerting)

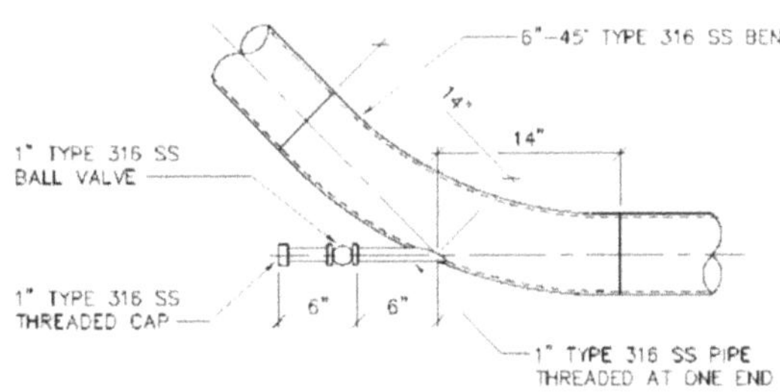

Figure 8-20. Flushing connection on a GAC transfer pipe (courtesy Malcolm Pirnie)

To minimize corrosion and plugging of GAC transfer lines, it is desirable to flush them with water after each use so that prolonged contact with GAC, especially partially dewatered GAC, is avoided. In addition, special taps for introduction of supplemental transfer water under pressure can be provided to allow complete removal of GAC from idle pipelines (Figure 8-20).

Avoid the use of throttling valves on GAC slurry lines; instead, install throttling valves on the water supply lines ahead of the point of GAC introduction. Knife-gate valves work well on GAC slurry lines because they have a full port design that allows a high flow capacity with low head loss. Knife-gate valves are also resistant to clogging.

Pretreatment

Chapter 5 demonstrated that as the background DOC concentration increases bed life of the GAC media decreases. Thus, optimizing pretreatment for the removal of

DOC will lower GAC operation and maintenance costs, which should be evaluated against increased pretreatment costs. For systems targeting the removal of DOC, lowering pH improves GAC performance and should also be considered in pretreatment.

Retrofit Limitations

Installing GAC treatment units in place of existing filters poses additional design challenges. The following challenges may need to be overcome during design and construction of a filter retrofit:

1. *The existing backwash pump station design does not allow fine-tuning of the backwash rate, and pumps are oversized for proper backwash (bed expansion) of GAC.* An entirely new backwash supply pump station may be constructed as part of the project.
2. *The existing filter underdrains require a layer of sand and gravel to support the filter media.* The existing underdrains may be replaced with low-profile underdrains to maximize the GAC bed depth in the filter structures. The existing troughs and surface wash system may be raised to accommodate greater media depth.
3. *The design of a retrofit is constrained by the existing space available on-site.* Existing plant piping can also be a limitation. New piping and valves required for the GAC facility should be installed in existing filter galleries, which may already be congested.
4. *Construction of a retrofit at an existing plant is more complicated than with construction of a new facility because some existing piping cannot be moved or taken off-line.* If the plant must remain in service at all times, tie-ins to existing pipes should be carefully coordinated with plant operations and are likely subject to time limitations (e.g., the plant can be shut down for only 4 hr while the tie-in is completed).

Operation and Maintenance

Operation and maintenance of a GAC adsorber mainly consist of the following operations: filling and emptying a unit, offline storage, backwashing, and water quality monitoring.

Filling a GAC Treatment Unit

For GAC fill lines, slurry is typically conveyed in pipes that are located along the sides of the GAC treatment unit perpendicular to the pipe gallery and are attached to the wall or to the bottom of the backwash troughs above the planned installation depth of the media. GAC outlets can be spaced uniformly along the length of the treatment unit. The angle of repose of GAC in water is about 45°, so there will be some mounding upon discharge into the treatment unit (Figure 8-21).

GAC slurry exits the first fill line outlet in the GAC treatment unit. Once the GAC mound blocks the first outlet, the slurry starts flowing out of the second outlet. Uniform distribution and leveling of the GAC within the treatment unit can be accomplished by operating the backwash flow at a low flow rate (approximately 4 gpm/ft^2) for a short time.

GAC media are purchased by mass, and adsorbers that are filled dry will typically initially appear to have insufficient depth. The denser packing of the media results from the bed not being stratified. The initial backwash to remove the fines will result in media stratification and an increase in bed depth that should achieve the design EBCT. The increase in depth may be critical if filters are filled dry to the depth mark where, after the initial backwash the bed can expand about 10 percent, which could interfere with a surface wash system.

Fines Removal

Operating facilities have reported that fines removal during filling operations is very important (Scottsdale CAP WTP, Case Study No. 7; Cincinnati Richard G. Miller WTP, Case Study No. 1; Lake Pleasant WTP, Case Study No. 8; Cholla WTP, Case Study No. 9). Inadequate backwashing after virgin or reactivated GAC has been placed into GAC treatment units has resulted in premature termination of operations due to excessive head loss and accumulation of fines in GAC effluent turbidimeters and downstream finished water reservoirs. These facilities have reported alleviation of such issues by backwashing at frequent intervals during the filling process (e.g., after every 50,000–80,000 lb of GAC are added; after the treatment unit is one quarter, half, three quarters, and completely filled).

Figure 8-21. Accumulation/mounding of GAC during treatment unit filling

Emptying a GAC Treatment Unit

The most common approach used for emptying a GAC treatment unit is by means of an eductor. Eductors have been used primarily in two ways:

1. *Fixed installations* (Figure 8-22): With a fixed installation, each treatment unit has a dedicated eductor hard-piped to the service water, treatment unit, and either a transfer station to transfer the spent GAC out of the facility or an on-site reactivation facility (if present).
2. *Mobile installations* (Figure 8-23): Some facilities have used

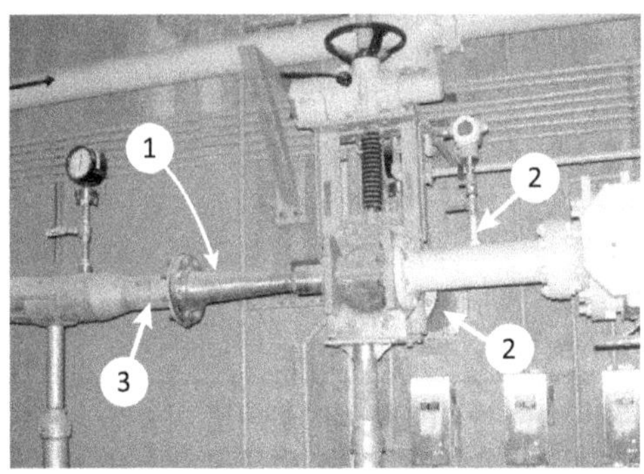

Figure 8-22. Permanently installed eductor with service water (1), wall connection to gravity treatment unit through knife-gate valve (2), and discharge piping to transfer station (3)

Figure 8-23. Eduction of GAC from a GAC treatment unit with temporary piping and a mobile eductor

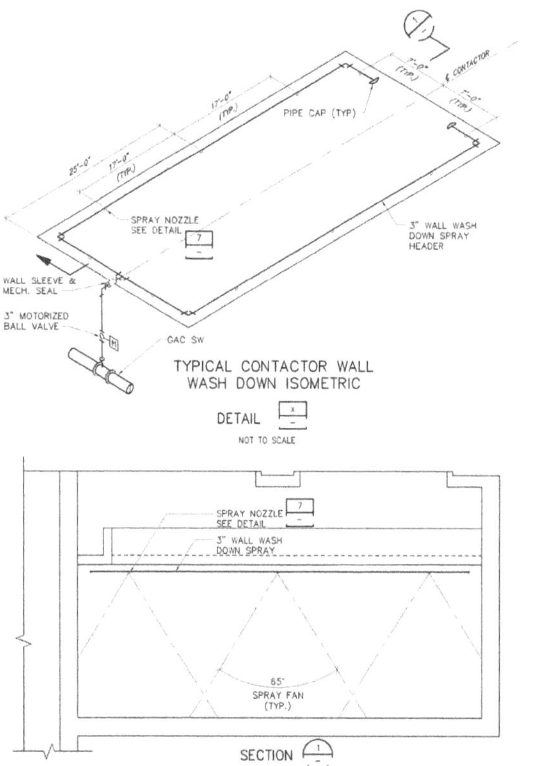

Figure 8-24. (a) GAC treatment unit wall washdown schematic (courtesy Malcolm Pirnie)

a mobile eductor with temporary piping to reduce the capital cost of the facility. The portable eductor can be moved around within the GAC treatment facility to remove GAC from any one of the GAC treatment units. Flexible hose is used to make the connections of the eductor to the plant service water and spent GAC piping. To remove GAC from each treatment unit, the eductor is either in the pipe gallery connected to a wall fitting or at the operating level with a vacuum suction line lowered into the treatment unit. In the latter case, a person must be lowered into the treatment unit to manipulate the portable suction line as shown in Figure 8-23. (NOTE: proper safety guidelines need to be followed at all times. See more information at the end of this chapter.)

Backwash water is typically added to the GAC treatment unit at a low rate (approximately 4 gpm/ft^2) during the eduction process to facilitate fluidization of the media. However, operations staff should experiment with low rate backwash to optimize removal operations staff at the Richard G. Miller WTP in Cincinnati, Ohio (Case Study No. 1), have observed better GAC transfer out of the treatment units with the media submerged but without concurrent backwash water addition (Waymeyer 2003).

Flushing/Wash Systems

Water spray systems inside a gravity treatment unit can facilitate removal of GAC. A wall washdown system, as shown in Figure 8-24a, is installed around the perimeter of the treatment unit. The system is supplied by the service water system and is equipped

Figure 8-24. (b) Model FF spray nozzle (left); (c) spray pattern of the Model FF spray nozzle (right) (BETE Fog Nozzle, Inc.)

with special spray nozzles (Figure 8-24b and c). The nozzle angle and spray pattern can be selected to cover the entire wall surface prior to the point it joins the slab to improve the flow of GAC to the treatment unit outlet. In addition, including hoses and hose connections allows flexibility for maintenance staff to direct washwater as needed during the media removal process.

Some GAC treatment units incorporate sloping the floor of the structure to facilitate GAC removal. A trench sloped from the far end of the treatment unit to the pipe gallery end can be equipped with spray nozzles at each end to help wash the spent GAC down the trench. A plain pipe with a duckbill check valve has been used for washdown (Figure 8-25).

Figure 8-25. Post-filter contactor with center trench and center flushing line

On-Site Storage of GAC

The need for on-site storage of GAC should be considered in the design of a facility. On-site storage can provide operational flexibility. Purposes of storage may include the following:

1. Virgin or reactivated storage to replace the GAC in one or more treatment units.
2. Storage of spent GAC from one or more treatment units to be sent to on-site or off-site reactivation or disposal. Media can also dewater and dry more to reduce transportation costs.

3. Storage of virgin makeup GAC for addition to GAC treatment units on an as-needed basis (e.g., to replace media lost during backwashing or due to attrition during a reactivation process).

On-site storage of GAC can be a logistical challenge. On-site storage can amount to any of the following:

1. *Pallets of 50-lb sacks.* Sacks are a more likely alternative for PAC feed systems because GAC treatment systems typically use larger quantities than would be practically stored in 50-lb sacks. However, it may be an option for storing makeup GAC for facilities that use reactivated GAC.
2. *A staging area of 1,000- or 2,000-lb bags.* GAC can be stored on pallets in 1,000-or 2,000-lb bags. Storing spent GAC in bags is relatively inexpensive because no additional structures or equipment are needed, but it does require a large amount of space.
   a) Dry virgin GAC can be stored indoors or outdoors, but bags stored outdoors need to be covered by tarpaulins to prevent degradation of the sack material by UV light.
   b) Wet GAC can also be stored in 1,000-lb bags. The staging area would include means to manage drainage from the wet GAC prior to removal from the site.
3. *Silos or other permanent structures for storage of virgin, reactivated, or spent GAC* (Figure 8-26). Silos do not require much space but do require additional systems for moving the GAC such as hoppers, rotary screw feeders, and eductors or pumps for transferring the GAC to or from the GAC treatment units.
4. *A spare offline GAC treatment unit that can be filled or emptied at any time.* A spare unit can reduce operating costs by eliminating transfer of GAC to/from intermediate facilities. Instead, delivery trucks can empty their loads of virgin or reactivated GAC into the offline treatment unit and then receive a load of spent GAC from another treatment unit that is being taken offline. Spent GAC can also be stored in an extra offline treatment unit, where feasible.

Figure 8-26. Reactivated GAC storage silos

Storage can affect the long-term quality of the GAC. GAC may be stored dry or wet in a treatment unit. Wet storage can only be for short periods of time (on the order of a week) without flushing the GAC with freshwater to avoid stagnation and biological growth. When GAC is stored in a dry or moist environment for long periods of time (months to years), the internal surfaces (pore walls) can become coated with oxygen-containing functional groups (Radovic 1999). Functional groups are formed when gaseous oxygen enters the GAC pores and interacts with highly reactive sites found on the pore walls. These sites are a natural consequence of the heating process employed as part of typical activation or reactivation protocols (Menendez et al. 1996). GAC adsorption performance for a variety of organic compounds has been found to be inversely correlated with the presence of oxygen-containing functional groups (Karanfil and Kilduff 1999, Leng and Pinto 1997, Pendleton et al. 1997). Thus GAC stored for long periods often does not perform as well as fresh GAC of the same type. It is recommended that the adsorption performance of "aged" GAC be determined prior to utilizing that material in a full-scale adsorber (so as to avoid the negative consequences associated with poor adsorption performance).

Offline Storage of GAC Media

When gravity GAC treatment units are out of service for extended durations, they should be drained while leaving the GAC in place. To return to service, the contactor or filter is filled with water so that the GAC can soak, allowing any air that has adsorbed on the GAC to be displaced by the water. Facilities surveyed have used soaking times of 24 hr up to 1 week. After soaking, the media should be backwashed until biological water quality objectives can be met (that is, heterotrophic plate counts [HPC]), followed by a filter-to-waste or contactor-to-waste period, then production.

Backwashing

Frequent backwashing of post-filter contactors is considered undesirable because mixing of the GAC media destroys the mass transfer zone, which could lead to earlier breakthrough of the target adsorbate(s). Media mixing during backwashing is managed by the larger uniformity coefficients that result in stratification of the GAC bed. Once backwash is complete, the flow is slowly ramped down so that the largest GAC particles settle first; then progressively smaller particles settle until the backwash flow is completely stopped. Stratification results in particles that were near the bottom of the adsorber before backwash returning to the bottom of the adsorber after backwash, approximately preserving the mass transfer zone.

For adsorbers targeting DBP control, about 90 percent of DOC does not desorb, so maintenance of the mass transfer zone is not a concern. Corwin and Summers (2010) showed that even complete mixing of the GAC media had no effect on breakthrough curves of target organics because desorption was hindered by the accumulated organic matter. Thus, backwashing may not be as critical to performance as once thought. Consequently, if maintenance of the mass transfer zone is not deemed critical for a particular application, the need to specify a GAC with a high uniformity coefficient is not as great. The higher uniformity coefficients result in a much wider range of fluidization during backwash, which can lead to blowout of the smallest media before fluidization of the largest-size fraction occurs.

Filter Adsorbers

Filter adsorbers are regularly backwashed to remove accumulated particles. Backwash is initiated when (1) maximum time, (2) low flow, or (3) maximum head loss has been reached in the same manner as traditional filters.

Post-Filter Contactors

Post-filter contactors do not require frequent backwashing because they follow an upstream turbidity removal process. Head loss will not accumulate rapidly in a post-filter contactor if the GAC has been properly installed (fines are thoroughly washed out of the media) and the upstream filtration process is working properly. Post-filter contactors do need to be backwashed occasionally, however, to overcome compaction of the media that occurs during normal operations. GAC media compaction may reduce the flow paths through the bed, causing water to flow preferentially through some portions and not through others, which is often referred to as *filter cracking*. To prevent short circuiting through filter cracks, post-filter contactors may be backwashed at a low rate for a short period (~5 min) to "fluff" the bed. The Cincinnati Richard G. Miller WTP (Case Study No. 1) has reported backwashing each post-filter contactor every 1–2 weeks to reduce the effects of GAC media compaction (Waymeyer 2003).

Water Quality Monitoring

Monitoring requirements for adsorbers depend on application, target adsorbate, and operator preference. Adsorbers must be monitored for turbidity, and most plants will also monitor particle counts on filter adsorbers as with traditional filters. If GAC is used to meet an MCL (e.g., atrazine), the SDWA and permit determine the monitoring required. Most plants periodically monitor UVA and DOC to track performance, particularly for DBP control. However, most target organic compounds are more strongly adsorbed than DOC, so the GAC will still have excess capacity when the DOC has completely broken through. For these compounds, filtrate samples should be analyzed for the compound(s) of interest. Many utilities also monitor capacity of the GAC media, most commonly with the iodine test. While not a direct measurement of the remaining capacity for the target adsorbate, in some instances it may be able to be related to the treatment objective (e.g., MIB or MTBE) as a surrogate.

Chlorination

Free Chlorine

Allowing chlorine residuals in the GAC adsorber is typically avoided. Free chlorine is reduced to chloride in a reaction at the GAC surface. The reaction is very fast and occurs in the first few minutes of EBCT. Continuous exposure to chlorine results in degradation of the GAC structure and will produce more fines over time. The chlorine may also react with organics adsorbed on the GAC surface creating potentially toxic by-products that may desorb into the finished water.

Pilot testing of GAC reactivation at the Richard G. Miller WTP (Case Study No. 1) found that dioxin was formed during the GAC reactivation process when prechlorination was practiced upstream of the GAC post-filter contactors. When prechlorination was not used, dioxin did not form. Although the mechanism of dioxin formation during reactivation was not fully understood and levels were much less than regulated levels, prechlorination was stopped to prevent the formation from occurring (DeMarco et al. 1988).

If chlorination upstream of the GAC treatment process is practiced prior to installation of GAC, it is recommended that use of the upstream chlorination points be limited as described below for limiting biological growth. New chlorination points downstream of the GAC treatment process should be installed if not already present.

Biological growth through the treatment facility can occur without chlorine present, especially in the summer months. Excessive biological growth can present a maintenance issue. Although biologically active GAC treatment processes are used for TOC removal, excessive growth can increase head loss through the GAC treatment process and reduce its performance. Biological growth in these areas can be addressed by the following approaches:

1. *Add chlorine at the head of the plant.* In such cases, upstream chlorine doses should be limited so that the residual reaching the GAC treatment units is nondetect or as close to nondetect as possible.
2. *Use alternative oxidants, such as potassium permanganate, ozone, or chlorine dioxide in lieu of chlorine.* Contact between chlorine dioxide and GAC should also be avoided; chlorine dioxide decays faster than free chlorine, allowing it to be considered an alternative.
3. *Use chlorinated backwash water for the GAC treatment process.* Contact with backwash water is much less than contact with water during the treatment process, and this limited chlorination can help reduce head loss due to biological growth.

Combined Chlorine

Combined chlorine also reacts with GAC in a manner similar to free chlorine; however, the ammonia is not destroyed, and free ammonia will be released into the system. Free ammonia may affect downstream chlorination and possibly result in nitrification in the adsorber.

GAC Safety

Activated carbon products have safety issues that users should acknowledge. Some information from material safety data sheets for GAC manufacturers listed below illuminates the issues:

Activated carbon (especially when wet) can deplete oxygen from air in enclosed spaces, and dangerously low levels of oxygen may result (Norit Americas Inc. 2007). Before entering such an area, sampling and work procedures for low oxygen levels should be taken to ensure ample oxygen availability, observing all local, state or provincial, and federal regulations (Calgon Carbon Corporation 2008).

As with most organic solids, fire is possible at elevated temperatures or by contact with an ignition source. Activated carbon is difficult to ignite (the ignition temperature is greater than 220°C) and tends to burn slowly (smolder) without producing smoke or flame. Carbon monoxide and carbon dioxide gas may be generated if combusted. Contact with strong oxidizers such as ozone or liquid oxygen may cause rapid combustion (Calgon Carbon Corporation 2008).

Airborne dust (i.e., fines) from GAC can be a respiratory hazard. Manufacturers recommend GAC handling procedures be conducted to minimize production of dust and respirators be worn if necessary to protect workers.

Exposing GAC to hydrocarbons (oils, gasoline, diesel fuel, grease, paint thinners, etc.) can cause spontaneous combustion (AWWA B605).

References

American Water Works Association. AWWA Standard B605, Reactivation of Granular Activated Carbon. Denver, Colo.: AWWA.

Amirtharajah, A., N. McNelly, G. Page, and J. McLeod. 1991. Optimum backwash of dual media filters and GAC filter-adsorbers with air scour. Denver, Colo.: AWWA [now Water] Research Foundation.

Calgon Carbon Corporation. 2008. Filtrasorb 400: Granular activated carbon for municipal specifications. Pittsburgh, Pa.

Corwin, C. J. and R.S. Summers. 2010. Scaling trace organic contaminant adsorption capacity by granular activated carbon. *Environ. Sci. and Technol.*, 44(14):5403–5408.

Corwin, C.J. and R.S. Summers. 2011. Controlling trace organic contaminants with granular activated carbon adsorption. *Journal AWWA*, 104(1):43–44.

DeMarco, J., R. Miller, and D.J. Hartman. 1988. Discovery and elimination of dioxins from a carbon reactivation process. *Journal AWWA*, 80(3):66–72.

ITT/F.B. Leopold Company. 1999. Type S underdrain brochure. Zelienople, Pa.

Karanfil, T., and J.E. Kilduff. 1999. Role of granular activated carbon surface chemistry on the adsorption of organic compounds. 1. Priority pollutants. *Environ. Sci. Technol.*, 33(18):3217.

Kawamura, S. 2000. *Integrated design and operation of water treatment facilities*, 2nd ed. New York: John Wiley & Sons, Inc.

Kearns, J. 2004. Interview by Brian Leto. Greenway WTP operations.

Leng, C.C., and N.G. Pinto. 1997. Effects of surface properties of activated carbons on adsorption behavior of selected aromatics. *Carbon*. 35(9):1375.

Menendez, J.A., J. Phillips, B. Xia, and L.R. Radovic. 1996. On the modification and characterization of chemical surface properties of activated carbon: In search of carbons with stable basic properties. *Langmuir*, 12:4404.

Naylor, B. 2003. Interview by Brian Leto. Norit Americas Inc.

Norit Americas Inc. 2007. Norit 1240 MSDS. Marshall, Texas.

Pendleton, P., S.H. Wong, R. Schumann, G. Levay, R. Denoyel, and J. Rouquerol. 1997. Properties of activated carbon controlling 2-methylisoborneol adsorption. *Carbon*, 35(8):11–41.

Radovic, L.R. 1999. Surface chemistry of activated carbon material: state of the art and implications for adsorption. In *Surfaces of nanoparticles and porous materials*. J.A. Schwarz and C.I. Contescu, eds. New York: Marcel Dekker.

Sanks, R.L., ed. 1989. *Pumping station design*. Boston, Mass: Butterworth Publishers.

Scott, R. 2006. Interview by Brian Leto. Water Operations Superintendent.

USEPA (US Environmental Protection Agency). 2001. Filter backwash recycling rule. *Fed. Register*, 66(111):31085–31105.

Waymeyer, J. 2003. Interview by Brian Leto. Operator, Richard G. Miller WTP (ret.).

Chapter 9

Pressure GAC Treatment System Design

This chapter discusses the use of GAC in pressure treatment systems for liquid-phase treatment. Pressure GAC treatment systems are usually selected over gravity systems in groundwater treatment applications for removal of volatile organic compounds (VOCs) and/or synthetic organic compounds (SOCs). More recently, prefabricated units are being considered and used for T&O and organics removal for DBP control in smaller surface water treatment plants. The design flow rate for groundwater applications is usually less than a surface water treatment application, as the size of the wellhead is limited. Additionally, wellhead groundwater treatment systems may require flexible installations to fit treatment within a small footprint, which makes a modular pressure treatment system a better fit than a gravity system.

Design and Construction

Pressure GAC System Overview

In a pressure system, GAC is contained in pressure-rated vessels as small as 55-gal drums containing approximately 200 lb of activated carbon, to vessels as large as 60,000 lb. Pressure systems can use permanently or temporarily installed vessels for containing GAC. While much of the content of this chapter is applicable to the range of vessel sizes, the chapter's focus is on the larger vessel systems because those are usually required for drinking water applications. Because of their size, large vessel systems have greater implications on site utilization and coordination of operations and maintenance activities. An example of a large pressure vessel installation is shown in Figure 9-1.

The standard pressure vessel system consists of two free-standing vessels connected by a common center piping rack or tier manifold. The dual-vessel system may be referred to as a skid, pair, or system and contains the typical features shown in Figure 9-2.

Pressure GAC systems are typically configured for downflow operation in either series or parallel. Series operation describes the mode in which the vessels operate in

Figure 9-1. Dual GAC pressure vessel installation (large vessel)

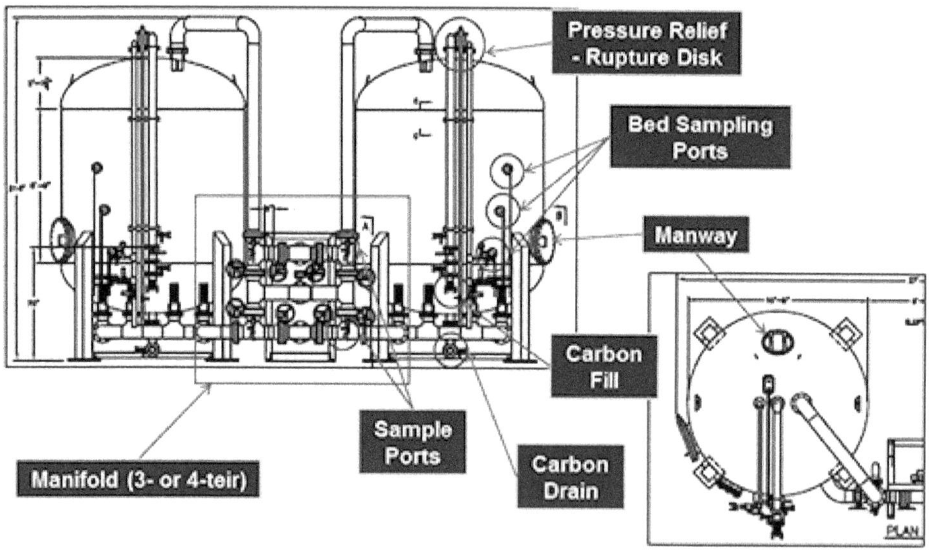

Figure 9-2. Typical pressure GAC system features

a lead/lag configuration. The lead vessel first processes the flow followed by the lag vessel as shown in Figure 9-3.

Parallel operation describes the mode in which each vessel processes flow independently, as shown in Figure 9-4. As with all applications of GAC, pressure GAC systems undergo a backwash event after a GAC replacement. The backwash supply

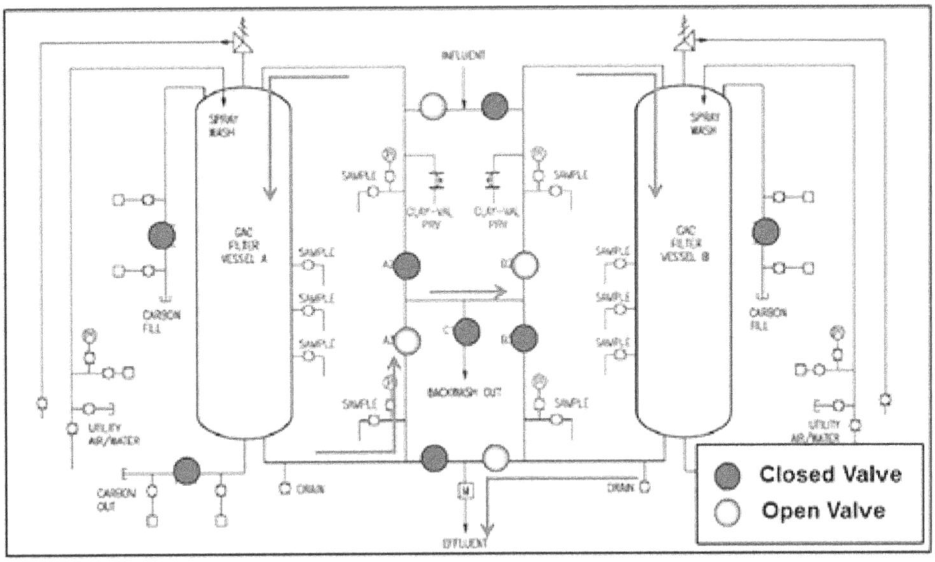

Figure 9-3. Operation of vessels—series flow

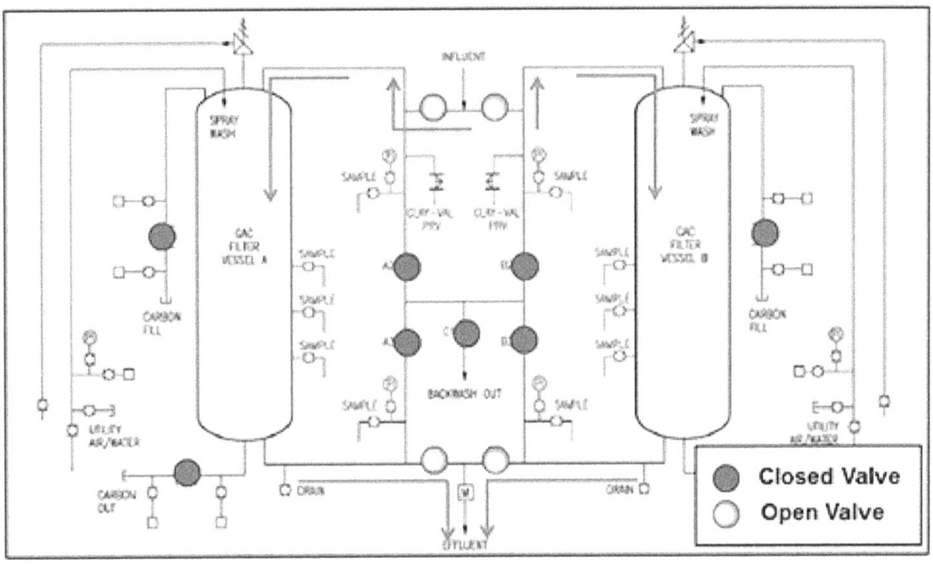

Figure 9-4. Operation of vessels—parallel flow

enters the vessel through the bottom, flows through the underdrain system, and exits through the top of the vessel. The backwashing operation for a three-tier manifold system is shown in Figure 9-5. The tier manifold design allows either mode of operation (series or parallel); however, unless all valves are manually operated, the system design typically only allows for one mode of operation to simplify the instrumentation and controls setup. Selecting the mode of operation is discussed in more detail in following sections.

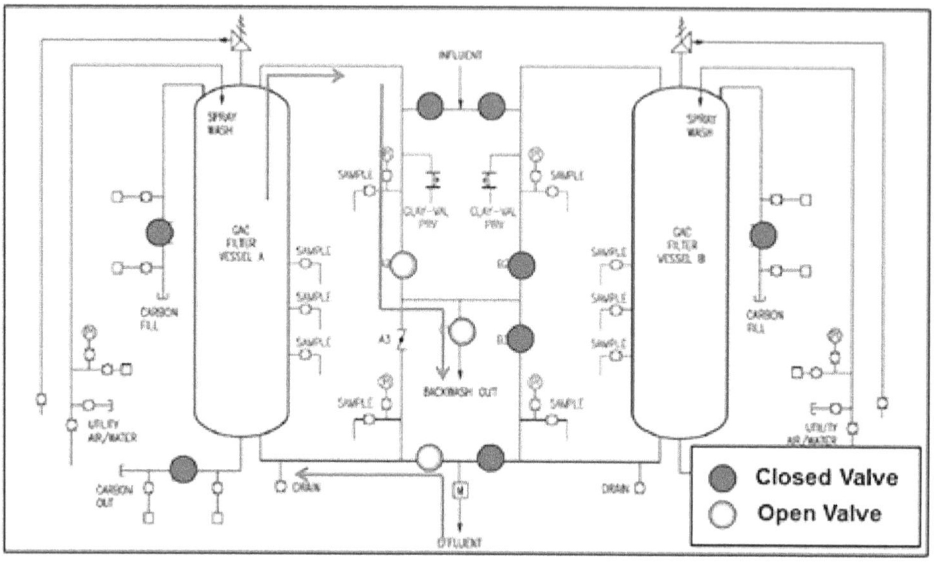

Figure 9-5. Operation of vessels—backwashing

Review Key Design Parameters

The conceptual design phase for the design of pressure GAC treatment systems consists of determining the key design parameters, verifying regulatory requirements, and obtaining specific vessel manufacturer information.

When a pressure GAC system is selected, the first step of the design phase is a review of the key design parameters from the master plan:

1. Treatment objectives
2. Influent concentrations of target and background DOC
3. GAC type
4. Operational mode
5. Empty-bed contact time (EBCT)
6. Design and vessel flow rate
7. System configuration

Review of the treatment objective, influent concentrations of target and background DOC, GAC type, operational mode, and EBCT are similar to the key design parameters determined for gravity GAC treatment systems and are discussed in chapter 8, Review Key Design Parameters.

Design and Vessel Flow Rate

Once the EBCT has been determined, the minimum and maximum design flow rates for the entire GAC process should be determined to select the size and number of vessels required. The maximum design flow rate is typically equal to the plant design capacity, but it may be determined based on future design capacity. The minimum design flow rate is typically equal to the minimum plant flow rate, but it may differ if GAC bypassing is used. It is important to note that bypassing is not allowed in certain locations, including California, and should be coordinated with the regulatory agency during the conceptual design phase. Once the design flow rates have been determined, the minimum and maximum vessel flow rates should be determined.

Acceptable flow rates are provided by GAC manufacturers and should be obtained to determine the size and number of vessels required.

System Configuration

Because pressure systems are made up of modular components (vessels), a range of possibilities exists for system configuration. Units may consist of one pressure treatment unit (vessel), several vessels configured in series, several vessels configured in parallel, or a combination of multiple parallel-series treatment trains. Examples are shown in Figure 9-6.

The system configuration may be a function of the purpose for GAC treatment. For example, in a remediation application (e.g., removal of TCE from a contaminated groundwater source), the regulatory agency having jurisdiction may require that a GAC treatment system consist of two to three treatment units/stages in series. A series or combination treatment scenario provides the opportunity to monitor the first treatment unit/stage for breakthrough of the contaminant, schedule GAC replacement once it is detected, and make the replacement while still not having detectable concentrations at the outlet of the last treatment unit/stage. The piping and valve

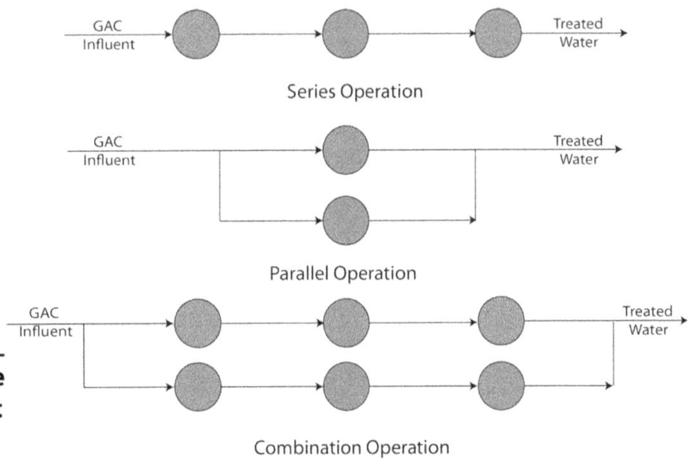

Figure 9-6. Example of modular treatment unit configurations

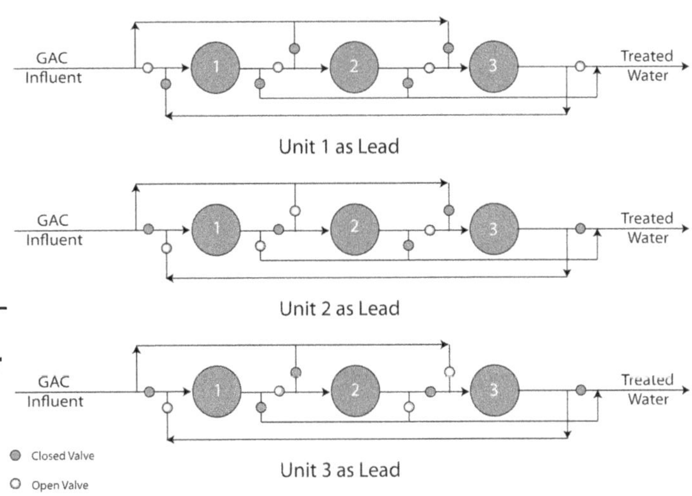

Figure 9-7. Series operation of modular treatment system with varying lead-treatment units

arrangements for series or combination systems are often designed to allow the treatment units to be operated in any order so vessels that are not being replaced can move up in the treatment series order without having to be physically moved. A conceptual schematic is shown in Figure 9-7.

In contrast, an application designed for removal of organic matter to reduce DBP formation in treated water will not require the same monitoring and discharge requirements. For DBP control, parallel treatment units can be operated in staged parallel mode as described in chapter 4.

Facility Layout

Before the layout of the facility can be integrated into the existing site, the pressure GAC treatment facility components should be considered in detail. Determination of whether to house the pressure GAC treatment system in a building should be made early in the design process because it directly affects the layout of the facility.

Obtain Vessel Manufacturer Information

Once the key design parameters have been determined, the vessel manufacturers should be contacted to obtain additional information on specific products for the application. Vessel manufacturers can typically provide the following information that can be used to design pressure GAC treatment systems:

- Detailed GAC information, including recommended GAC for the application, EBCT, downflow head loss, and backwash rate curves.
- Detailed vessel product information, including dimensions, clearance requirements, pipe connections, operating weights, interior/exterior coating, and temperature/pressure ratings.
- System pressure-drop information.

Vessel Sizing

To begin site design for the GAC facility, the size of the modular treatment units (vessels) must be considered. If the GAC treatment facility is large enough, utilizing vessels that contain 20,000 dry lb of GAC (10- to 12-ft diameter) simplifies site operations and GAC replacement logistics. Because wet, drained GAC weighs approximately twice as much as dry GAC, and one GAC delivery truck can haul 40,000 lb of wet cargo (Calgon 1993a), GAC replacement operations are simplified to one truck per delivery in or out of the treatment facility, and the modular units are sized accordingly. However, if footprint is limited on an existing site or if the number of valves and other efficiencies are a concern, then larger vessels can be investigated. Manufacturers fabricate larger 30,000-, 40,000-, and 60,000-lb vessels as well. The 30,000- and 40,000-lb vessels are designed with a 12-ft diameter, whereas the 60,000-lb vessel has a 14-ft diameter. Benefits of the larger vessels include reducing the following:

1. *Number of vessels required.* Using 40,000-lb vessels reduces the number required by half.
2. *Footprint required.* Footprint is often a significant concern when adding additional treatment processes to existing treatment plants. Footprint restrictions are often the case for GAC facilities installed for TOC reduction. Because fewer vessels are required when utilizing larger vessels, the footprint required

Table 9-1. Number of GAC delivery trucks required

Vessel Size (lb)	Number of GAC Trucks Required to Deliver Fresh GAC	Number of GAC Trucks Required to Remove Spent GAC	Total Number of GAC Trucks Required
20,000	1	1	1–2
30,000	1	2	2–3
40,000	1	2	2–3
60,000	2	3	3–5

is reduced. However, larger vessels are taller and require taller buildings, if a building is provided for the GAC process.

3. *Number of valves and instruments required.* Again, using 40,000-lb instead of 20,000-lb vessels reduces the number required by half. Reducing the number of valves and instruments decreases the risk of maintenance issues, especially if automated valving is used.

However, significant drawbacks to designing and using larger vessels include the following:

1. *Increased complexity of GAC replacements.* Using vessels larger than 20,000-lb capacity requires more GAC delivery trucks to replace the GAC in one vessel. Table 9-1 summarizes the number of trucks required to remove and replace the GAC in one vessel depending on vessel size. The logistics of more delivery trucks is a concern because deliveries require operational resources during the replacements, and deliveries should be coordinated with other traffic occurring at the treatment plant. For some plants, a spare unit is used for storage so that a delivery truck can fill the spare unit upon arrival and fill the truck with spent GAC before leaving the site, resulting in fewer total truckloads.
2. *Increased backwash supply rate and equalization volume.* The backwash supply rate is based on the surface area of the vessel. Because the 30,000- and 40,000-lb vessels have the same diameter (12 ft) as the 20,000-lb vessels, there is no increase in the backwash supply rate. However, 60,000-lb vessels are fabricated with a 14-ft diameter; therefore, the backwash supply rate and subsequently the backwash supply volume and equalization volume increase with the larger vessel size. For example, a 12-ft-diameter vessel with a backwash rate of 13.3 gpm/ft^2 requires 1,500 gpm or 45,000 gal for a 30-min backwash. A 60,000-lb vessel with a 14-ft diameter under the same conditions requires approximately 2,050 gpm (an additional 550 gpm), resulting in a total of 61,500 gal, or an additional 16,500 gal required from the backwash supply source. The increased backwash supply rate should be considered to determine if adequate supply and space for the equalization basin are available.
3. *Lowered EBCT during GAC replacements.* When the process is operating under maximum design capacity and one vessel is out of service, processes with larger vessels will be operating at a lower EBCT than those with 20,000-lb vessels because a larger percentage of the GAC is out of service. The lower EBCT may cause the in-service vessels to become exhausted more quickly, and/or treated water contaminant levels may not be met. The importance of reducing the EBCT during these periods depends on the specific water quality goals,

contaminants, and system flexibility. The EBCT achieved with one adsorber out of service needs to be considered in detail during the design.

4. *Higher pressure drop.* Larger vessels have deeper beds than the 20,000-lb vessels, which results in a higher pressure drop across the bed. However, because most of the pressure drop is in the piping and underdrain, the variation is not considerably different. The pressure drop of a 40,000-lb vessel would conservatively be approximately 20 percent higher than the 20,000-lb pressure drop (Calgon 2008a). Pressure drop curves should be obtained from the vessel manufacturers to verify pressure drops for 20,000-lb and larger vessels.
5. *Increased building height.* Although using larger vessels results in a smaller footprint, the building (if provided) will be taller. Taller buildings may be a concern depending on the location and zoning of the existing treatment plant or wellhead.
6. *Complex fabrication and shipping.* Currently, 30,000- and 40,000-lb vessels can be manufactured at the vendor facility and shipped to the construction site, which results in a uniform and tested product. However, a 60,000-lb vessel is designed with a 14-ft diameter, which requires special fabrication to ensure quality assurance of vessel lining and painting. Special fabrication shops may not be available in all parts of the country, which may limit the availability of large-diameter vessels. Also, the size of the vessels may require special permits for transportation due to the large diameter. These factors result in additional costs.

Vessel Number and Loading Rate

Once the vessel size has been determined, the number of vessels required can be determined. The number of vessels needed depends on vessel size, system flow rate, EBCT, and allowable vessel loading rates. Although vessels are classified based on a weight of GAC (20,000 lb, etc.), the number of vessels is determined by their volume in order to achieve the desired EBCT. Once the minimum number of vessels is determined, then the loading rate should be verified. Each manufacturer's product will have a specific minimum and maximum loading rate defined and should be verified with the manufacturer. However, a loading rate of 2–6 gpm/ft^2 is a standard acceptable range if not otherwise specified.

Once the minimum number of vessels required has been determined, other factors must be considered before selecting the final number of vessels. These factors include design symmetry with vessel skid design, operation during GAC replacements, and future plant expansions. Typically, if more than one vessel is required for an application, pressure vessel systems are arranged in pairs that consist of two freestanding vessels sharing one common center piping network or tier manifold. The standard design configuration usually allows for piping efficiencies for both series and parallel operations.

The design should also consider operations while replacing GAC. During a GAC replacement, at least one vessel is out of service for a period of time. As aforementioned, the amount of time depends on the number and availability of delivery trucks required. For example, a 40,000-lb vessel could be out of service for approximately 20–36 hr, assuming the following:

- 8 hr to complete the GAC replacement
- 12–24 hr to allow the GAC to soak
- 1–2 hr required to complete a backwash and vessel-to-waste event

During the period when one vessel is out of service and the GAC process is operating under full design capacity, the GAC process will be operating at a lower-than-designed EBCT. The importance of the lower-than-designed EBCT depends on the specific water quality goals, contaminants, and system flexibility. For some utilities during certain portions of the year, operating at a lower EBCT for a short duration will not affect overall treatment goals; for others it will. If the lowered EBCT is a concern, the design should consider adding one or more additional vessel(s) that would be placed in service as others are taken out of service. The spare unit allows for the design EBCT to be met at all times, but it has an associated additional capital cost that should also be considered.

During a GAC replacement for remediation applications where the vessels are operated in series, two vessels will be out of service. The EBCT and loading rate with two vessels out of service should be considered with great care.

Another consideration in determining the number of vessels desired pertains to possible future plant expansions or future regulations. The design should consider procurement of additional vessels now or in the future to meet expected future expansions or regulations. Vessels can be installed during the construction phase, but not be placed in service until the plant requires. The design could also consider replacing vessels with larger ones in the future to meet expansions or regulations. Those planning for future larger vessels need to consider the additional space required for the larger vessels so that the vessel arrangement can accommodate them. Ultimately, the design should take all of the above considerations into account to choose the correct number of vessels for the application, which will vary from project to project. The number of vessels determined can then be confirmed by the vessel manufacturer.

EXAMPLE 12-1. Determining the number of treatment units

A 10-mgd plant plans to control DBP formation with prefabricated GAC vessels. In working with the manufacturer, the utility has decided on the following vessel:

Vessel Size	Diameter	Surface loading Rate	Units per skid
40,000 lb	12 ft	2–6 gpm/ft^2	2

The treatment study report recommends a 20-min EBCT operated in staged parallel mode and a bituminous 12 × 40 GAC (π = 0.45 kg/L). How many units are needed?

SOLUTION: Start by calculating the volume of the unit by converting the reported weight of GAC into a volume using the apparent density of the GAC:

$$\text{Bed volume} = \frac{40{,}000\ \text{lb}}{0.45\ \text{kg/L}} \times \frac{1\ \text{kg}}{2.205\ \text{lb}} \times \frac{1\ \text{gal}}{3.785\ \text{L}} = 10{,}650\ \text{gal}$$

Next, determine the flow rate to achieve the design EBCT:

$$Q_f = \frac{V}{\text{EBCT}} \times \frac{10{,}650\ \text{gal}}{20\ \text{min}} = 532\ \text{gpm}$$

Check the filter velocity:

$$v_f = \frac{Q_f}{A_f} \times \frac{532 \text{ gpm}}{\pi/4\ (12 \text{ ft})^2} = 4.7 \text{ gpm/ft}^2$$

Calculate the minimum number of units:

$$n = \frac{Q_t}{Q_f} \times \frac{10 \times 10^6 \frac{\text{gal}}{\text{day}} \times \frac{1 \text{ day}}{1{,}440 \text{ min}}}{532 \frac{\text{gal}}{\text{min}}} = 13.05 \text{ units}$$

Thirteen (13) units are required, but because units come in skids of 2, 14 units will be used. All 14 units may be used for production ensuring that the design conditions can be met with one unit out of service. Alternatively, an additional unit can be used as a spare so that GAC delivery trucks can fill one unit and empty another on a single trip. In this case, 12 units will be left for production with one empty and one unit out of service, resulting in an EBCT of 18.4 min and a filter velocity of 5.1 gpm/ft^2. These are likely acceptable conditions for short periods of time in this plant.

Construction Details of Treatment Units

As discussed earlier, pressure GAC treatment facilities can be designed to remove a wide array of contaminants including VOCs, SOCs, and DBP precursors. The design of the facility and associated components is directly related to the contaminant, facility capacity, and available footprint in the existing treatment plant or wellhead site. Depending on the size and GAC replacement frequency of the facility, the design will include all or some of the following typical facility structures, equipment, and resources:

- GAC treatment units (i.e., vessels)
- Process piping and valves
- Sampling ports
- GAC feed pumps and/or pump station (if required by hydraulic profile)
- Backwash system (supply source, pumps, waste removal)
- Vessel-to-waste piping and/or pumping
- GAC equalization basin and pumps
- Plant service water
- Plant service air
- Power supply and controls for pumps and automatic valves
- Instrumentation and controls
- GAC building
- Access for GAC delivery trucks
- Possible HVAC requirements (if new building constructed)
- Site security
- Expansion space for future units

Vessel Shell and Lining

The vessel shell is constructed of carbon steel and is supported by four structural steel legs and baseplates. Each vessel is then equipped with manways, overpressure protection, an underdrain system, plant air and water lines, and media loading and unloading lines (Calgon 1993b, Siemens 2006). Each vessel is also lined with an abrasion-resistant vinyl ester resin to prevent corrosion of the vessel. The vessel lining is applied after the vessel is manufactured and has a specific curing time. After curing, the lining should not be compromised in any way because corrosion will occur. At the time of publication, Carboline Plasite® 4110 was the standard lining specified for pressure vessels by vessel manufacturers Calgon Carbon Corporation and Siemens Water Technologies (Calgon 2008b, Siemens 2006). The specific type of lining desired should be selected and specified for the application and should be confirmed with the vessel manufacturer.

Vessel Pressure Rating

The standard pressure vessel carries the ASME-coded pressure rating of 125 psi at 150°F. Vessels can be manufactured and coded for higher pressures; however, it is not typically necessary for water treatment applications and increases the vessel cost. Each vessel is equipped with an overpressure protection system. Typically a relief rupture disk or valve mounted to the top of the vessel will recycle flow to the plant headworks to protect the vessel from pressures greater than 125 psi (Calgon 1993b, Siemens 2006).

Vessel Underdrains

The vessel underdrains, located in the bottom of the vessel, collect the treated water and support the media without allowing the media to exit the vessel through effluent piping. There are two main types of underdrain systems for pressure vessels: false bottom type and external header type. Vessel manufacturers typically offer one of the underdrain types as their standard; however, most manufacturers can provide either type if specified.

False Bottoms. The false-bottom underdrain consists of approximately 80 to 120 underdrain intake nozzles that are mounted on an internal false-bottom cone, as shown in Figure 9-8. The false-bottom cone is constructed of carbon steel and is welded to the inside of the vessel along the entire circumference, increasing the vessel height. Because the cone requires welds inside the vessel, the vessel lining should be applied in four sections, one above and below the cone, and one on the top and bottom of the cone. The nozzles, which are polypropylene (PPL) or stainless steel, are mounted on the false bottom with a gasket that seats each nozzle on the cone. The nozzles and the false bottom maintain the GAC above the cone. When the nozzles or the false bottom require maintenance, the vessels require confined-space entry (Calgon 2008a). It should be noted that appropriate safety measures for confined-space entry and work in a dry GAC environment should be followed. For more information about GAC safety, refer to chapter 8.

External Header. The external-header underdrain consists of eight nozzles arranged on an external header located beneath the pressure vessel, as shown in Figure 9-9. The proprietary nozzles are stainless-steel v-wire-type screens. When the nozzles require maintenance, the external header can be removed by dropping

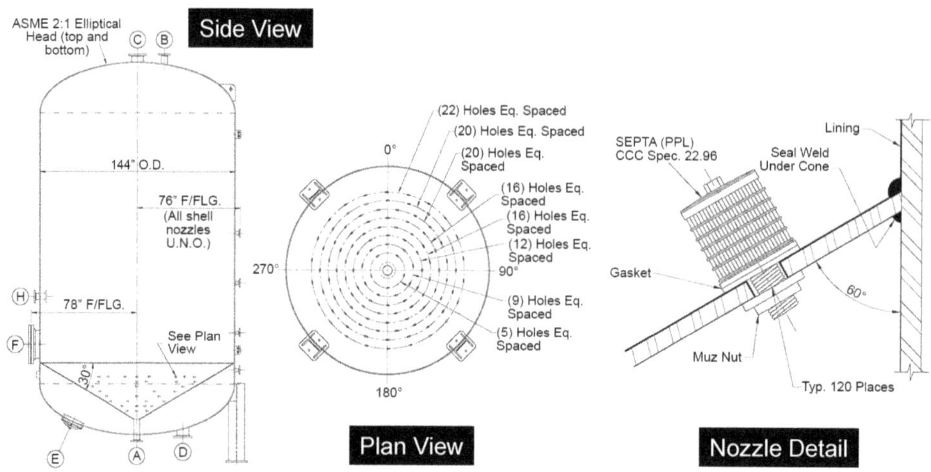

Figure 9-8. False bottom underdrain system

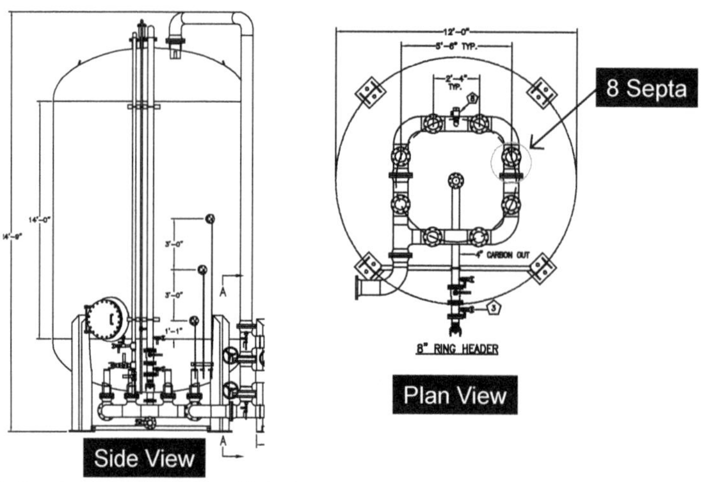

Figure 9-9. External header underdrain

the header out of the vessel, which does not require confined-space entry. Although confined-space entry is not required, care should be exhibited near any opening of the vessel during header maintenance, as the environment adjacent to the opening could be oxygen depleted. For more information about GAC safety, refer to chapter 8. There are few welds inside the vessel for sample ports and manways, which allows for a consistent vessel lining (Siemens 2008). Both underdrain types should be considered in detail and discussed with maintenance personnel to select the appropriate design for the application.

Vessel Accessories

Manufacturers provide additional vessel accessories that can enhance performance and use of the vessels, if specified, including sampling ports and additional access.

Bed sampling ports, shown in Figure 9-10, are optional for most manufacturers and can be specified to allow sampling of the bed to determine the rate of GAC exhaustion or breakthrough. The ports are located on the exterior of the vessel at various heights of the bed. If used, the design will specify the desired heights of the ports. Additional sample ports can be specified on the piping as well. These sample ports allow for more flexibility in testing water quality along various portions of piping.

Additionally, the standard vessel package typically does not include ladders or platforms to access the top manway on each vessel. Vessel access ladders and platforms can be specified to provide a permanent means of accessing the vessel and appurtenances, as shown in Figure 9-11. Depending upon the frequency of access required and maintenance personnel preferences, permanent vessel access may be desired.

Figure 9-10. Bed sampling ports (Celina)

Piping and Valving

Piping and valving beyond the prefabricated unit piping may be required.

Manifold Design. The standard pressure vessel system consists of two freestanding vessels connected by a common center piping rack or tier manifold. The tier manifold may be a three- or four-tier design as shown in Figure 9-12.

The four-tier manifold consists of piping and valves for backwash waste, vessel influent, vessel effluent, and backwash supply. The four-tier manifold allows one vessel to perform a backwash while the other continues to process water. Therefore a backwash does not disrupt the normal operation of vessels in parallel. For vessels in series, the lag vessel will become the lead vessel during the backwash so that only one vessel will be out of service during a backwash. The four-tier design should be selected if more than one vessel out of service is a concern due to the lower-than-designed EBCT during GAC replacements and backwashing.

Figure 9-11. Ladder and platform vessel accessories

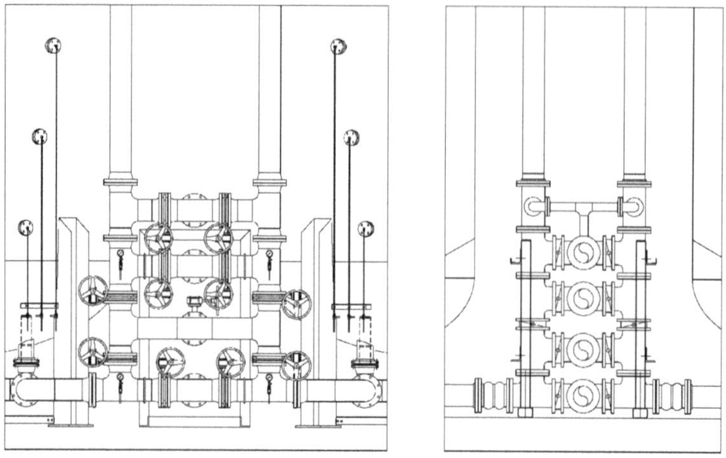

Figure 9-12. Three- and four-tier manifolds

With the three-tier manifold design, the backwash supply uses the same piping as the treated water; therefore the system cannot process water while either vessel in the system performs a backwash for both parallel and series operations. If series operation is required for a remediation system, both vessels in the pair would be shut down during GAC replacements and backwashing. Series systems do not require a four-tier design and should be designed with a three-tier design. The three-tier design consists of fewer valves and less piping, which should be weighed against the effects of two vessels out of service during GAC replacements and backwashing. The intended use of the vessel systems should also be considered.

Pipe Types and Velocities. For each vessel system, vessel manufacturers provide the piping on the manifold (GAC influent, GAC effluent, backwash supply, and backwash waste) and the GAC fill, GAC drain, plant service water, and plant service air lines. Although the standard piping systems and types can be modified, vessel manufacturers typically provide the following pipe sizes and types for the vessel systems (Calgon 2008b, Siemens 2006):

- Process and backwash piping: 8-in. Schedule 40 carbon steel
- GAC fill lines: 4-in. Schedule 40 carbon steel or Schedule 10 stainless steel
- Spent GAC drain lines: 4-in. polypropylene-lined carbon steel or Schedule 10 stainless steel
- Plant service water and air lines: Schedule 80 carbon steel

The design engineer should account for and specify any piping not provided by the manufacturer, including the process, backwash waste, and backwash supply lines to connect the vessels to the existing processes. These pipes can be specified as carbon steel, stainless steel, polyvinyl chloride (PVC), ductile-iron piping (DIP), or other pipe types as applicable. The type of piping selected depends on the specific design application, water quality, and cost. While PVC enhances the schedule for fabrication and provides a less expensive alternative to DIP, the lack of rigidity requires additional pipe supports. The design should investigate the various pipe types and select the one appropriate for the application. If the vessels will not be housed within a temperature-controlled building, heat tracing and/or insulation should be investigated to protect piping from potential freezing conditions.

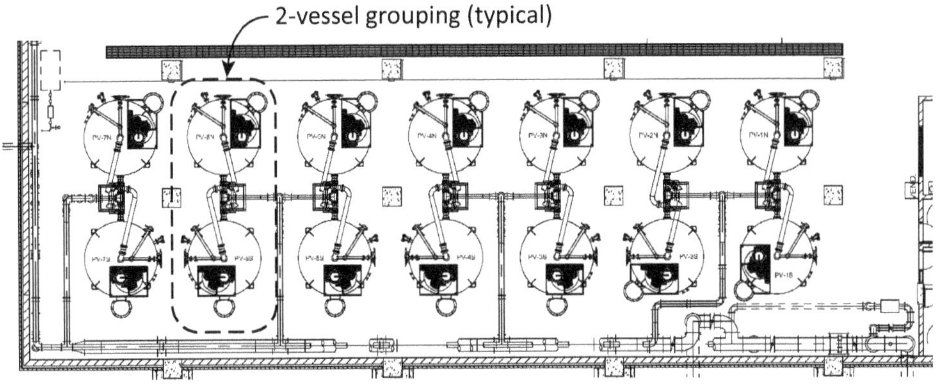

Figure 9-13. Pressure vessel system with 2-vessel grouping (courtesy Malcolm Pirnie)

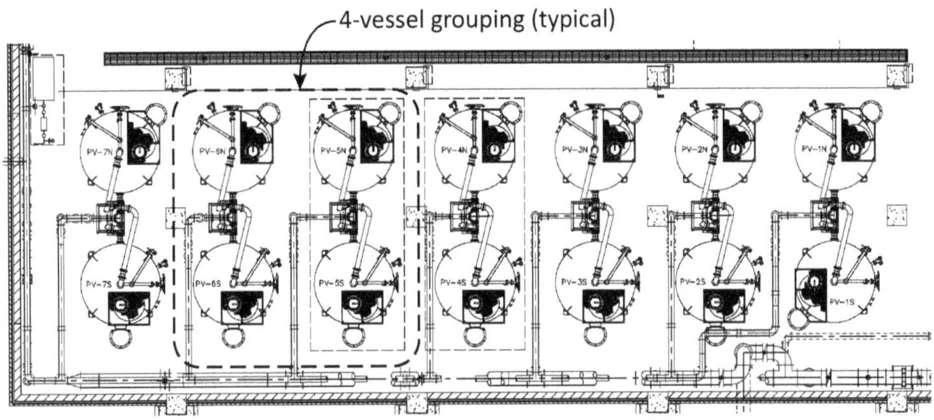

Figure 9-14. Pressure vessel system with 4-vessel grouping (courtesy Malcolm Pirnie)

Process and backwash piping should be sized for velocities of 4 to 8 ft/sec. To avoid clogged piping, long radius bends should be considered for piping that will carry backwash waste, GAC fill, or GAC drain.

Piping Arrangements. When multiple vessel systems are considered, it is important to investigate grouping vessel systems to gain piping efficiencies, save space, and provide adequate room for GAC replacements. For example, when seven dual-vessel systems are required, each vessel system pair could be connected to the main header via separate piping, as shown in Figure 9-13. However, by investigating grouping vessel systems, less piping can be used by grouping two systems or four vessels to the main header via the same piping, as shown in Figure 9-14. Depending on the available footprint, vessel configuration, and building layout, various grouping options can be investigated to use less piping, resulting in a lower overall project cost.

Valving. Pressure vessel systems can require a large number of valves. A typical two-vessel system will include the following valves on the tier manifold (Calgon 2008b, Siemens 2006):

- Two backwash waste valves
- Two supply (water inlet) valves

- Two treated-water valves
- Two isolation valves
- Two backwash supply valves (four-tier manifold only)

All the manifold valves are typically cast-iron wafer-type butterfly valves. Additionally, other valves (Calgon 2008b, Siemens 2006) on the pressure vessels include:

- GAC fill and drain valves: To perform GAC replacements, each vessel is equipped with one GAC fill valve and two GAC drain valves. These are typically stainless-steel full-port ball valves.
- Sample port valves: If sample ports are specified, each is typically equipped with a ½-inch bronze ball valve.

The type and material of all valves should be verified with the vessel manufacturer and detailed in the design specifications.

Flow Control. The rate of flow through each vessel should be controlled to provide an equal flow rate to all parallel treatment trains or groupings of vessels. Flow control can either be manual or automated.

- Manual flow control: Facilities with fewer numbers of vessels and longer time between GAC replacements will typically be manually operated because of fewer valves and fewer changes to valve positions.
- Automated flow control: In organics removal applications where GAC replacements occur frequently or a large number of vessels are present, automated valves should be considered.

Vessel manufacturers typically provide manually operated valves as part of their standard equipment packages; however, options are available to provide automated valves. Automated valves can be specified as open/close or modulating. If automated valves are specified, open/close valves can be used for all process valves (backwash supply, influent, backwash waste) with the exception of the effluent valves. Treated-water valves may be designed as modulating, where the valve will adjust in response to a flow meter located on the vessel supply line(s).

Although automated systems require additional instrumentation and controls, signals can be wired to control stations through SCADA for ease of operation. Control stations can be located at each vessel system, in a centralized GAC control room, or in the main control room of the treatment plant. The location of the controls is specific to each application and should consider operations personnel and vessel operations.

Hydraulic Considerations

Once the location of the GAC facility is determined, the hydraulic profile should be investigated to verify if additional pumping facilities will be required to supply the vessels and to deliver treated water to the clearwell or distribution system.

To determine if the GAC facility requires an influent or effluent pump station, the pressure drop across the vessels should be determined. The majority of the pressure drop (or head loss) for the pressure vessel system is due to the piping and underdrain system. However, head loss also occurs through the GAC media. Figure 9-15 shows series pressure drop curves for individual 8-ft-diameter 10,000-lb, 10-ft-diameter 20,000-lb, and 12-ft-diameter 20,000-lb pressure vessels with piping and 8×30 mesh GAC installed. The manufacturer pressure drop curves should be requested from the specific manufacturers during the conceptual or preliminary design phase to verify pressure drop through the selected vessel and media.

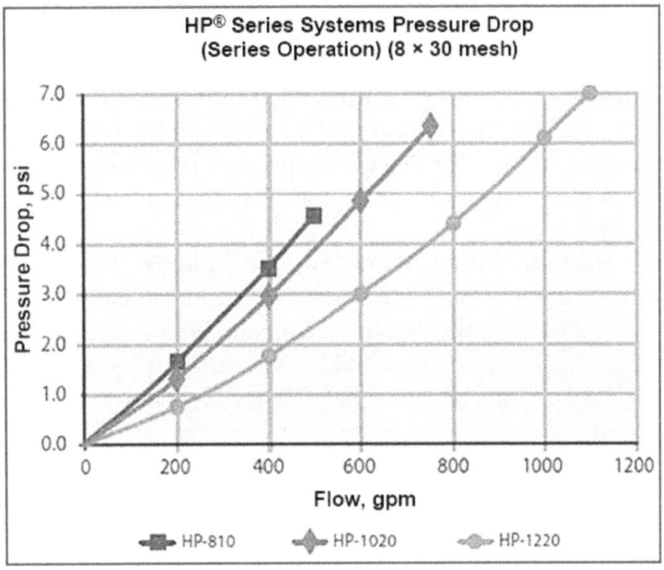

Figure 9-15. HP® systems pressure drop curves (courtesy of Siemens Industry, Inc.)

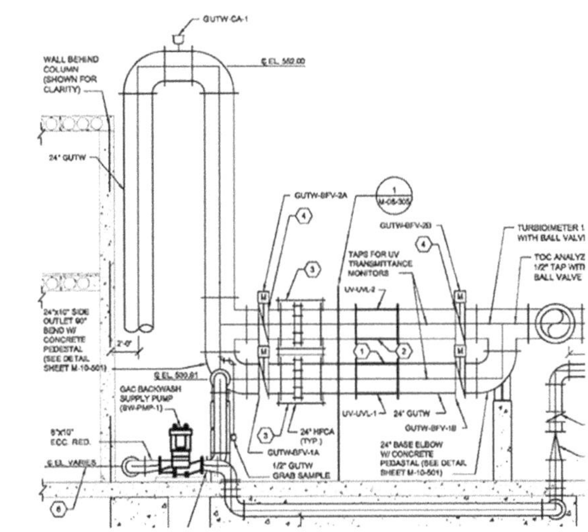

Figure 9-16. Positive head loop example (courtesy Malcolm Pirnie)

In order to ensure that the pressure vessels are fully submerged at all times, the design of pressure vessel systems may incorporate a positive head loop located on the GAC-treated water line, much like a seal well. An example of a positive head loop is shown in Figure 9-16. The centerline of the treated water line should be higher than the vessel supply line. The positive head loop elevation is the controlling water elevation for the new GAC facility.

Once the pressure drop through the vessels and the controlling water elevations are determined, the updated hydraulic profile can be calculated to determine if additional pumping facilities are required. Pressure vessel systems located on wellhead sites may not require additional pumping facilities because pumping from the well through the vessel and to the system may be sufficient. The existing system hydraulics should be considered to ensure that adequate pressure is maintained in the distribution system. Other hydraulic considerations include backwash supply water and plant service water supply. These considerations are discussed specifically in the following sections.

Backwash System

The backwash system provides backwash supply water to perform backwashes and provides a mechanism for disposal of the backwash waste once the backwash is complete. Backwashes are performed for two functions: to remove GAC fines after GAC transfers and to relieve GAC compaction. Depending on the type of system designed, the system can consist of backwash pumps, a backwash wetwell, backwash piping, and an air gap.

Backwash Rate, Duration, and Volume. The first function of the backwash system, removal of GAC fines, occurs immediately after GAC is transferred into a vessel. Although backwashes during long-term operations are not expected to generate a significant mass of solids, backwashing should be performed periodically to remove any fines that may be present to meet turbidity requirements and reduce the potential for plugging of the media. The backwash flow rate requirements when GAC is initially loaded into the vessels are related to fine particle removal. The fluidization required for fine removal varies with water temperature and GAC media characteristics. To determine the backwash flow rate required for the application, first the bed expansion curves and recommended bed expansion should be requested from the GAC manufacturer. The maximum bed expansions for backwash are typically 25 to 30 percent for pressure vessels (Siemens 2008, Calgon 2008a). An example of a bed expansion curve is shown in Figure 9-17. Once the bed expansion curve is obtained and the range of water temperatures is known, the minimum and maximum flow rates can be calculated as gpm/ft^2. The time for backwashing is equal to the amount of time required to remove the majority of fines, which can be as long as 20 to 30 min, but varies depending on GAC type and transfer methods. The specific backwashing time should be decided upon before designing an equalization basin, which is discussed in further detail in following sections.

The second function of the backwash system, relieving GAC compaction between GAC replacements, should occur when the pressure differential and flow modulation indicate that the GAC bed has compacted. Backwashing the GAC bed for a short period (5 to 6 min), also known as "fluffing" the bed, allows the GAC to expand without redistributing the GAC bed. The frequency that relieving GAC compaction will be needed will vary with each application.

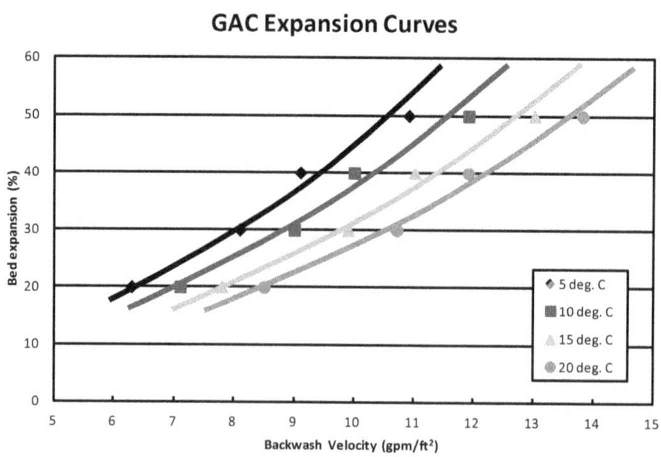

Figure 9-17. Effect of temperature and flow rate on bed expansion (courtesy of Siemens Industry, Inc.)

Now that the rate and time of the backwash have been determined, the volume of water required for backwashing can be determined. For example, if a backwash rate of 13.3 gpm/ft^2 for a 12-ft-diameter vessel is desired for 30 min, then 45,000 gal of backwash supply water is required.

Backwash Supply Source. With the volume of supply water determined, the design can begin to investigate possible sources for the backwash supply water. Possible supply sources include the following:

- GAC influent
- GAC effluent
- Outside water source (hydrant or water truck)

GAC influent water can be contaminated, especially in remediation applications. In applications where using the GAC influent would require disposing of contaminated water, it should not be used. GAC effluent is treated water and can be used for applications where drawing water from the effluent still allows for the demand needs of the plant to be met. Lastly, in some cases, using either GAC influent or effluent may not be feasible, and an outside water source such as a hydrant or water truck will be necessary. No matter which source is chosen, the design should consider the implications of the contaminants in the water as well as the chlorine concentration.

Once the backwash supply source is determined, how the backwash water is supplied to the GAC process should be determined. Options for supplying the GAC process include, but are not limited to, the following configurations:

Option 1. Dedicated backwash supply pumps in GAC influent pump station
Option 2. Dedicated backwash supply pumps in finished-water clearwell
Option 3. In-line booster backwash supply pumps drawing from GAC effluent
Option 4. Plant high service main
Option 5. Hydrants or water trucks

The first option for the backwash supply source consists of using the GAC feed pump wetwell with dedicated backwash supply pumps. Option 1 is not feasible for facilities that do not require a feed pump station. The feasibility of option 1 for facilities with a feed pump station depends upon the location of the feed pump station with respect to the vessels. If the distance is significant, typically one of the other options will be more favorable. If the pump station is relatively close, then this option may be feasible; however, other considerations may still favor other options. These considerations include possible control issues that would arise during backwashing events when the rate of flow to the GAC process would be reduced by the flow required for the backwash, the increase in wetwell size to accommodate the backwash supply volume, and the increase in the size of the influent pump station to house the additional pumps. These considerations should be investigated in detail, weighing the cost considerations and the other possible options for the supply source.

The second option for the backwash supply source consists of using the finished-water clearwell with dedicated backwash supply pumps. Option 2 can only be considered if the regulatory agency allows backwash pumps to be housed in the clearwell or reservoir. Approval of this setup should be confirmed prior to investigating option 2. If approval from the regulatory agency is granted and the clearwell is not a significant distance from the GAC facility, then this option may be feasible. Also, it may be necessary to find an alternative source of backwash water during start-up and construction unless the clearwell can be backfilled from the distribution system.

The main advantage of Option 2 is the quantity of water that would be available for backwashing events.

The third option for the backwash supply source consists of using in-line booster pumps located in the GAC facility taking suction from the GAC treated-water line prior to the positive head loop. By taking suction from the GAC effluent line, the pumps generally require less horsepower to achieve pumping requirements than pumps located in a feed pump station or clearwell. Option 3 also typically requires less footprint and piping than other options mentioned. The line pressures and feed requirements should be considered in detail with respect to the needs of both the backwash system and the treatment process. One major disadvantage to this option is backwashing under start-up conditions. When the vessels are filled with GAC for the first time, backwashing is required prior to the GAC process being placed in service. Selecting Option 3 will require another means of backwash supply during start-up of the GAC facility. Initial backwash can be performed by a water truck, hydrant, or other temporary source for the one-time condition. Also, because this option depends on availability of GAC-treated water, there may be times when distribution system demands dictate that a backwashing event could not be performed.

The fourth option for backwash supply consists of using the plant high-service discharge. Pressure-reducing valves may be required to reduce the pressure from a high-service source. Option 4 could be feasible depending on the distance from the GAC process. If distance is not a concern, the design should verify that the distribution system can handle the reduced pressures and capacity during backwashing events, which can be verified by the utility or through the use of distribution modeling software. If the option is feasible, it has the advantage of having a small cost and small footprint because no pumps or wetwells are required. However, if the distance from the GAC process is far, piping costs could be high.

The fifth option for backwash supply consists of using hydrants or water trucks to perform backwashes. Option 5 is not feasible for facilities that will perform frequent backwashing events and is typically only considered for small one-vessel systems that are located on wellhead sites a significant distance from the treatment plant. Facilities with larger numbers of vessels should be designed with permanent backwashing systems to facilitate long-term operation.

Backwash Supply Dechlorination. Once the backwash supply source has been determined, the amount of chlorine in the supply water should be considered. Because chlorine can react with the GAC and any adsorbed organics on the GAC surface, some designs may elect to dechlorinate the supply source. Using dechlorination efficiency curves obtained from the GAC manufacturer, the amount of GAC lost to chlorine per gallon of water used can be calculated. The cost of the GAC lost can then be compared to the cost of the dechlorination facility required. Typically, the cost of the lost GAC is much less than the cost of the dechlorination facility, and chlorinated water is accepted as the backwash supply source.

Backwash Waste Discharge Location. Once the backwash supply source has been determined, the location to send the backwash waste can be determined. The backwash waste will typically be returned to the head of the plant or be discharged to the sewer. Both options may require an equalization basin so that backwash waste can slowly be recycled to the head of the plant or to allow for low-flow discharges that will not overload the sewer. Sizing an equalization basin is discussed more in following sections.

When returning backwash waste to the head of the plant, operational concerns should be considered. These include the amount or percentage of flow and limitations on times when flow can be recycled. These should be discussed and decided upon with the utility. The main advantage of recycling backwash waste is the recuperation of water that has already been treated.

When discharging flows to the sewer and ultimately to a publicly owned treatment works (POTW), pretreatment sampling and flow restrictions compliant with the requirements of the POTW's permitting program must be considered and incorporated into the design. Discharging backwash waste to sewer may waste water that has already been partially or fully treated, but it does not spend additional resources to pump water to the head of the plant and then retreat the water.

The backwash system design should include an air gap on the backwash waste piping to prevent backflow from the backwash process from siphoning back into the GAC vessel. An air gap can also provide a visual indication for the operator to indicate when the majority of the fines have been removed and backwashing can be terminated.

Vessel-to-Waste System

Vessel-to-waste refers to the action of sending the GAC treated-water flow to waste for a period of time after filling or backwash. Performing vessel-to-waste operations gives the vessel time to clear any remaining fines and to verify that water quality targets are met prior to returning the vessel to normal operation. The vessel-to-waste flow stream can be sent to the same location as the backwash waste. The duration of the operation varies from application to application and is flexible to change throughout the life of the facility depending on operational needs. A typical vessel-to-waste operation could last anywhere from 30 min to 2 hr or longer. The design duration should be determined prior to sizing an equalization basin to ensure that waste flows can be handled. The vessel-to-waste stream can use the same piping as the backwash waste stream, provided that the connection is made downstream of the air gap. In most cases, there is no need for additional piping unless the stream is being sent to a different location than the backwash waste stream.

Washwater Equalization Basin

Equalization basins are included in pressure GAC facilities so that backwash waste can slowly be recycled to the head of the plant or to allow for low-flow discharges that will not overload the sewer. The following sections describe how an equalization basin is sized and located and what types of basins are used.

Equalization Basin Sizing. When determining the equalization volume required, the flow streams that will be stored should first be determined. The following various flow streams could be stored in the equalization basin:

- *Drain water from spent GAC truck loads.* When spent GAC is transferred to GAC trucks, the truck should drain excess water prior to leaving the GAC facility to meet transportation weight limits. The amount of water drained per 20,000 lb of wet spent GAC is approximately 10,000 gal (Calgon 2008a).
- *Drain water from GAC filling operations.* Prior to transferring fresh GAC to the vessel, the vessel is filled with approximately 3,000 gal of water used as a cushion for the incoming fresh GAC. As the vessel fills with fresh GAC, a portion of the cushion is drained from the vessel, approximately 1,000 gal. The drain volume

can be sent to an equalization basin or the sewer. Because it is typically a very small volume and is unlikely to overload the sewer, the design may select to discharge the flow directly to the sewer by gravity, when possible (Calgon 2008a).
- *Truck runoff.* Truck runoff, including any water used to wash down GAC trucks, stormwater, or snowmelt, may be sent to the equalization basin as well. However, these flows are typically routed directly to the sewer or a drain with a sump pump because the water will likely contain oil, gasoline, or other substances. These substances contaminate the treated water in the equalization basin.
- *Vessel backwash waste.*
- *Vessel-to-waste.*

Once the flow streams requiring equalization are identified and the associated volumes are calculated, the final size of the equalization basin can be calculated. The basin should be designed for the worst-case scenario, which occurs when the plant is operating at minimum design flow and the backwash event is immediately followed by a vessel-to-waste event. Low plant flow corresponds to the worst case because the equalization pumps should limit pumping to the head of the plant at a percentage of the plant flow per recommendations in the guidance manual for the Filter Backwash Recycling Rule (USEPA 2002). Once the volume of the equalization basin is determined, adding a safety factor to the final volume should be considered.

Equalization Basin Location. Determining if a new or existing equalization basin should be used depends on the volume required. Existing basins used to equalize vessel-to-waste or other flow streams may have the volume required to store GAC waste streams. Using existing basins saves capital and should be considered if possible. If a new equalization basin is required, it should be located as near to the GAC facility as possible.

Equalization Basin Type. The type of equalization basin designed depends on both the volume and the available footprint of the facility. Types of basins that could be designed include aboveground concrete basins, belowground concrete basins, or metal tanks. The type of tank selected will vary based on application, but all equalization basins should be designed with proper venting and overflow capabilities and should be capable of housing the equalization basin pumps. Because the backwash waste contains GAC fines, the design should consider a sloped bottom basin to facilitate removal of settled fines.

Equalization Basin Pumps. Once the location and type of equalization basin have been determined, the destination of the equalization basin discharge should be determined. Typically, the backwash waste is recycled to the head of the plant or sent to the sewer. Determining where the waste is sent is discussed in the section above. Whichever option is selected, equalization basin pumps are required and are to be located in the basin. The pumps are selected to be capable of meeting the flow and head conditions to send the waste to the desired location.

GAC Handling

GAC is typically delivered to GAC facilities with pressure vessels via tanker trucks. The GAC transfer process requires less time than gravity systems because pressure can be used. Vessels can also be filled with GAC via 1,000- to 2,000-lb bags and hoppers; however, this requires additional time and on-site facilities. For designing facilities for GAC transfers with bags, refer to chapter 8. GAC transfers via tanker trucks have specific plant water and air requirements as discussed below. It is important to communicate with potential GAC suppliers during design to verify the needs of the delivery trucks and to ensure that the necessary connections and services are provided at the site. Working with the GAC manufacturer regarding GAC shipments helps minimize operations and maintenance costs for GAC facilities.

Plant Service Water. Plant service water needs to be supplied to the GAC facility in some fashion to supply water for the various water requirements of the GAC transfer process as follows:

- *GAC spray wash during spent GAC transfer.* During spent GAC transfers, approximately 500 gal at 100–200 gpm is introduced into the top of the vessel for 5 min to wash down the sides of the vessel to ensure that all the spent GAC is removed (Calgon 2008a).
- *GAC filling operations.* During fresh GAC transfers, approximately 3,000 gal at 100–200 gpm is introduced into the vessel prior to adding GAC to provide a cushion of water for the entering GAC (Calgon 2008a).
- *GAC truck preparation.* Prior to fresh GAC transfers, approximately 5,000 gal per 20,000 gal dry GAC is added to the GAC truck to create a GAC slurry to prepare for the GAC transfer (Calgon 2008a).

Multiple sources are available for plant service water in the GAC facility:

- Existing plant service water system.
- Plant service water pumps and a hydropneumatic tank, which can use water treated by the GAC process or water prior to GAC treatment.
- High service discharge.

Plant Air. Depending on the size of the vessels selected and the frequency of GAC replacements, plant air may need to be provided. Air compressor systems located on GAC delivery trucks can be rented from the GAC provider if GAC replacements are infrequent. For applications where replacements occur more regularly, it may be more cost effective to design a permanent air system for the GAC facility. The air requirements of the GAC transfer process are as follows:

- *Vessel preparation.* During spent GAC transfers, the vessel is pressurized to allow GAC to transfer from the vessel to the GAC truck. Typically 100 scfm at 30 psig is required for vessel preparation, but this should be confirmed with the vessel manufacturer (Calgon 2008a).
- *Spent GAC truck drain procedure.* Once the spent GAC has been transferred to the GAC truck, the transfer water needs to be drained before the truck leaves the site. To facilitate dewatering, the GAC trucks are pressurized with 100 scfm at 15 psig maximum.
- *Truck preparation.* During fresh GAC transfers, the GAC truck is pressurized to allow GAC to transfer from the truck to the vessel. Typically 100 scfm at 15 psig maximum is required for truck preparation, but this should be confirmed with the GAC provider.

Because the air is being provided for potable water, the design should consider specifying food-grade oil for air compressors as well as a desiccant dryer and air filters to avoid contamination of the potable drinking water. Oil-less compressors can also be considered; however, they have higher capital costs. A cost analysis can be completed to select the ideal compressor for the specific application.

GAC Storage. On-site GAC storage can be provided in the form of silos, 1,000- or 2,000-lb bags, or extra empty vessels. Pressure GAC treatment systems are typically small enough that one facility would not warrant the additional cost of extensive on-site storage, but it should be considered to weigh the benefits versus the costs:

- Including a silo(s) would not require much space on-site, but installing the silos and related equipment, such as hoppers and eductors, could be a significant cost that may outweigh the benefit.
- 1,000- to 2,000-lb bags require additional storage space, and transferring GAC from the bags to vessels requires additional equipment as described earlier.
- Extra vessels are sometimes included to provide a means of storing GAC and maintaining the facility's EBCT during GAC transfers.

Each of these storage options can be considered to fit the specific needs of the application.

Site Design

When all the components of the GAC facility have been selected, sized, and designed appropriately, the facility layout can be determined and integrated into the existing plant site. First, a GAC building should be considered. For small facilities located on wellhead sites, a building is typically not used. However, heat tracing or piping insulation needs to be provided in climates where freezing temperatures are encountered. Under normal conditions, the small-diameter piping may contain sitting water, and the backwash lines are full and susceptible to freezing. During a backwash, the influent and effluent lines are full and susceptible to freezing if no freeze protection is provided. Heat tracing and insulation should also be provided for GAC loading and unloading lines. For larger facilities in colder climates located on existing treatment plant sites, a building should be considered. Providing a building for the GAC facility allows the vessels to be protected from the elements in a temperature-controlled environment. Buildings eliminate the need for heat tracing and piping insulation. Also, providing a building on an existing treatment plant site allows the new facility to be integrated into the existing site by utilizing similar architectural features, especially if the plant site is in a residential area. If a building is selected, the addition of a control room(s), an electrical room(s), restroom facilities, and possibly a conference room can be considered based on the specific application.

Regardless of whether a building is provided for the GAC facility, the logistics of GAC replacements should be considered in laying out the final facility. The vessels should be configured in such a fashion to ease GAC replacements. Delivery trucks typically carry two 20-ft sections of flexible hose to perform the transfer. Ideally the vessels are located so that the trucks can be within 20 to 40 ft from the vessel. Additional distance from the vessel requires additional hose lengths and adds time to the GAC replacement process. If vessels must be greater than 40 to 60 ft from the delivery trucks, the design may consider connecting the vessel GAC fill and drain lines via hard piping to GAC transfer stations. The number of stations required would depend on the arrangement of vessels, number of vessels, and location of the truck loading and unloading. However, hard piping the GAC fill and drain lines should be avoided if possible, as they are highly susceptible to clogging. If it is unavoidable, long radius bends should be specified to reduce pipe clogging.

Depending on the climate of the facility location, vehicle access may want to be provided indoors, outdoors, or covered. For locations where freezing concerns are present, providing a facility with indoor vehicle access should be considered. Indoor access with adequate draining facilities avoids icy conditions and promotes a safer working environment for operations staff. Because providing indoor access results in a higher building capital cost, an outdoor, covered facility may be considered. Completely outdoors facilities with no covering provided may be adequate for facilities expecting very infrequent replacements or where freezing concerns are not present. Whichever vehicle access is provided, adequate drainage for the GAC trucks should be provided. When transferring spent GAC, the trucks drain approximately 10,000 gal of water for every 20,000 lb of spent GAC, which should be adequately drained to avoid icy or muddy conditions.

Additionally, the facility layout should consider the location of the supply of plant air and water. As mentioned in the sections above, plant air and water should be supplied to both the vessels and GAC trucks for GAC replacements to occur. The number and location of air and water stations depends on the number and arrangement of vessel systems as well as the location of the GAC truck loading and unloading. The design should include adequate stations to provide both water and air to the vessels and GAC trucks as needed.

Vehicle access to the GAC facility should also be considered and should account for the sizes and turning radii of vehicles expected to service the site. Tanker trucks are the standard truck that delivers GAC to pressure vessels. For larger facilities where loading and unloading of multiple vessels may occur simultaneously, additional driveway width and length can be provided adjacent to the facility if footprint is available. Depending on site constraints, access design should consider providing a loop drive around the facility or a driveway through the GAC building that allows truck drivers to navigate the site without reversing. Installing a truck scale can be considered as part of the design so that drivers can verify the shipping weight of trucks prior to leaving the site. Providing flexibility at the site to accommodate various delivery vehicles should be weighed against the additional costs of design and construction.

The site paving should be designed to handle the load of a full GAC delivery truck. The design of paving in the area adjacent to the GAC transfer stations is especially important because trucks loaded with 20,000 to 40,000 lb of GAC will have approximately 4,000 to 6,000 gal of water, or an additional 30,000 to 50,000 lb, added to them while trucks are parked to slurry the GAC into the vessels. High-strength asphalt paving or reinforced concrete paving is recommended for the transfer area to prevent excessive wear and tear on the pavement.

Pretreatment

In addition to the potential modifications discussed in chapter 8, GAC facilities, particularly those located on wellheads, may require pretreatment for arsenic, iron, and/or manganese removal. For applications where additional facilities are required for arsenic removal, the GAC and arsenic treatment processes may be able to share the equalization basin for backwash waste, backwash supply piping, backwash air gap, additional piping, and possibly various instrumentation and controls. Iron and manganese removal may be required to prevent fouling of GAC and any reductions in adsorptive capacity. Iron and manganese removal is often achieved though greensand filtration.

On wellhead sites where sand is introduced into the system through the well pumping, the design may consider removal of sand prior to pre-GAC treatment using

Figure 9-18. Sand separators

sand separators, shown in Figure 9-18. If sand is not removed, it will be filtered out by the GAC, but because it has a higher density than GAC, it will settle at the bottom of the vessel during backwashing rather than being removed, resulting in an increase in the head loss.

References

Calgon Carbon Corporation. 2008a. Information request memorandum. Pittsburgh, Pa. (Unpublished)

Calgon Carbon Corporation. 2008b. *Modular Model 12 adsorption system (40,000 lbs) using granular activated carbon.* Specification. Pittsburgh, Pa.

Calgon Carbon Corporation. 1993a. Operating manual: Model 12 granular carbon adsorption system. Pittsburgh, Pa.

Calgon Carbon Corporation. 1993b. Model 10 modular adsorption system, equipment bulletin. Pittsburgh, Pa.

Siemens Water Technologies. 2006. HP® series liquid phase adsorption systems (ASME Code). Warrendale, Pa.

Siemens Water Technologies. 2008. Information request memorandum. Warrendale, Pa. (Unpublished)

USEPA. 2002. *Filter Backwash Recycling Rule technical guidance manual.* Washington, D.C.: USEPA Office of Ground Water and Drinking Water.

Chapter 10

Vapor-Phase GAC Treatment Design

This chapter addresses the use of GAC in vapor-phase treatment systems associated with potable water treatment. Emissions of contaminated air from water treatment technologies can result in air quality issues or permit noncompliance. In instances where emission rates exceed air quality regulations or where health and safety concerns exist, vapor-phase GAC treatment systems may be used to mitigate the release of contaminants into the atmosphere. Examples of water treatment technologies that may require vapor-phase GAC treatment include air stripping and multi-stage diffused bubble aeration (MSDBA). Schematics of both water treatment technologies are shown in Figures 10-1 and 10-2, respectively.

Numerous examples exist of using air stripping to remove volatile organic compounds (VOCs) from contaminated groundwater sources and using the treated groundwater as a potable supply. Air stripping is recognized by the USEPA as a BAT for removing VOCs (AWWA 2012) from water, and MSDBA is also used for some applications.

Vapor-phase GAC treatment systems are designed to remove VOCs from vapor streams. This chapter focuses on providing vapor-phase GAC treatment for air-stripping systems, although the design is similar for MSDBA.

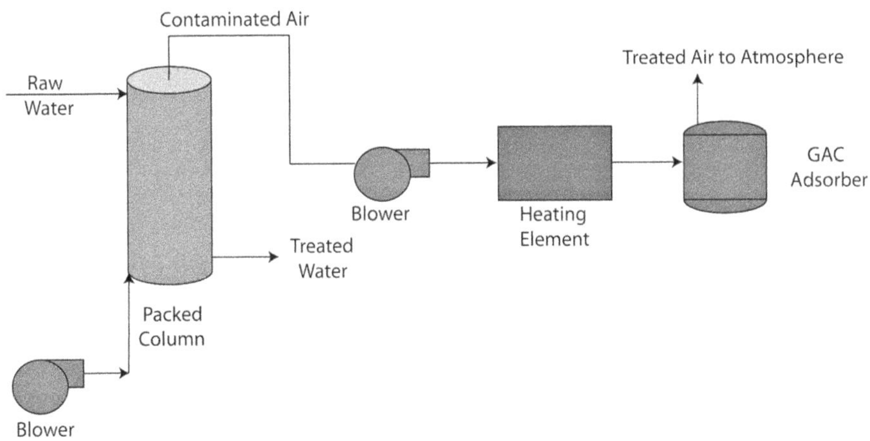

Figure 10-1. System schematic for packed column air stripper followed by vapor-phase GAC

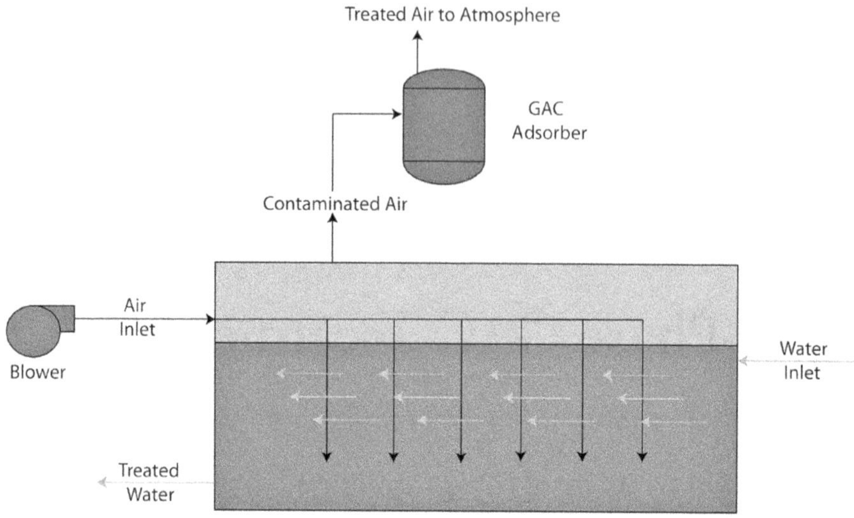

Figure 10-2. System schematic for multi-stage diffused bubble aeration followed by vapor-phase GAC

Design and Construction

Vapor-Phase GAC Treatment System Overview

Air stripping is a treatment process that removes a gas from water via aerators or packed columns. *Packed columns* are also known as *air strippers* or *packed towers* and are commonly used in water treatment plants to remove VOCs from contaminated groundwater. The air-stripping system consists of a column, packing material, and a blower. The blower forces air upward through the column while water flows downward through the column by gravity, stripping the contaminants from the water phase into the vapor phase. The concentration of contaminants in the air increases as the air passes through the column. The contaminated air stream is then released into the atmosphere or enters vapor-phase treatment. If vapor-phase GAC treatment is provided, the system may consist of some or all of the following components:

- Ductwork and dampers
- Duct heaters
- GAC treatment units (adsorbers)
- Flowmeters
- Temperature and humidity probes
- Blowers

A schematic of an air-stripping system with vapor-phase GAC treatment is shown in Figure 10-1. As shown in the schematic, clean air is forced through a packed column by a blower, and the contaminated air is forced through the GAC adsorber by a second blower. The typical design can be varied as follows:

- Placing the second-stage blower downstream of the heating element.
- Having air flow downward through the adsorber rather than upward.
- Only having one blower for the system, located either before the packed column, after the packed column, or after the GAC adsorber.

Once the air exits the blower, it enters a heating element. The heating element increases the temperature of the air stream to reduce the relative humidity, which prevents condensation from occupying GAC pore sites thereby increasing the adsorption capacity of the media to remove the desired contaminant(s) (AWWA 2012).

The GAC adsorber is the main component of a vapor-phase GAC system. Standard prefabricated adsorbers range in size from 180 to 20,000 lb GAC, which can process air from 100 to 20,000 scfm. Larger, custom-made vessels can also be used. Adsorbers may be cylindrical or rectangular and may have a single or a dual bed. Figure 10-3 shows a photo of a single-bed adsorber.

Figure 10-3. Vapor-phase single-bed adsorber (CARBTROL Corporation, Bridgeport, CT)

Determine Key Design Parameters

Once a treatment process requiring vapor-phase treatment is selected for use, the first step of the design phase is determining the key design parameters. The most important design criteria for a vapor-phase treatment system are as follows:

1. Vapor stream flow rate and superficial velocity
2. Vapor stream relative humidity and temperature
3. Vapor-phase adsorption characteristics of contaminants and GAC
4. GAC selection
5. System components and configuration

Vapor Stream Flow Rate and Superficial Velocity

The vapor stream flow rate is determined by the water treatment process. For example, an air-stripping facility will have target minimum and maximum air-to-water ratios for optimal process performance. The air-to-water ratios and the water flow rates set the capacity of the vapor-phase treatment system.

In considering the system vapor flow rate, the standard modular adsorbers available from GAC manufacturers also need to be considered. The adsorber manufacturers have limits on recommended flow rates depending on their size. For example, one manufacturer's standard product line includes vapor-phase adsorbers that hold 180 to 18,000 lb of GAC, resulting in design air flow rates ranging from 100 to 10,000 scfm. Air flow rates are typically based on standard conditions with sea-level properties. Depending on the specific location and temperature of the vapor-phase GAC treatment system, the designer may need to calculate the actual flow rate (acfm) from the standard flow rate (scfm). Eq 10-1 below summarizes the conversion from scfm to acfm.

$$\text{acfm} = \text{scfm}\left[\frac{P_{std}}{P_{act} - P_{sat}\Phi}\right]\left[\frac{T_{act}}{T_{std}}\right] \qquad \text{(Eq 10-1)}$$

Where:

acfm = Actual cubic ft/min
scfm = Standard cubic ft/min
P_{std} = Standard pressure, 14.7 psia
P_{act} = Actual pressure (psia)
P_{sat} = Saturated pressure at the actual temperature (psi)
T_{act} = Actual temperature (°R)
T_{std} = Standard temperature, 530°R
Φ = Actual relative humidity

The superficial velocity, which is also referred to as the surface loading rate, is the flow rate divided by the cross-sectional area of a GAC adsorber. In considering the superficial velocity, GAC manufacturers recommend that systems be designed to achieve a superficial velocity of approximately 50 ft/min. The treatment units should also be designed for a maximum of 100 ft/min superficial velocity (Shepherd 2001).

Vapor Stream Relative Humidity and Temperature

The vapor stream from an air-stripping process is saturated with water when exiting the packed column, and the vapor stream is at the temperature of the water present in the air stripper. The relative humidity needs to be reduced to 40–50 percent for optimal adsorption onto the vapor-phase GAC media (AWWA 2012). The water temperature and vapor flow rate are used to size in-line duct heaters, which reduces the relative humidity by heating the vapor stream.

Vapor-Phase Adsorption Characteristics of Contaminants and GAC

The concentration of contaminants requiring treatment in the vapor stream is determined by the water-phase concentration and the efficiency of the water treatment process. Once the vapor-phase contaminant concentration is known, the adsorption capacity can be estimated. The GAC adsorption capacity depends on the contaminant concentration, the affinity of GAC to adsorb the contaminants, and the total GAC surface area. The adsorption capacity is typically presented as an isotherm. Although many isotherm equations could be used to represent the equilibrium between the vapor and solid phases, the Freundlich isotherm is the most widely used. Eq 10-2 presents the Freundlich equation in vapor-phase form.

$$\frac{x}{m} = KC^{\frac{1}{n}} \qquad \text{(Eq 10-2)}$$

Where:

x = Amount of contaminant adsorbed (µg, mg, or g)
m = Mass of GAC (mg or g)
K = Empirical constant determined for each combination of contaminant, GAC type, and temperature (mg/g)(L/mg)(1/n)
C = Concentration of contaminant in vapor stream at equilibrium (mg/L)

$\frac{1}{n}$ = Empirical constant determined for each combination of contaminant, GAC type, and temperature (dimensionless)

The empirical constants are derived from the log-log plot isotherm with concentration on the x-axis and x/m on the y-axis. On this plot, the slope is represented by 1/n, and the intercept is represented by K, as shown in Eq 10-3.

$$\log \frac{x}{m} = \log \mathrm{K} + \frac{1}{n} \log \mathrm{C} \qquad \text{(Eq 10-3)}$$

An example of a vapor-phase isotherm for TCE using Carbtrol® GAC at 32 °F and 140 °F is shown in Figure 10-4. Isotherms or theoretical GAC usage rates can be obtained from literature or from GAC manufacturers. Software programs such as ETDOT, developed by Michigan Technological University, can also be used to help estimate the adsorption capacity. However, the amount of information available for vapor-phase GAC is not as extensive as that available for liquid-phase GAC (USACE 2001). Also, adsorption capacities obtained from isotherms should be carefully reviewed because isotherms are typically developed for clean water or vapor and do not account for competitive adsorption. Once the GAC usage rate is determined, a factor of safety can be added to help account for the variability of adsorption. Factors of safety may be as little as 20 percent up to 100 percent for conservative design (USACE 2001). Additionally, monitoring the GAC effluent air quality either continuously or intermittently can help to determine when to perform GAC replacements.

Once the theoretical adsorption capacity is determined, GAC replacement frequency projections can be made by determining the GAC usage rate. Below is an example of how to calculate the GAC replacement frequency projection for a vapor stream with two contaminants.

Example 10-1. GAC Replacement Frequency Projection Calculations.

A facility located at sea level has an air-stripping treatment process with a vapor-stream flow rate of 2,000 cfm at 60 °F. The vapor stream contains 10 ppm by volume

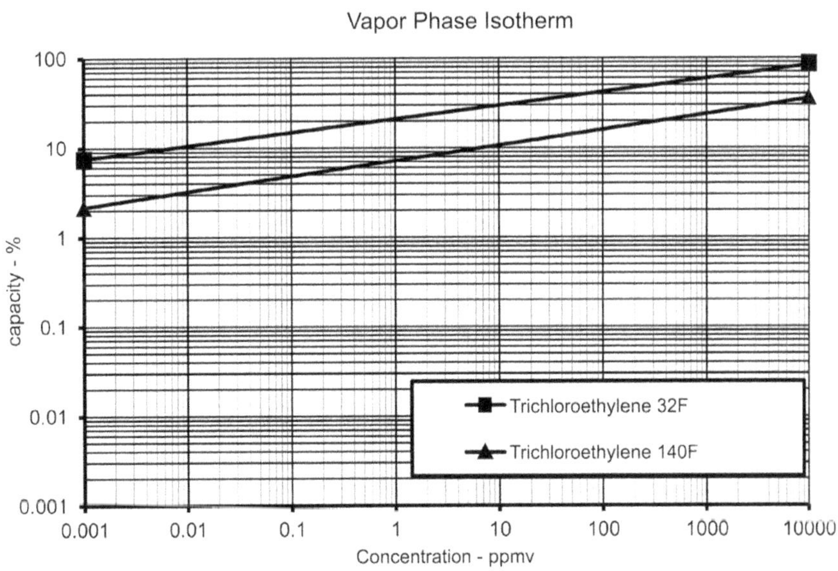

Figure 10-4. TCE example vapor-phase isotherm (Shepherd 2001; CARBTROL Corporation, Bridgeport, CT)

(ppmv) of PCE (MW = 165.8) and 12 ppmv of TCE (MW = 131.4). The adsorption capacities at the treatment objectives of each were derived from isotherm data to be 0.35 lb PCE/lb GAC and 0.21 lb TCE/lb GAC.

SOLUTION: To calculate the total GAC usage rate per day, first calculate the partial pressure of the contaminant in the vapor stream:

$$\text{Partial Pressure of Contaminant} = \text{Total System Pressure} \times \frac{\text{Contaminant Concentration}}{\text{Air Volume}}$$

$$\text{Partial Pressure of PCE} = 14.7 \text{ psia} \times \frac{10 \text{ ppmv}}{10^6 \text{ ppmv air}} = 1.47 \times 10^{-4} \text{psia}$$

Use the ideal gas law to determine the amount of contaminant (lb) adsorbed per day:

$$PV = nRT$$

Pressure × Volume = Number of Moles × Universal Gas Constant × Temperature

Solve for n:

$$n\left(\frac{\text{lb mol}}{\text{min}}\right) = \frac{P(\text{psia}) \times V(\text{cfm})}{R\left(\frac{\text{cf} \times \text{psia}}{\text{lb mol} \times {}^\circ\text{R}}\right) \times T({}^\circ\text{R})}$$

$$n\left(\frac{\text{lb mol}}{\text{min}}\right) = \frac{1.47 \times 10^{-4}(\text{psia}) \times 2{,}000(\text{cfm})}{10.73\left(\frac{\text{cf} \times \text{psia}}{\text{lb mol} \times {}^\circ\text{R}}\right) \times (60 + 460)({}^\circ\text{R})} = 5.26 \times 10^{-5} \frac{\text{lb mol PCE}}{\text{min}}$$

Convert to contaminant concentration (lb) per day using molecular weight:

$$5.26 \times 10^{-5} \frac{\text{lb mol}}{\text{min}} \times \frac{1{,}440 \text{ min}}{\text{day}} = 0.076 \frac{\text{lb mol PCE}}{\text{day}}$$

$$0.076 \frac{\text{lb mol PCE}}{\text{day}} \times 165.8 \frac{\text{lb}}{\text{lb mol}} = 12.6 \frac{\text{lb PCE}}{\text{day}}$$

Convert to GAC usage rate using the theoretical adsorption capacity:

$$12.6 \frac{\text{lb PCE}}{\text{day}} \times \frac{\text{lb GAC}}{0.35 \text{ lb PCE}} = 36 \frac{\text{lb GAC}}{\text{day}}$$

Following the same steps for TCE yields:

$$11.7 \frac{\text{lb TCE}}{\text{day}} \times \frac{\text{lb GAC}}{0.21 \text{ lb TCE}} = 56 \frac{\text{lb GAC}}{\text{day}}$$

For a total GAC usage rate of:

$$36 \frac{\text{lb GAC}}{\text{day}} + 56 \frac{\text{lb GAC}}{\text{day}} = 92 \frac{\text{lb GAC}}{\text{day}}$$

A factor of safety can then be added to account for adsorption capacity variability. Assume a 20 percent factor of safety:

$$\text{Total projected GAC usage rate} = 110\ \frac{\text{lb GAC}}{\text{day}}$$

If the design selects a prefabricated 3,000-lb adsorber with a maximum flow rate capacity of 2,000 cfm, adsorber replacement will occur approximately every:

$$3{,}000\ \text{lb GAC} \times \frac{\text{day}}{110\ \text{lb GAC}} = 27\ \text{days}$$

GAC Selection

Different types of GAC are available for vapor-phase applications. GAC can be classified by the source of its manufacture (examples include bituminous, lignite, coconut shell) or by its size (examples include 4×6 mesh and 4×8 mesh). Vapor-phase applications typically use 4×6, 4×8, or 4×10 mesh GAC (USACE 2001). In addition to size differences between liquid-phase and vapor-phase GAC, the mass transfer zone (MTZ) will also differ. Vapor-phase GAC systems have very short MTZs, meaning that effluent concentration goes from no breakthrough to full breakthrough very quickly. The short breakthrough for a vapor-phase GAC is shown in Figure 10-5 compared with a typical liquid-phase breakthrough curve. The steep breakthrough curves make series operation of vapor-phase adsorbers useful to avoid out-of-compliance operation.

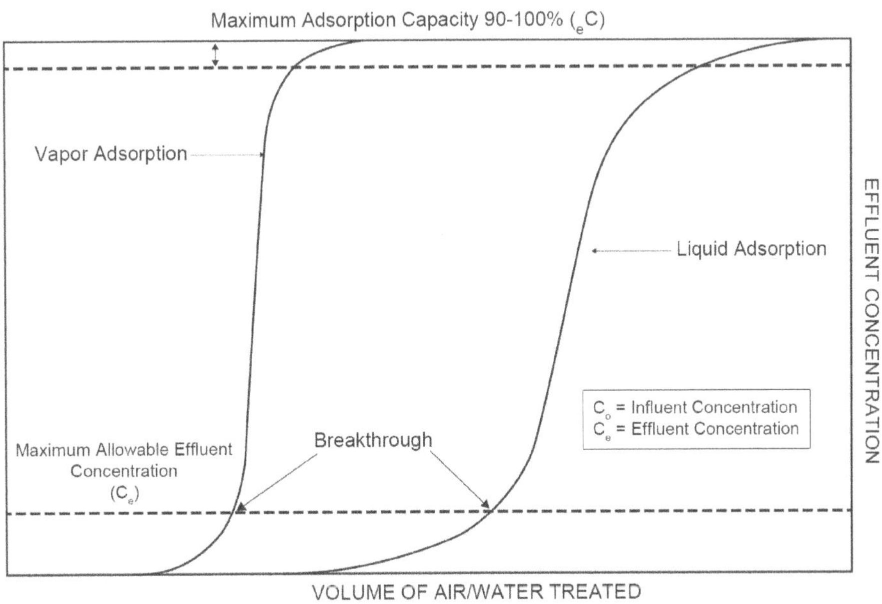

Figure 10-5. Comparison of vapor-phase (left) and liquid-phase (right) breakthrough curves (USACE 2001)

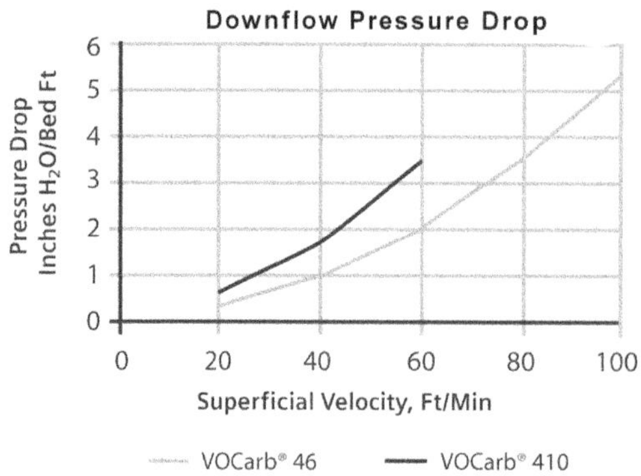

Figure 10-6. Example head-loss curves for VoCarb® 46 and VoCarb® 410 GAC (courtesy of Siemens Industry, Inc.)

GAC manufacturers can help determine which GAC type is best suited for the vapor-phase application because different contaminants have greater affinities for specific GAC types. GAC manufacturers can also be consulted for head-loss information to aid design of the blower sizing. An example of GAC head-loss information from one manufacturer is shown in Figure 10-6. In Figure 10-6, the pressure drop through the GAC bed is shown for various superficial velocities for two types of GAC. Manufacturers may provide pressure drop curves for both densely and loosely packed GAC. Densely packed pressure-drop curves should be used when available for sizing blowers, because GAC settles with use.

System Components and Configuration

Figure 10-7 presents a schematic of an air-stripping facility with vapor-phase GAC treatment. The system configuration will likely be driven by the water treatment process design and the type of monitoring and redundancy required. In Figure 10-7, three air-stripping towers operate in parallel to remove VOCs from the contaminated groundwater. Each air-stripping tower has a dedicated vapor-phase GAC treatment unit, shown as dual-bed adsorbers. Single-bed adsorbers could also be used. The blower is located between the air-stripping tower and the GAC treatment unit in each treatment train. Air is drawn through the air-stripping tower and discharged through the vapor treatment unit.

Vapor-phase treatment may include parallel treatment units as indicated in Figure 10-7. Other vapor-phase treatment configurations may also be possible such as (a) combining the vapor streams for a single vapor-phase treatment train (Figure 10-8), (b) splitting individual vapor streams for parallel vapor-phase treatment units (Figure 10-9), or (c) adsorbers in series (Figure 10-10).

Apply Regulatory Requirements

Any regulatory requirements that apply to the project location should be considered during the conceptual design phase. Some states may require redundant adsorbers so monitoring can be conducted after the lead adsorber. Regulatory requirements affect the facility layout and operation of the vapor-phase treatment system.

Obtain Manufacturer Information

Once the key design parameters have been determined, the adsorber manufacturers should be contacted to obtain additional information on specific products that are applicable. Manufacturers can typically provide the following information that can be used to design vapor-phase treatment systems:

- Detailed GAC information, including recommended GAC for the application and downflow head loss.

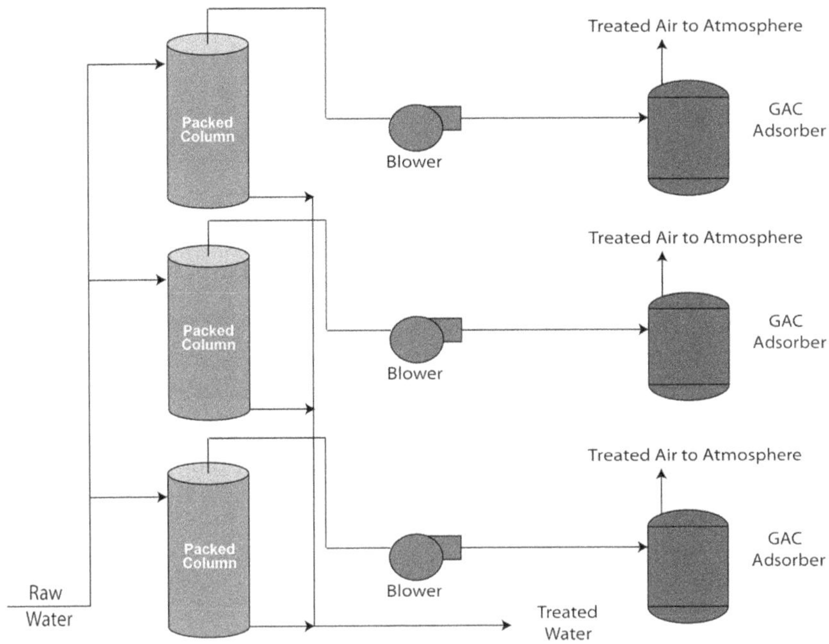

Figure 10-7. Example schematic: Packed column air-stripping facility with vapor-phase GAC treatment

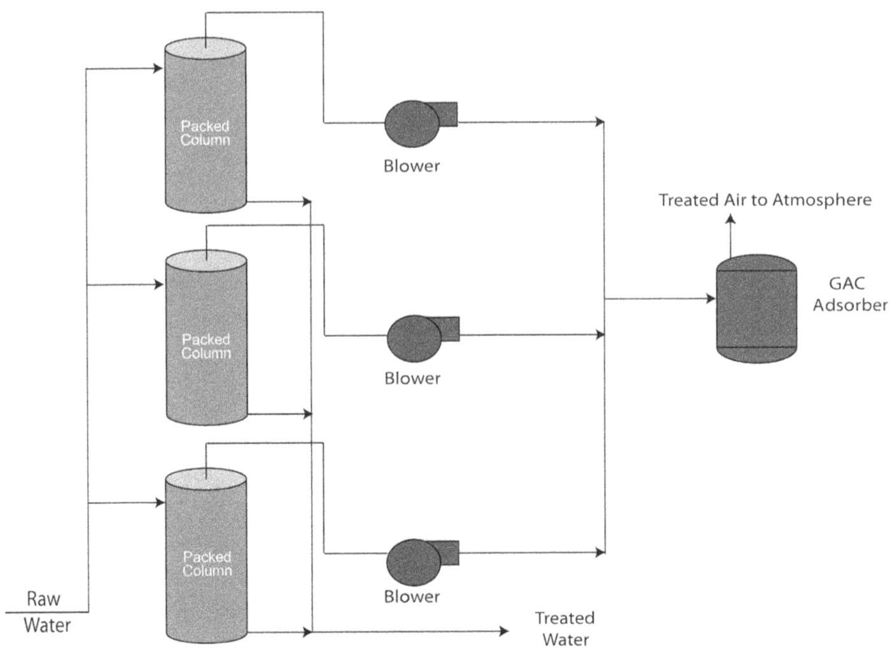

Figure 10-8. Example schematic: Packed column air-stripping facility with combined vapor-phase GAC treatment

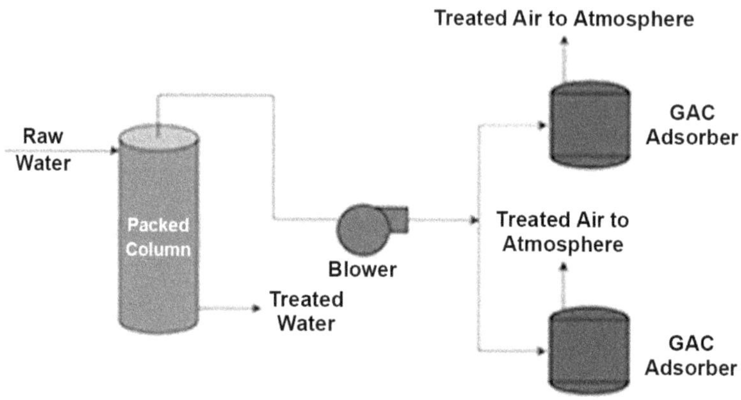

Figure 10-9. Example schematic: Packed column air-stripping facility with split, parallel vapor-phase GAC treatment

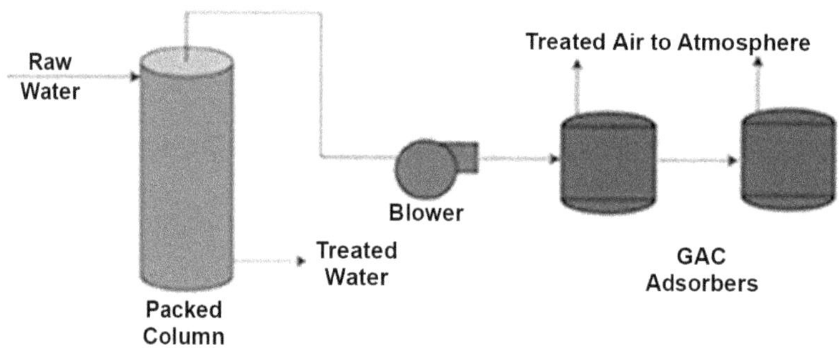

Figure 10-10. Example schematic: Packed column air-stripping facility with series vapor-phase GAC treatment

- Detailed product information, including dimensions, operating weights, interior/exterior coating, temperature/pressure ratings, and materials of construction.
- System pressure-drop information.
- Loading rate criteria.
- GAC transfer procedures.

Construction Details of Treatment Units

As mentioned previously, vapor-phase GAC treatment systems are designed to remove VOCs from vapor streams. The design of the system and associated components is directly related to the contaminant, the water treatment process, and the available footprint in the existing treatment plant or wellhead site. The design will include all or some of the following typical vapor-phase GAC system structures, equipment, and resources:

- GAC treatment units (i.e., adsorbers)
- Blowers

Figure 10-11. Transportable adsorber (courtesy of Siemens Industry, Inc.)

- Heating elements
- Ductwork
- Flowmeters
- Temperature and relative-humidity probes
- GAC truck access

Before the layout of the system can be integrated into the existing site, the vapor-phase GAC treatment system components should be considered in detail.

Adsorber Unit

Selection of the adsorber unit includes choosing an adsorber type, bed arrangement, and other standards and options.

Adsorber Type. Adsorbers are designed to be either permanent or modular units. Permanent installations require on-site GAC replacements similar to those provided for liquid-phase gravity contactors or pressure vessels. Permanent installations are also hard-piped for continuous use. The modular units, also known as *transportable*, *disposable*, or *roll-off*, are designed to be removed and replaced when the GAC becomes exhausted. Transportable-type adsorbers require piping that can be disconnected easily and adequate vehicle access to perform the adsorber replacement. Figure 10-11 shows an example of a transportable adsorber on a trailer.

Either type of adsorber may be owned by the facility or may be rented from the manufacturer. The design should select the appropriate adsorber type depending on the frequency of replacements required, adsorber access, and a cost analysis.

Bed Arrangement. Both permanent and transportable-type adsorbers can have either single or dual beds. Figure 10-12 shows a

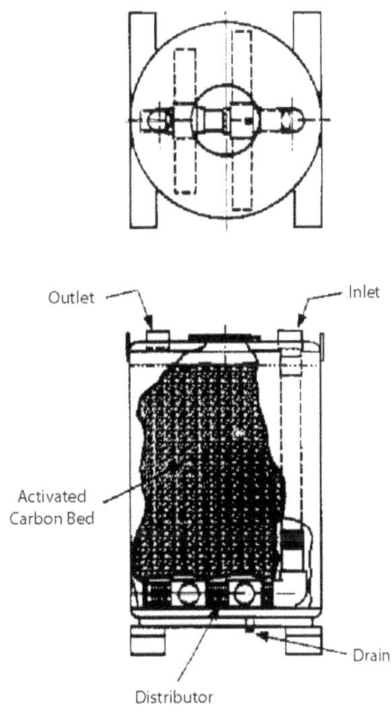

Figure 10-12. Single-bed adsorber cross section (CARBTROL Corporation, Bridgeport, CT)

Figure 10-13. Dual-bed adsorber installation

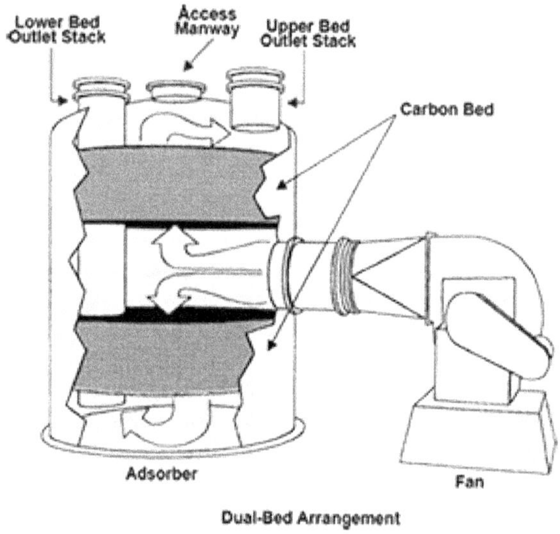

Figure 10-14. Dual-bed adsorber cross section

cross section of a single-bed adsorber. A typical adsorber consists of influent piping that connects to a distributor. The distributor, located in the bottom of the adsorber, distributes the airflow evenly across the adsorber bed. The air then exits through the effluent piping, where a discharge stack may be located. Dual beds, as the name states, have two beds that are separated. Dual-bed adsorbers are flexible because they can handle a range of flows and contaminant concentrations, can both be used simultaneously or separately, and can flow in series or parallel modes of operation. Figure 10-13 shows an installation of a cylindrical dual-bed adsorber. Figure 10-14 shows a cross section of a dual-bed adsorber.

Adsorber Standards and Options. Standards for vapor-phase adsorbers vary more widely than for liquid-phase pressure vessels depending on manufacturer and size. Depending on the size and application, the adsorber shell can be constructed of carbon steel, stainless steel, polypropylene, or fiberglass-reinforced plastic. Adsorbers typically are equipped with a manway access, inlet and outlet connections, interior distributor piping, and a condensate drain. Smaller adsorbers will be equipped with forklift guides for ease in moving. Vapor-phase adsorbers may include the following options:

- Influent/effluent sample and pressure taps
- Discharge stacks
- Built-in ladder access
- Saturation indicators (discussed in Instrumentation section, below)
- Flanged inlet/outlet connections
- Epoxy internal/external lining or coating

Adsorber Sizing

To begin site design for the GAC facility, the size of the vapor-phase treatment units (adsorbers) should be considered. The size of the adsorbers chosen is directly related to the vapor flow rate and the superficial velocity. Prefabricated adsorbers are rated for a maximum flow rate and range in sizes from 100 to 20,000 cfm. If the prefabricated sizes are not adequate, it is not unusual for manufacturers to custom design and fabricate adsorbers for larger flow rates. Because the vapor flow rate determines the size of adsorbers chosen, the system configuration should be determined first. Vapor-phase treatment may include treatment for each vapor stream in the system, treatment for a combined vapor stream, or split of one vapor stream into multiple streams for parallel treatment. The design is largely based on splitting or combining the flow to allow for the minimum number and footprint of adsorbers. For example, for a treatment process with a vapor flow rate of 40,000 cfm, the design could split the stream into two equal 20,000-cfm streams using prefabricated adsorbers, or the design could select a custom adsorber for 40,000 cfm. Once the configuration has been determined and the vapor flow rate finalized, the superficial velocity should be verified to be less than 100 ft/min.

Once the initial minimum size of the adsorber required has been determined, the frequency of GAC replacements is estimated. In some instances, the minimum-size adsorber required for the flow rate may require very frequent GAC replacements due to the contaminant GAC usage rate. By increasing the adsorber size, GAC replacements can be decreased, if desired. Referring to the example calculation in the section Vapor-Phase Adsorption Characteristics of Contaminants and GAC, with a GAC usage rate of 110 lb GAC/day and a 3,000-lb adsorber, GAC replacements occur approximately every 27 days (or about once a month). If a longer time between replacements is desired, the design could select a 10,000-lb adsorber requiring replacements approximately every 90 days (or about once every 3 months). The design should weigh the frequency of GAC replacements with the amount of available facility footprint to determine the optimum design for the vapor-phase system.

Adsorber Number

Once the system configuration has been determined, the number of adsorbers required can also be determined. The number of adsorbers will first be determined by the number of flow streams in the system configuration. Then the design should consider adding redundant adsorbers under average or maximum design conditions to account for GAC replacements or other circumstances where one or more adsorbers may be out of service. The design may also add additional adsorbers in series to meet regulation requirements or to allow for operations to monitor the lead adsorber breakthrough. By monitoring the lead adsorber breakthrough, the system wastes less GAC than performing replacements based on the theoretical adsorption capacity. Adsorbers in series can be replaced when the lead adsorber experiences a certain percentage of breakthrough, allowing adsorption capacity to be maximized. The regulations in the locality of the project along with the toxicity of the contaminants aid in determining whether additional adsorbers in series are required.

To reduce the number of adsorbers required, dual-bed adsorbers can also be used. Dual-bed adsorbers are flexible in that they can handle a range of flows and contaminant concentrations, can be used simultaneously or separately, and can flow in series or parallel. Operational flexibility can help to decrease the number of adsorbers used.

Figure 10-15. Example blower (CARBTROL Corporation, Bridgeport, CT)

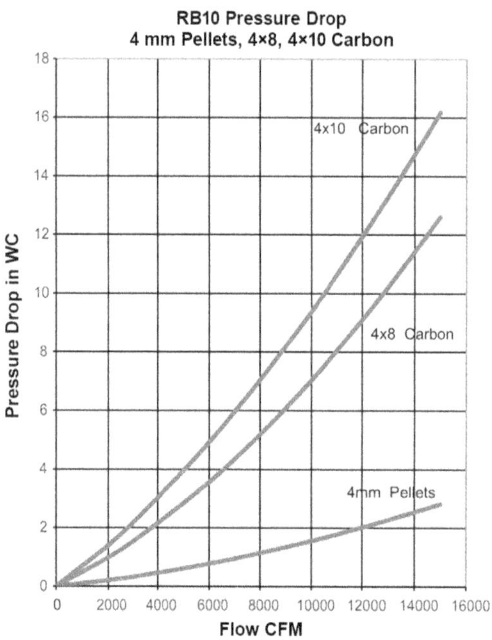

Figure 10-16. Example vapor-phase pressure drop curves (courtesy of Siemens Industry, Inc.)

Adsorber Appurtenances

Blowers, heaters, and ductwork are all required for successful performance of vapor-phase GAC systems.

Blowers. Blowers are required to force air through the GAC adsorbers. High-pressure centrifugal blowers are most often used (USACE 2001). Figure 10-15 shows an example of a blower for a vapor-phase installation, and Figure 10-16 shows an example pressure-drop curve. The manufacturer pressure-drop curves should be requested from the specific manufacturers during the conceptual design phase to verify pressure drop through the selected adsorber and media. The pressure-drop curves for the adsorber with GAC media are then used in conjunction with blower performance curves to size the blower, which should overcome the pressure drop in the piping, adsorber, and GAC media. The blower on each treatment train could also be located on the inlet of the air-stripping tower, so the tower and vapor treatment unit are both under positive pressure, or on the outlet of the vapor treatment unit, so the tower and vapor treatment unit are both under negative pressure. In these configurations, the blower would also need to overcome the pressure drop through the packed tower.

Heaters. Duct heaters, shown in Figure 10-17, are required to increase the temperature of the vapor stream and to reduce the relative humidity. The vapor stream leaving the air stripper is saturated with moisture (100 percent relative humidity). Heaters should be designed so that the relative humidity can be reduced to 40–50 percent or less. In direct heating, adsorption capacity is not affected as long as 115 °F is not exceeded. The temperature range for the duct heater can be determined by reading a psychrometric chart, shown in Figure 10-18. For example, if the ambient temperature of the vapor stream exiting the air stripper is 77 °F with 100 percent relative humidity, the 50 percent relative humidity range is reached by heating the vapor stream to 99 °F, as shown in Figure 10-18.

Figure 10-17. Gas-fired duct heater

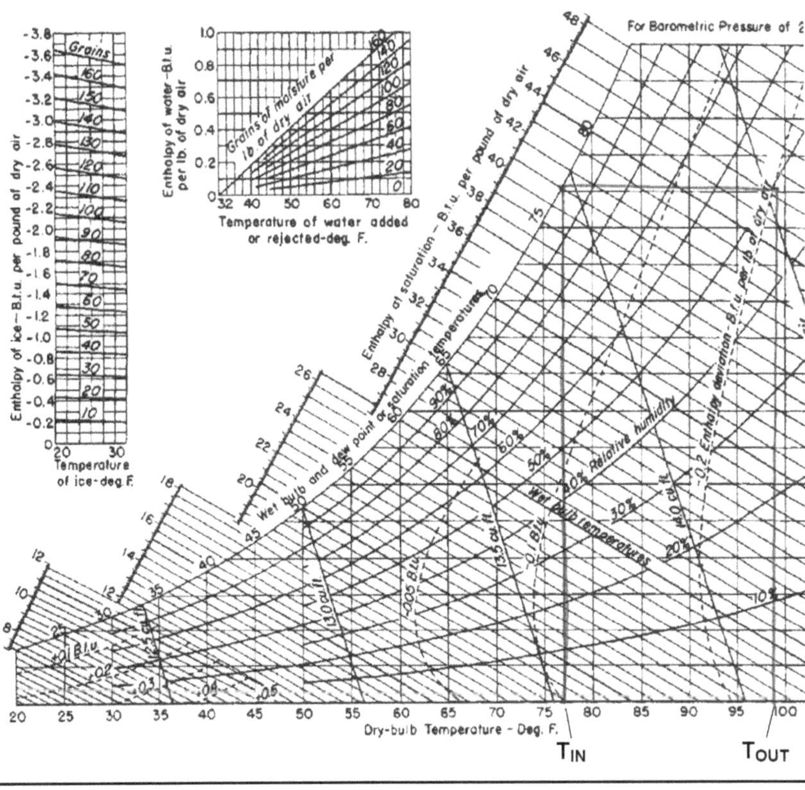

Figure 10-18. Example psychrometric chart (courtesy Ingersoll-Rand)

For installations outdoors, the off-gas may require additional heating during the winter. Duct heaters may be designed as gas-fired or electric. The type chosen depends on the specific application (how much heat is needed), availability of power sources, and a cost analysis to select the most efficient design.

Ductwork. Ductwork is required to connect the water treatment system to the vapor-phase treatment system. Ductwork materials and sizing should be designed based on the appropriate air velocities and contaminants expected. It should also be insulated to prevent loss of heat and increase in relative humidity of the vapor stream between the heater and the GAC adsorber.

Instrumentation

Vapor-phase GAC systems require instrumentation for monitoring vapor stream temperature, humidity, and VOC concentrations. Temperature probes can be located along the ductwork upstream and downstream of the heating element to adjust heating as necessary. Relative-humidity probes can also be located upstream of the GAC adsorbers to monitor the humidity entering the units. Additionally, monitoring VOC levels to aid in determining when to perform GAC replacements is necessary for vapor-phase GAC systems. VOC monitoring can be performed in a number of ways including, but not limited to, the following techniques:

- *Saturation indicator.* Indicators are located on GAC bed to determine initial breakthrough of GAC. The indicator changes color once it detects VOCs and must be checked manually (Calgon 2007).
- *Tedlar bags.* Samples can be collected manually for air exiting the adsorbers. The sample is then sent to a laboratory to be tested for air quality.
- *Summa canister.* Samples can be collected manually for air exiting the adsorbers. The sample is then sent to a laboratory to be tested for air quality.
- *Continuous online analyzer.* Continuous VOC analyzers can be installed on discharge stack to continuously monitor and report the air emissions from adsorbers. A total hydrocarbon analyzer could be installed to indicate a rise in outlet concentration of total hydrocarbons. When hydrocarbons are detected, an analysis would be completed to determine whether breakthrough is occurring for the contaminants of interest. Various types of analyzers are available depending on the contaminant. Continuous VOC monitoring is more expensive than manual monitoring and is not normally required.

GAC Handling

GAC Transfer

Vapor-phase adsorbers (prefabricated up to 20,000 lb) typically arrive at the site prefilled with GAC, so the initial GAC fill is complete prior to start-up. The subsequent GAC replacements for the permanent-type adsorbers require the following steps:

1. Isolate adsorber from vapor stream by closing influent dampers.
2. Open manway and remove GAC manually by buckets or by vacuuming.
3. Inspect the adsorber for damage.
4. Refill the adsorber with fresh GAC through the manway using GAC loaded in 1,000- to 2,000-lb bags.

5. Rake the top of the GAC down to create a flat surface.
6. Close the manway and reopen the influent dampers to place the adsorber back in service.

The subsequent GAC replacements for transportable-type adsorbers require a roll-off trailer arriving at the site to deliver an adsorber with fresh GAC. The truck removes the adsorber with the spent GAC and delivers it to the reactivation facility where the adsorber is emptied, cleaned, inspected, and refilled with fresh GAC. The transfer process for a transportable-type adsorber requires adequate access to the adsorber location for the truck to replace the spent adsorber. The use of the roll-off trailer eliminates the need for on-site loading/unloading equipment.

Transfer operations may vary slightly depending on the specific adsorber type and manufacturer. The GAC transfer process should be discussed with the adsorber manufacturer during design to ensure that adequate equipment and footprint are available and appropriate safety measures for confined-space entry and work in a dry GAC environment are followed. For more information about GAC safety, refer to chapter 8.

GAC Storage

On-site GAC storage can be provided for permanent-type adsorbers in the form of silos, 1,000- or 2,000-lb bags, or redundant adsorbers. Vapor-phase GAC treatment systems are typically small enough that one facility would not warrant the additional cost of extensive on-site storage, but it could be considered to weigh the benefits versus the costs:

- Including a silo(s) would not require much space on-site, but installing the silos and related equipment, such as hoppers, could be a significant cost that may outweigh the benefit.
- The 1,000- or 2,000-lb bags require additional storage space that may or may not be available given site layout.
- Maintaining an off-line, redundant adsorber would provide a means of storing GAC and allowing the facility to operate during GAC transfers.

Each storage option can be designed to fit the specific needs of the application.

Facility Layout

Now that all the components of the GAC facility have been selected, sized, and designed appropriately, the facility layout can be determined and integrated into the existing plant site. The logistics of GAC replacements should be considered in laying out the final facility. The adsorbers should be configured in such a fashion to allow for ease and speed of GAC replacements. The importance of replacement efficiency depends upon the GAC replacement frequency. If replacements are expected to occur often, then the importance increases. For facilities with permanent-type adsorbers, the layout should allow for 1,000- or 2,000-lb GAC bags to be delivered, staged, and manually transferred into the adsorbers. For the transportable-type adsorber facilities, the layout should allow adequate access for the delivery trucks to replace the spent GAC adsorbers. Vehicle access should account for the sizes and turning radii of vehicles expected to service the site as well as the expected loading of the trucks on site pavement.

Operation and Maintenance

GAC Replacements

Vapor-phase GAC treatment unit system configurations can be optimized to allow for more flexibility during GAC replacements as follows:

- Incorporate redundant train(s)
- Rotate adsorbers in series
- Perform simultaneous shutdowns

Redundant treatment trains can be incorporated into the design so that continuous treatment of the vapor stream is provided. For example, two treatment trains are designed to treat the required flow rate. A third treatment train would be added so that flow could be diverted to the redundant train when GAC replacements or maintenance occur. Interconnects between the three treatment trains allow for seamless transition of GAC replacements and continuous treatment.

For systems with treatment units in series, the piping and valve arrangements can be designed to allow units to be operated in any order so that units that are not being replaced can move up in the treatment series order without being physically moved. The GAC in the exhausted treatment units can be replaced one at a time or once all treatment units are exhausted. A conceptual schematic is shown in Figure 10-19.

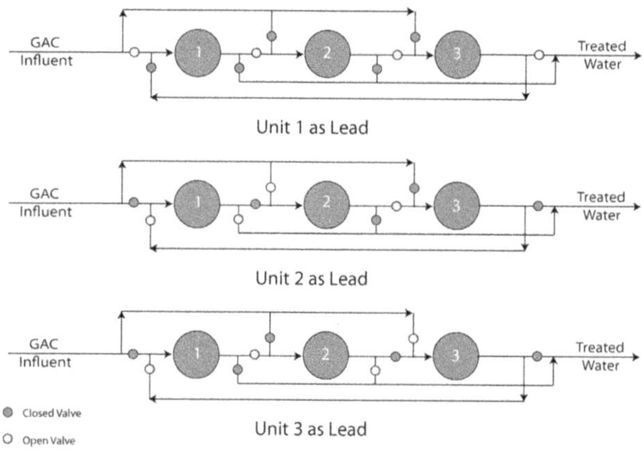

Figure 10-19. Series operation of treatment units with varying lead treatment unit

Another way to optimize GAC replacements is to perform simultaneous shutdowns for replacements. All parallel treatment trains can be shut down simultaneously once GAC replacement is needed to economize on GAC replacement costs.

References

AWWA and ASCE. 2012. *Water treatment plant design.* 5th ed. S.J. Randtke and M.B. Horsley, eds. New York: McGraw-Hill Professional.

Calgon Carbon Corporation. 2007. Carbon saturation indicator, information brochure. Pittsburgh, Pa.

Shepherd, A. 2001. Activated carbon adsorption for treatment of VOC emissions. EnviroExpo, Boston, Mass.

U.S. Army Corps of Engineers (USACE). 2001. Adsorption design guide no. 1110-1-2.

Chapter 11

On-Site GAC Reactivation Design

When a GAC adsorption process is used to treat water, organic compounds in the water adsorb to sites on the GAC. Over time, more and more adsorption sites become occupied by the organic compounds, and the adsorptive capacity of the GAC is reduced. Eventually, the GAC can no longer produce the desired treated–water quality; at this point, the GAC is described as "spent" (AWWA B605). To restore the adsorption capacity of the GAC treatment unit(s), the spent GAC can either be replaced with virgin GAC, or the adsorptive capacity of the spent GAC can be restored through reactivation.

The term *regeneration* refers to removal of contaminants from GAC without destroying them, and can be accomplished by exposing the GAC to steam or hot gas (USACE 2001). *Reactivation* refers to the process of removing the adsorbed compounds from GAC and destroying them at high temperatures (i.e., greater than 1,500°F), such as in a rotary kiln or multiple-hearth furnace (USACE 2001). In referring to GAC use in water treatment, reactivation is used, although the terms *regeneration* and *reactivation* are often used interchangeably. Because much of the adsorbed organic material is nondisplaceable, regeneration is usually ineffective at restoring the adsorption capacity of the GAC.

Reactivation of spent GAC may be accomplished by thermal, chemical, hot gas, solvent, or biological methods. Chemical and biological regeneration are discussed briefly, but this chapter focuses primarily on thermal regeneration, which is the technique most commonly used for large-scale regeneration.

Reactivation of GAC can be performed in an on-site facility or it can be part of a turnkey service provided by a GAC manufacturer. The capital cost of constructing a reactivation facility will not be practical for most small GAC installations such as wellhead treatment systems. On-site reactivation becomes more cost-effective for large water treatment plants:

- Records from the 220-mgd Richard G. Miller WTP in Cincinnati, Ohio (Case Study No. 1) indicate the cost of on-site reactivation of GAC is approximately 33 percent of the cost of virgin GAC (Moore et al. 2001).
- The cost for reactivating GAC off-site is approximately 60 to 70 percent of the cost of virgin GAC (Scott 2009).

Some WTPs contract with GAC manufacturers to have their spent GAC removed and either reactivated or landfilled. When GAC is reactivated off-site, the reactivated GAC can be returned to the original WTP or sent to an industrial facility (it is not reused at a different WTP). GAC that is reactivated off-site may be certified by the manufacturer with a Certificate of Destruction, which relieves the WTP from liability for the contaminants that were adsorbed to the GAC. The relief of liability is an advantage of reactivation over landfilling, where the WTP retains liability for the GAC indefinitely.

For on-site reactivation, the WTP retains responsibility for the contaminants adsorbed to the GAC through the reactivation process, including through any air-treatment processes downstream of the reactivation furnaces.

Not many utility-managed reactivation facilities exist. The majority of the information presented in this chapter not explicitly cited was developed for a few utilities/facilities:

- The 220-mgd Richard G. Miller WTP in Cincinnati, Ohio (in addition to numerous published studies of the facility's operation, preceding conceptual and preliminary design studies were done).
- 80-mgd Lake Pleasant WTP in Phoenix, Ariz.
- A centralized reactivation facility for several City of Phoenix water treatment plants.
- Metropolitan Water District of Southern California.
- City of Columbus, Ohio.

Reactivation Methods

Thermal Reactivation

Thermal reactivation is the most common method of GAC reactivation. Thermal reactivation is typically carried out in five steps (Figure 11-1) (Knappe et al. 1992, DeMarco et al. 1988):

1. *Dewatering.* GAC slurry is dewatered to 40 to 50 percent moisture by draining. It is more efficient to dewater the GAC by draining than to evaporate the extra water in the furnace.
2. *Drying.* The GAC enters the low-temperature (300°F to 700°F) section of the furnace for removal of additional water.
3. *Baking or pyrolysis (charring).* Some adsorbed compounds are volatilized and removed from the GAC at medium temperatures (930°F to 1,600°F). Other organic adsorbates within the GAC structure are converted to char.
4. *Reactivating.* The char is removed from the GAC via incineration at high temperatures (1,300°F to 1,800°F).

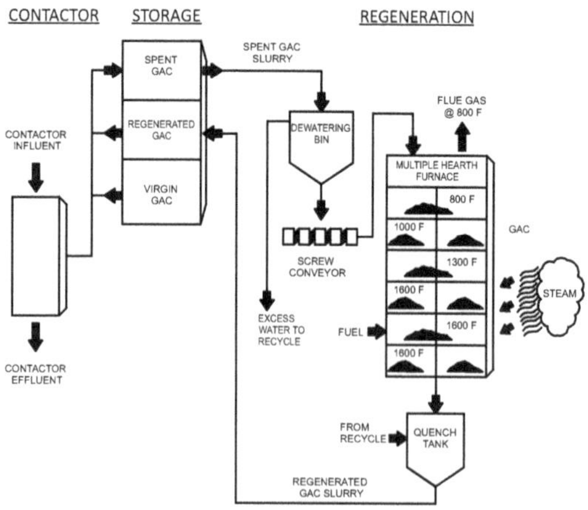

Figure 11-1. GAC thermal reactivation

5. *Quenching.* The reactivated GAC is quenched (cooled), which can be accomplished by placing it in a water bath.

The typical retention time for a GAC reactivation furnace is ½ to 2 hr. The GAC spends approximately one-third of the retention time in the hot zone, where the char is converted to a gas (Step 4, Reactivating, above). Regeneration of one pound of spent GAC typically requires 0.5 to 2.0 lb of steam, 0.13 to 0.33 lb of natural gas, and 1.2 lb of oxygen from air (DeMarco et al. 1988, Adams et al. 1988, Eberle 2009).

The thermal reactivation process creates an exhaust gas that is a mixture of organic and inorganic oxidation by-products and particulates resulting from the combination of adsorbate, GAC, and reactivation system fuel. Air-pollution control systems such as high-temperature afterburners, scrubbers, and other techniques are usually required to satisfy air-quality regulations.

Four types of furnaces can be used for reactivation of GAC:

- Electric infrared oven
- Rotary kiln
- Fluidized bed furnace
- Multiple hearth furnace

A brief overview of each type of furnace is provided in the following paragraphs. Multiple-hearth furnaces and rotary kilns are the most common methods of regenerating spent GAC (USEPA 2000).

Electric Infrared Oven

Infrared light is located between visible light and radio waves on the electromagnetic spectrum, with wavelengths between 0.8 and 10 microns (USDOE 2007). Infrared energy can be transmitted and absorbed and can be used in many manufacturing applications for heating, drying, thermal bonding, and sterilizing.

Spent GAC passes through the electric infrared oven on a conveyor belt and is exposed to the infrared energy (Figure 11-2). Electric infrared systems typically include an emitter, a reflector, and controls (USDOE 2007). An electric infrared system used for reactivation of GAC would also require a materials handling system and a ventilation system.

Electric infrared ovens are best used for fairly flat products because reactivation is effective only at the surface of the product but does not penetrate deeply into the

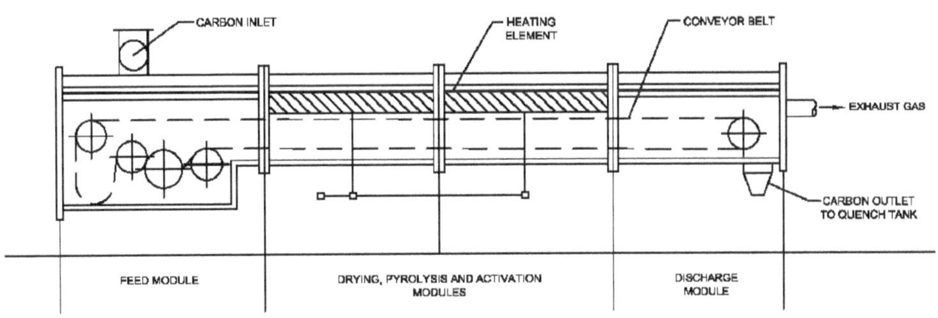

Figure 11-2. Infrared oven schematic (Inhoffer and McGinnis 1980, referenced in Sontheimer et al. 1988)

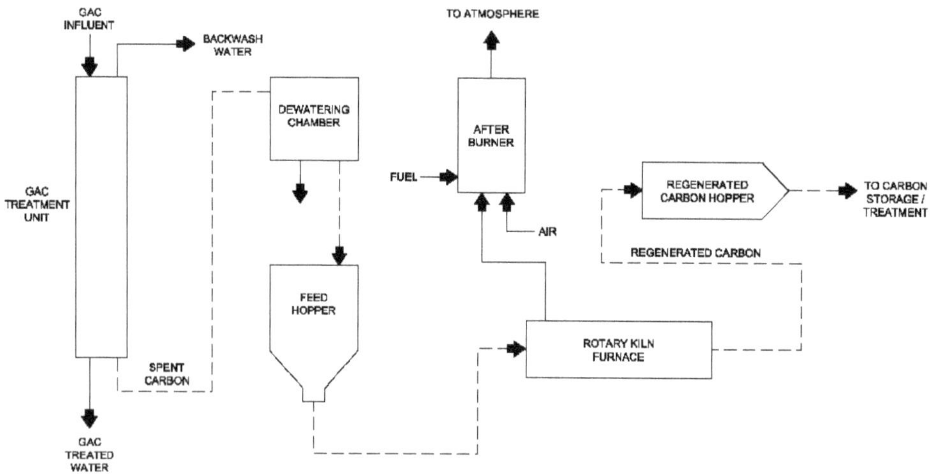

Figure 11-3. Rotary kiln schematic

product (USDOE 2007). Therefore, the depth of GAC on the belt should be approximately 2 cm (Sontheimer et al. 1988).

Rotary Kiln

A rotary kiln consists of a tilted, rotating cylinder through which spent GAC and hot gases flow (Figure 11-3). The spent GAC and hot gases can flow in the same direction or countercurrent to each other. The kiln can be designed either with only hot gases flowing through the kiln or with an additional flame located at one end of the kiln. These types of rotary kilns are called direct fired because the heat source is in direct contact with the media being treated.

Figure 11-4. Rotary kiln at Lake Pleasant Water Treatment Plant

Exhaust gases from the kiln are captured and treated prior to release into the atmosphere. In cases where the adsorbates on the GAC are considered hazardous, the use of an indirect-fired rotary kiln can reduce the volume of exhaust gas that must be treated. With indirect-fired rotary kilns, the heat sources (i.e., tangential burners) remain outside the kiln, and heat is transferred through the wall of the kiln to the material inside. The heating gases do not combine with the adsorbate-laden exhaust gas, thus reducing the volume of air that must be treated to remove the hazardous compounds. Zones can be created within the kiln by strategically arranging the burners on the outside of the kiln. Figure 11-4 shows a rotary kiln at the Lake Pleasant WTP in Phoenix, Ariz.

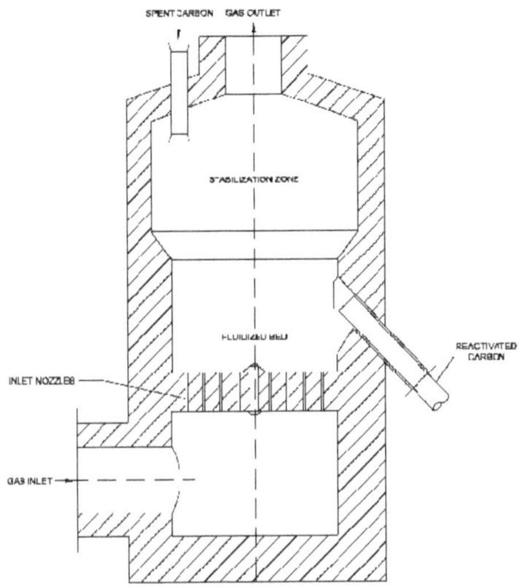

Figure 11-5. Example of fluidized bed furnace (Strack 1976, referenced in Sontheimer et al. 1988)

Figure 11-6. Fluidized bed furnace (photo by Metropolitan Council, Minneapolis, MN)

Fluidized Bed

The fluidized-bed furnace is a vertical process in which the GAC progresses downward through the furnace in the opposite direction of rising hot gas, which carries off volatiles as it dries the spent GAC and pyrolyzes the adsorbates (Miller and Hartman 1982). A schematic of a fluidized-bed furnace is shown in Figure 11-5. The airflow is sufficient to keep the bed of GAC fluidized through the furnace. Burners are placed along the height of the furnace to adjust the temperature in different zones. Water connections can also be provided for addition of steam. Figure 11-6 shows a fluidized-bed furnace installed to incinerate solids at a WWTP in St. Paul, Minn.

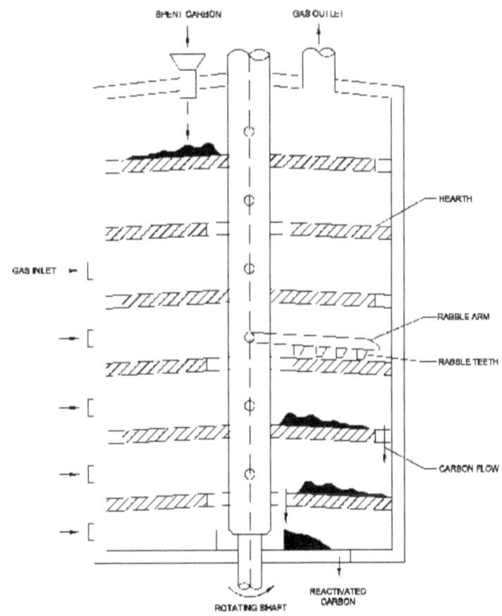

Figure 11-7. Multiple hearth furnace (von Dreusche 1978, referenced in Sontheimer et al. 1988)

Multiple Hearth

In a multiple-hearth furnace, dewatered spent GAC is introduced into the top of the furnace through inlet ports (Figure 11-7). Radial arms with rabbling teeth move the GAC downward through a number of hearths; the GAC drops through holes in the hearths from level to level. Figure 11-8 shows a multiple-hearth furnace.

Hearths can be equipped with automatic natural-gas–fired burners that are controlled as a function of hearth temperatures. Steam injectors can also be provided.

Figure 11-8. Example of a multiple-hearth furnace

Ultraviolet scanners can be provided to shut down individual burners in case of poor combustion or flame failure. Other safety features that shut down the furnace or alert the operator's attention may include:

- High gas pressure
- Low gas pressure
- High combustion-air pressure
- Low scrubber-water pressure
- High stack-gas temperature
- Draft fan not operating
- Shaft cooling air fan not operating

An example of staged reactivation of GAC using a multiple-hearth furnace is as follows:

- Hearth No. 1: GAC drying and volatilization of organic compounds at 800°F
- Hearth No. 2: Temperature increase to 1,000°F to pyrolyze adsorbates
- Hearth No. 3: Temperature increase to 1,300°F to pyrolyze adsorbates
- Hearth No. 4: 1,600°F
- Hearth No. 5: 1,600°F
- Hearth No. 6: 1,600°F

In Hearths 4, 5, and 6, steam is added to reactivate the GAC by removing the char. The heat vaporizes and drives off the impurities that are adsorbed on the GAC, and the injection of steam provides improved control of the reactivation portion of the process and aids in controlling the apparent density and the iodine number of the reactivated GAC.

Rotary Kiln vs. Multiple-Hearth Furnaces

Rotary kiln and multiple-hearth furnaces are the most commonly used methods of thermal regeneration of GAC. Each has its own advantages and disadvantages.

Rotary kilns generally have a lower capital cost than multiple-hearth furnaces. Rotary kilns also have a shorter shutdown and startup cycle, which can usually be completed within one day, compared to a week for multiple-hearth furnaces. As a result, rotary kiln systems are more nimble in response to operational variations.

Multiple-hearth furnaces contain more internally heated moving parts and kilned bricks that can crack via thermal fatigue, which may require more preventive maintenance. However, multiple-hearth furnaces allow more precise control of the reactivation process, which can result in a more consistent quality of the reactivated GAC.

Chemical Regeneration

Chemical regeneration with sodium hydroxide (caustic soda), acids, or solvents might be more cost-effective than thermal regeneration for some facilities (Norit Americas 2001). Chemical regeneration can be accomplished without removing the spent GAC from the contactor, which can reduce the media losses that occur with thermal regeneration. When chemical regeneration is used, the chemical flow is typically in the opposite direction of normal forward water flow through the GAC bed. Chemical regeneration is most commonly used for industrial wastewater applications with relatively high concentrations of a specific compound or where it is possible to recover the chemical used for regeneration (Sontheimer et al. 1988).

Biological Regeneration

Biologically active GAC may partially "regenerate" itself because biodegradation consumes organic matter that is adsorbed on the GAC, thus creating a greater capacity for organic matter than otherwise achieved strictly through adsorption. Biological regeneration has the potential to remove previously sorbed organic chemicals and increase the GAC service life. Speitel and DiGiano (1987) achieved significant bioregeneration when using biologically active GAC to remove phenol and paranitrophenol (PNP) from drinking water. Biodegradation of the adsorbates ranged from 8 to 22 percent. Bioregeneration rates may be affected by the concentration of adsorbates in the outer portion of the GAC particle where adsorption sites are readily available to microorganisms. Bioregeneration rates for adsorbates further inside the GAC particle are slower but could still extend the service life of the GAC (Speitel and DiGiano 1987).

Bioregeneration may be a constant process that takes place while the filter is in operation, or the filter may be shut down for regeneration and a water supply that is seeded with the desired microorganisms can be backflushed through the filter (Sontheimer et al. 1988). Continuous bioregeneration is most applicable to wastewater streams with readily biodegradable compounds that are adsorbed onto the GAC. Refer to chapter 8 for more information on biologically enhanced activated filters.

Indicators of Reactivation Efficiency

GAC can be reactivated multiple times, although it can also be damaged during the process due to rough handling or overreactivation, in which a portion of the GAC itself is charred and gasified. During reactivation, furnace conditions should be carefully controlled to achieve effective removal of the adsorbate without negatively affecting the adsorptive capacity of the GAC (Lambert et al. 2002). Reactivated GAC should be as similar to the virgin GAC as possible in terms of particle size distribution, effective size, and uniformity coefficient to ensure consistent filtering performance (AWWA B605).

Spent GAC is usually reactivated to a predetermined apparent density. If the apparent density is less than the predetermined value, the furnace temperature is likely too high or the feed rate is too low, which results in partial incineration of the GAC.

The adsorptive capacity of the reactivated GAC should also match the virgin GAC and can be evaluated using the iodine number. Progressive decreases in the iodine numbers of reactivated GAC indicate that the reactivation temperature is too low and that adsorbates are being carbonized and left in the original pore structure. The temperature of the GAC itself during reactivation should be at least 1,500°F to avoid deterioration, and the corrective action required is to raise furnace temperatures. Temperatures as high as 1,750°F to 1,800°F may be required to obtain full restoration of the iodine number. However, in an efficient reactivation system, the iodine number may decline by 10 to 20 mg/g through each reactivation cycle.

Effects of Inorganic Compounds on GAC Reactivation

Although GAC treatment units are designed to remove organic carbon from water, some inorganic compounds will also adsorb on the GAC. The types of inorganic compounds that accumulate on the GAC depend on the characteristics of the source water, but calcium is usually the most commonly found adsorbed metal (Lambert et al. 2002). Metals such as aluminum, iron, manganese, and magnesium may also accumulate on the GAC.

Calcium that adsorbs inside the GAC pores can catalyze the reactions of the oxidizing agents (steam and CO_2) with the GAC, leading to overreactivation of the GAC (Knappe et al. 1992). Overreactivation occurs when some of the GAC itself is pyrolyzed and gasified, and it results in mass loss and widening of the internal pore structure. Accumulated metals may also inhibit the adsorption of certain organic contaminants. The presence of metal ions on a GAC surface would presumably increase the polarity (charge) of that surface, and it has been shown that the adsorption of low-molecular-weight hydrophobic compounds decreases with increasing GAC surface polarity (Karanfil and Kilduff 1999).

Acid-washing the GAC prior to reactivation can remove accumulated metals and improve the reactivation process by reducing the rate of GAC gasification (Lambert et al. 2002). An acid-wash system could be used periodically to remove inorganic compounds from the GAC and would consist of an acid-wash vessel and a chemical feed and storage system. The GAC would be placed into the acid-wash vessel and allowed to soak for a few hours. After the GAC has soaked and been rinsed several times with water, it is sent to the reactivation furnace. One cycle of acid-washing may take up to 12 hr. The acid-bath solution needs to be neutralized and disposed of after the acid-washing operation. The spent GAC for a given site should be tested

Table 11-1. Recommended testing for reactivated GAC

Test	Used to Determine
Sieves	Particle size distribution, effective size, and uniformity coefficient
pH	pH of reactivated GAC slurry – can indicate whether inorganic compounds are adsorbed on the GAC
Apparent density	Correct rate of reaction during reactivation – if apparent density of reactivated GAC is low, some incineration of GAC is occurring
Abrasion number	Sensitivity of reactivated GAC to damage during transport between furnace and contactors
Iodine number	Relative mass of organic chemical that will adsorb to the reactivated GAC
Pore volume or surface area	Change in GAC pore structure over multiple cycles of reactivation

to determine the extent of calcium adsorption and whether acid-washing is necessary prior to design and installation of such a system.

Cannon et al. (1993) observed calcium-catalyzed reactivation of field-spent GAC but concluded that the presence of calcium did not justify the use of acid-washing prior to reactivation. The authors found that it was possible to reactivate two GAC samples (with and without acid-washing prior to reactivation) to near the characteristics of the virgin GAC and that acid-washing provided a slightly superior reactivation. Both GAC samples also experienced changes in their pore structure, with decreased micropore volumes, decreased surface area, and increased mesopore volumes. The authors concluded that the reactivation benefits did not outweigh the additional expense, labor, and safety issues associated with acid washing.

Reactivated GAC Quality Monitoring

The reactivated GAC should be tested to ensure that it meets the design parameters. The testing capabilities shown in Table 11-1 are recommended for a GAC reactivation facility.

Air Quality Considerations

A by-product of thermal reactivation is exhaust gas that may require treatment prior to release to the atmosphere. Air quality considerations for reactivation furnaces include criteria pollutants and air toxics, both of which are regulated to preserve ambient air quality.

Criteria Pollutants

The following criteria pollutants are regulated under the National Ambient Air Quality Standards (NAAQS) set by the 1970 Federal Clean Air Act and amendments:

- Ozone
- Particulate matter (PM10 and PM2.5)
- Carbon monoxide
- Nitrogen oxides
- Sulfur dioxide
- Lead

Table 11-2. Criteria pollutants found in reactivation furnace exhaust

Furnace Exhaust Component	Criteria Pollutant
Combusted GAC	CO,* particulates
Volatilized Adsorbate	Hydrocarbons
Combusted Natural Gas	Hydrocarbons, CO, particulates, NO_x,† SO_2‡

* CO = Carbon monoxide

† NO_x = A general category of compounds containing only nitrogen and oxygen

‡ SO_2 = Sulfur dioxide

Emissions associated with reactivation furnace exhaust gases are composed of GAC fluidized in the exhaust gas, adsorbates driven off the GAC, and combustion products of the natural gas or other fuel used to heat the furnaces. These three components generate the criteria pollutants shown in Table 11-2. The quantities of criteria pollutants can be estimated as indicated below.

Combusted GAC. Combustion of GAC in the furnace creates emissions of carbon monoxide (CO) and particulates. GAC attrition in the furnace can generally be assumed to be approximately 5 to 15 percent of the GAC feed rate (Waer et al. 1992). The emission factors provided in the USEPA's AP 42 for external bituminous coal combustion can be used to approximate CO and particulate emissions from the combusted GAC (USEPA 1995). Particulate emission factors are provided for total particulates, PM10 (particulates with diameter equal to or less than 10 μm), and PM2.5 (particulates with diameter equal to or smaller than 2.5 μm); emissions can be estimated from the particle size distributions provided in AP 42. PM10 can be assumed to constitute 50 percent of total suspended particulates by weight in the furnace exhaust (SBCAPCD 1991a).

Volatilized Adsorbates. GAC reactivation volatilizes adsorbates from GAC. A mass balance can be used to approximate the TOC loading on the spent GAC. Data on the composition of the adsorbate (i.e., percent TOC by weight) can provide a correlation between TOC and adsorbate loading. The TOC loading is then divided by the assumed weight fraction of TOC in the adsorbate to determine the adsorbate loading.

Combusted Fuel. GAC reactivation also produces combustion by-products of the natural gas or other fuel used to heat the furnace. A combustion analysis can be used to estimate the unit fuel consumption required per pound of GAC. Emission factors for combustion of the fuel can be used to convert fuel consumption to pollutant emission rates (USEPA 1995).

Emissions Evaluation

The results of one emissions evaluation are summarized in Table 11-3. Note how the contributions to emissions vary for the combusted GAC, volatilized adsorbate, and combusted natural gas.

Table 11-3 shows that hydrocarbons are generated primarily by volatilized adsorbate, while particulates are generated primarily by the combusted GAC. Combusted natural gas is the primary contributor to NO_x and SO_2 emissions. CO is generated almost equally by both combusted GAC and natural gas. The assumptions made to estimate these emission precursors (i.e., combusted GAC, volatilized adsorbate, and combusted natural gas) strongly affect criteria pollutant emission estimates.

Table 11-3. Emission contributions for drinking water facility with TTHM goal of 10 µg/L

Emission Constituent	Hydrocarbons	CO	Particulates	NO_x	SO_2
Combusted GAC	0%	47%	97%	0%	0%
Volatilized Adsorbate	99.9%	0%	0%	0%	0%
Combusted Natural Gas	0.1%	53%	3%	100%	100%

Air Toxics. Air toxics, or hazardous air pollutants (HAPs), are compounds that are known to cause cancer or other serious health effects, such as reproductive effects or birth defects, or cause adverse environmental effects (USEPA 2009). HAPs are regulated under the Clean Air Act and its amendments. GAC reactivation facility emissions may contain the following HAPs:

- Hydrochloric acid
- Ammonia
- Chlorinated organics (dioxins/furans)

Hydrochloric acid is produced in the afterburner by the combustion of chlorinated organics, a constituent of the adsorbate. Ammonia emissions result from the application of injections of ammonia into the exhaust gas to convert NO_x to N_2 via oxidation/reduction. Wet scrubbers achieve a high degree of control of emissions for both hydrochloric acid and ammonia.

Dioxins and furans may be formed from regeneration of municipal GAC during the combustion of chlorinated organics in the furnace. A previous evaluation of dioxin and furan formation found the incremental cancer risk associated with dioxin/furan emissions to be insignificant, especially when an afterburner was used to treat the reactivation system emissions (DeMarco et al. 1988). Elimination of prechlorination practices upstream minimizes total organic halides (TOX) formation prior to GAC adsorption and reduces the chlorine loading on the adsorbate, which in turn reduces the potential for the formation of dioxins and furans during GAC reactivation.

Addressing Public Perception

In addition to the regulatory issues affecting the implementation of GAC reactivation facilities, nontechnical issues also strongly affect project feasibility. The foremost nontechnical issue is public acceptance. Many communities are resistant to having an industrial facility, such as a GAC reactivation facility, constructed nearby, even though the project might otherwise be technically feasible and effective.

One proposed GAC reactivation facility in Flagstaff, Ariz., faced such strong public opposition to its construction that Flagstaff was abandoned in favor of a location in Utah. However, public opposition in Utah also prevented the facility from being sited there. The facility was eventually constructed in 1992 in Parker, Ariz.

Public acceptance is an important issue for on-site and off-site GAC reactivation facilities. It is advantageous to address public concerns early and continually through the implementation of a project that may be perceived as a health risk to the community. The following are focus points for generating a constructive dialogue with the public (Shovlin and Tanaka 1990):

- Public participation is the key to community acceptance of projects and positions. In order for people to support a project or position on an issue, they must be satisfied with the process by which the decision is made.
- The community should be involved early in the planning stages of the project.
- Possibly controversial information should be distributed quickly.
- As much as possible, requested information should be provided in a form that is easily digested.
- Subjectivity in risk judgments is not unique to nontechnical people; utility scientists and engineers should recognize that their risk judgments also contain subjectivity.
- Risk judgments are often based upon factors other than the degree of hazard, such as whether the risk is voluntary or coerced, natural or unnatural, morally relevant, culturally symbolic, etc.
- Broad-based groups should be involved in communication efforts whenever possible.
- Ongoing relationships with elected officials and the news media should be strengthened and maintained.
- Forging a strong cooperative bond between technical and public affairs staff is important.

The preceding list identifies the minimum efforts that a project proponent should undertake if a potentially controversial project such as construction of a GAC reactivation facility is to be sited successfully. However, no one particular prescribed process can ensure public acceptance.

Design of On-Site Reactivation Systems

The design of an on-site reactivation system is site-specific and should be developed for the individual facility's reactivation requirements. Locating a reactivation facility involves estimating air emissions from the facility, evaluating the current air quality in the proposed location, and developing a good relationship of communication with the surrounding neighbors. Design considerations that apply to most reactivation facilities are included in this section.

Facility Layout

Once the location of the GAC reactivation facility has been selected, the layout of the facility can be developed. An on-site reactivation facility should be located as close to the GAC treatment units as possible to minimize pipe runs. Proximity is beneficial for several reasons. First, minimizing pipe runs reduces the construction cost of the facility. Second, the chances of GAC accumulation or clogs occurring in the piping are reduced. Third, the energy needed to move the GAC, whether by water pressure or by slurry pump, is reduced for shorter pipe runs: For instance, at the Richard G. Miller WTP in Cincinnati, Ohio (Case Study No. 1), the GAC reactivation facility was installed in the middle of the building, central to all of the GAC treatment units. Fourth, shorter transfer distances mean less abrasion will occur during GAC transfer.

Another design issue is the height of the facility. Multiple-hearth furnaces can be quite tall. Zoning height restrictions should be reviewed during design to ensure that the buildings or structures housing the furnaces are of an acceptable height.

Vehicle access and truck loading/unloading stations should be provided for delivery of virgin (make-up) GAC to allow the GAC to be transferred from the delivery truck into the make-up storage facility, from where it is educted or pumped as a slurry to the GAC contactors.

For an on-site reactivation facility, spent GAC and reactivated GAC will likely be transported via slurry piping. However, if the reactivation facility is located remotely from the GAC treatment units similar to the Lake Pleasant WTP in Phoenix, Ariz. (Case Study No. 8), it might be more economical to transfer the spent and reactivated GAC using trucks instead of eductors. Vehicle access and transfer stations may be required for transfer of spent and reactivated GAC where remote reactivation facilities are used. Refer to chapter 8 for additional information on GAC transfer.

Process Design Criteria

The process criteria that have the greatest effect on GAC reactivation facility costs are:

- Furnace capacity
- Furnace loading rate (i.e., pounds of GAC per square foot of hearth area per day)
- Fuel consumption rate (i.e., Btu per pound of GAC)

Furnace loading rates vary widely. For example, loading rates for multiple-hearth furnaces can range from 20 to 100 lb of GAC per square foot of hearth area per day. The range reflects the diverse nature of the adsorbates removed at different installations. In general, spent GAC heavily loaded with industrial and wastewater adsorbates will require longer regeneration times and, in turn, lower furnace-loading rates. In comparison, spent GAC from a WTP will be relatively lightly loaded with NOM. The experience of large-scale reactivation operators indicates that NOM can be effectively removed from GAC at furnace loading rates of about 100 lb/ft^2/day (Zanitsch 1991). The furnace capacity, in pounds of GAC reactivated per day, can be calculated from the furnace-loading rate for a furnace with a given effective hearth area.

Fuel Consumption

Fuel consumption constitutes a significant portion of the operations and maintenance costs at a GAC reactivation facility. Reactivation energy requirements range from 2,500 to 10,000 Btu/lb (Eberle 2009). Multiple-hearth furnaces typically require 2,500 to 5,000 Btu/lb GAC. Furnaces are usually fired with natural gas, although some furnaces, such as infrared ovens, use electricity.

Waste heat from the furnace exhaust gas may be recovered to generate process steam required for GAC reactivation, reducing the need for additional fuel to fire a boiler. Further, additional recoverable heat is typically available following steam production and may provide an opportunity for cogeneration of power.

Number and Capacity of Furnaces

The furnaces should be sized to meet the maximum annual spent-GAC loading. Where feasible, provide enough furnaces to meet the design GAC-feed rate with the largest furnace out of service for maintenance. One or more furnaces can be taken out of service when loadings are lower or during seasonal periods when GAC reactivation is not required. A spare furnace provides a built-in preventive maintenance period for servicing the furnaces.

The Richard G. Miller WTP in Cincinnati, Ohio (Case Study No. 1), developed a furnace operation strategy that involved maximizing furnace use during the summer, when the plant typically saw high influent-TOC loadings and warm water temperatures, and minimizing reactivation in the winter, when low influent-TOC loadings resulted in low TTHM formation in the distribution system (Shrive et al. 1998). As a result, treated water TOC concentrations decreased during the summer and slightly increased in the winter, while still maintaining acceptable TTHM levels in the distribution system. The operational strategy allowed the plant to reduce TTHM formation during the summer months and also resulted in a cost savings because the furnaces were used less during the winter, when fuel costs were high.

Example 11-1. Furnace Sizing Calculations.

Assume a facility has six GAC treatment units, each containing 7,000 ft³ of GAC. The facility anticipates an average of one GAC replacement per treatment unit per year but could require up to two replacements per year. What size reactivation facility is required?

Solution. The total volume of GAC on-site is:

$$6 \text{ contractors} \times \frac{7{,}000 \text{ ft}^3}{\text{treatment unit}} = 42{,}000 \text{ ft}^3 \text{ GAC}$$

If the GAC is 12×40 mesh with an apparent density of approximately 37 lb/ft³, the total mass of GAC is:

$$42{,}000 \text{ ft}^3 \text{ GAC} \times \frac{37 \text{ lb}}{\text{ft}^3} = 1{,}554{,}000 \text{ lb GAC}$$

With one GAC replacement per contactor per year, that is the average amount of GAC that needs to be reactivated in one year. For maximum conditions (two replacements per year), the facility should be sized for 2 × 1,554,000 lb = 3,108,000 lb GAC per year.

Assume the entire reactivation facility is out of service for two months per year for maintenance. The furnaces then need to treat an average of:

$$\frac{3{,}108{,}000 \text{ lb GAC}}{\text{year}} \times \frac{1 \text{ year}}{10 \text{ months}} \times \frac{1 \text{ month}}{30.4 \text{ days}} = 10{,}224 \text{ lb GAC/day}$$

Assume a surface loading rate of 100 lb GAC/ft²/day for a multiple-hearth furnace. The furnace area required to reactivate the GAC is then:

$$\frac{10{,}224 \text{ lb GAC}}{\text{day}} \times \frac{\text{ft}^2 \times \text{day}}{100 \text{ lb GAC}} = 102 \text{ ft}^2$$

Assume three furnaces: two will operate normally, and one will be redundant. Each will have half of the area required to meet the design surface loading rate of 100 lb GAC/ft²/day, or 50 ft² per furnace. Furnaces with a diameter of 8 ft would be sufficient.

The energy usage for the furnaces can be estimated by assuming 5,000 Btu/lb GAC:

$$\frac{10{,}224 \text{ lb GAC}}{\text{day}} \times \frac{5{,}000 \text{ Btu}}{\text{lb GAC}} = 5.1 \text{ million Btu/day}$$

Steam Boiler

Steam use during the reactivation process requires a dedicated steam boiler. The purpose of the boiler is to heat water until it boils; the resulting steam is then withdrawn from the boiler and used for heating or other applications. Steam boilers can be designed to operate using one of several different energy supplies, including natural gas, propane, wood, coal, or electricity. In addition, heat recovered from the reactivation furnace is an excellent source of heat for the boiler.

The boiler should be sized to supply the amount of steam required, (0.5 to 2.0 lb for each pound of GAC reactivated). Where possible, provide a redundant unit so that boilers can be taken out of service for maintenance.

Boiler design standards are set by the ASME Boiler Code, which defines materials of construction and required pressure ratings. Appropriate materials for steam piping and valves should be selected. Steel piping is commonly used.

Scale buildup in the boiler is an important design consideration. As steam is removed from the boiler, compounds such as calcium or magnesium in the water remaining in the boiler become more concentrated. *Blowdown* refers to water that is discharged from the system to prevent excessive scaling on the surfaces within the boiler that are in contact with the water. The amount and frequency of blowdown for a given boiler depends on the source water characteristics. Blowdown needs should be evaluated during design to estimate make-up water requirements and to include blowdown flows when sizing the site sewers. Typically, a softener is used to reduce the hardness of make-up water. Scale-inhibiting chemicals can also be used to reduce scaling in the boiler, although inhibitors will not eliminate the need for blowdown.

Air Quality Controls

Air pollution controls (Table 11-4) are likely to be required to treat the furnace exhaust gases.

A waste heat recovery boiler can also be included in the air pollution control train to generate process steam and energy.

The total facility emissions after any air pollution control device should be compared to USEPA New Source Review (NSR) thresholds to determine regulatory requirements.

Table 11-4. Air pollution controls

Air Pollution Control	Removes or Reduces Concentration of:
Low-NO_x furnace burners	Hydrocarbons, CO, NO_x
Afterburner	Hydrocarbons and CO
Ammonia injection	NO_x
Wet scrubbers (e.g., Venturi scrubber)	Particulates and SO_2
Packed towers, spray towers, or dry scrubbers	Acid gases, including SO_2 and HCl
Baghouse	Particulates
Electrostatic precipitator	Particulates

Table 11-5. Low-NO_x burner emissions reductions

Criteria Pollutant	Potential Emissions Reduction
Hydrocarbons	20%
CO	20%
NO_x	40%

Table 11-6. Afterburner exhaust constituents

Afterburner Exhaust Component	Criteria Pollutant
Combusted GAC	SO_2
Volatilized adsorbate	Particulates, NO_x, SO_2
Combusted natural gas	Hydrocarbons, CO, particulates, NO_x, SO_2

Low-NO_x Furnace Burners. The use of low-NO_x burners results in more efficient combustion (Table 11-5). A low-NO_x burner could result in the following reductions in emission generation as shown in Table 11-5 (SBCAPCD 1991b).

Afterburners. Afterburners oxidize carbon monoxide and hydrocarbons into carbon dioxide and water. However, the afterburner generates additional pollutants as a result of the combustion of the sulfur impurities in the GAC combusted, the volatilized adsorbate, and the fuel used to heat the afterburner. In an afterburner, these three components would generate the criteria pollutants shown in Table 11-6.

VOC destruction efficiencies greater than 98 percent can be achieved by an afterburner. Emission factors and a combustion analysis can be used to estimate the emission rates in the afterburner exhaust. The afterburner requires fuel in addition to that required for reactivation: an additional 0.2 lb of natural gas per pound of GAC.

A high-temperature design afterburner (i.e., 1,800°F with a minimum residence time of 1 sec) may be used for destruction of any halogenated organics released in the reactivation process.

Afterburners typically require an air pollution permit, and the operating temperature, residence time, or both might be specified in the permit. Designers should consult with local regulatory agencies to determine any operational constraints for afterburners.

Ammonia Injection. Ammonia can be injected into exhaust gases from the afterburner to control NO_x emissions. Ammonia reacts with NO_x via oxidation and reduction to form nitrogen gas (N_2) and water. The temperature window under which the oxidation and reduction reactions are feasible is narrow (i.e., between 1,600°F and 1,800°F). A NO_x reduction efficiency of 25 to 50 percent can be achieved using ammonia injection without the use of a catalyst (Wood 1994). A catalyst, typically made of titanium and vanadium oxides, can increase NO_x removals to as high as 70 to 90 percent (Wood 1994); the effects of particulates and moisture from the exhaust gas on the catalyst should be evaluated in pilot facilities. In addition, depending on the form of ammonia used, ammonia storage facilities must comply with any applicable federal, state or provincial, and local accidental or catastrophic release requirements.

Waste heat recovery boilers can be required on either side of the ammonia injection process. The first boiler lowers the exhaust gas temperature to accommodate the kinetics of NO_x control with ammonia injection. The second boiler recovers additional waste energy prior to scrubbing.

Capital and operating costs can also be high for ammonia injection with catalytic reduction (Wood 1994). The design engineer should consider the regulatory implications of the projected NO_x emission rate versus the cost effect of increasing the NO_x destruction rate using ammonia injection.

Wet Scrubbers. Wet scrubbers control particulates, acid gases such as SO_2, and, to a slight extent, NO_x. Wet scrubbers include Venturi scrubbers and tray tower scrubbers (Figures 11-9 through 11-11), which can be used separately or in series. The Venturi scrubber provides turbulent conditions necessary for particulate removal. The tray tower scrubber provides an efficient air-liquid interface to allow the mass transfer of the acid gases from the gaseous phase to the liquid phase.

Baghouse Filters. Baghouse filters (Figures 11-12 and 11-13) are fabric-filter dust collectors and are one of the oldest and most

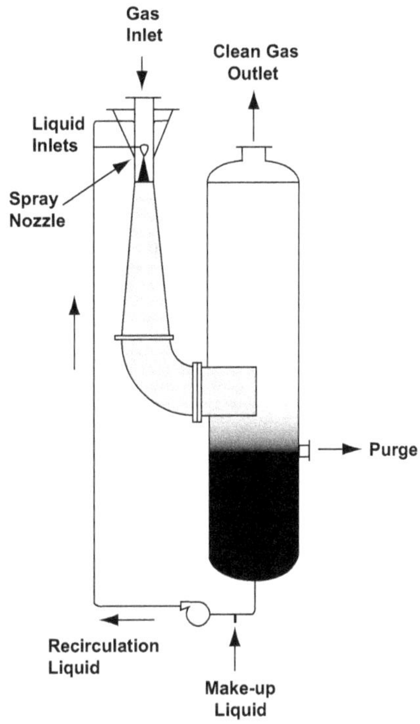

Figure 11-9. Venturi scrubber schematic (courtesy Schutte and Koerting)

Figure 11-10. Venturi scrubber (courtesy Schutte and Koerting)

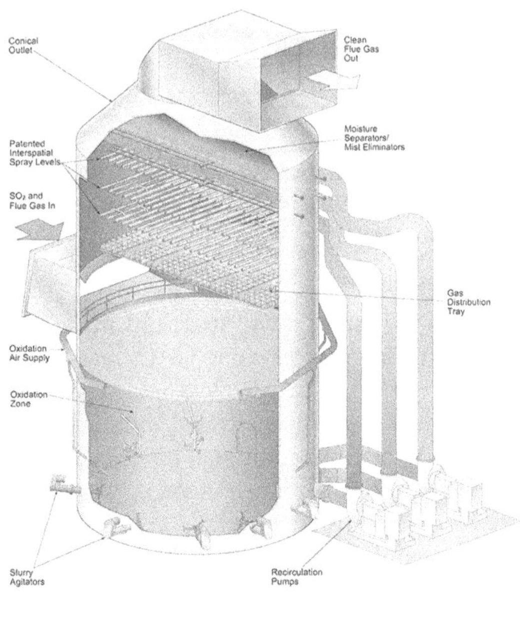

Figure 11-11. Tray-tower scrubber schematic (courtesy Babcock & Wilcox Company)

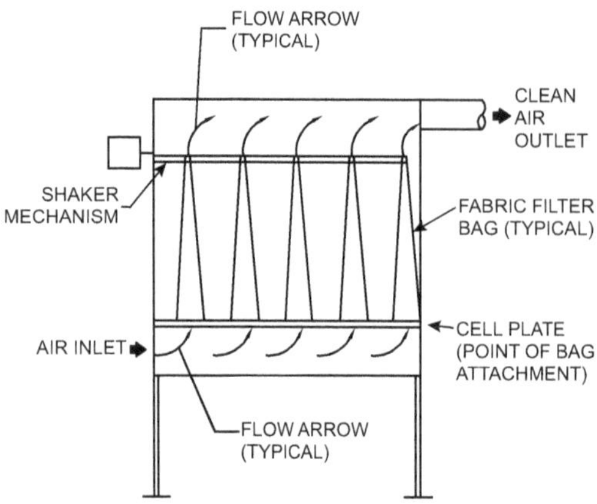

Figure 11-12. Baghouse schematic

Figure 11-13. Donaldson® Torit® Modular baghouse structure (405MBW10, courtesy Donaldson Torit)

common methods of particulate removal. At a GAC reactivation facility, baghouse filters can be used to collect particulates from the emissions leaving the furnaces. The temperature of the gas should first be lowered by quenching to prevent a fire in the baghouse. Baghouses can achieve PM10 control efficiency of 99.9 percent.

GAC Handling

The design of a GAC reactivation facility should take into account the storage and transfer of GAC within the facility. The spent, reactivated, and make-up GAC must be transferred between the treatment units, storage bins, and furnaces.

GAC Storage

Spent and reactivated GAC storage is provided to balance differences in GAC flows between the GAC treatment units and furnaces. The differences result from removal of GAC from and addition of GAC to the treatment units in batches, versus the continuous-flow operation of the furnaces. Storage should also accommodate temporary reductions in reactivation capacity as a result of unexpected mechanical problems with an active furnace.

Separate storage tanks can be provided for spent, reactivated, and make-up GAC. Each tank should have a total capacity equal to at least one GAC treatment unit plus sufficient freeboard for sedimentation to occur without loss of GAC over the overflow trough. Special care should be taken when sizing reactivated and make-up GAC storage tanks. More storage provides operators with additional capacity and flexibility but should be weighed against the cost of providing larger tanks.

Storage tanks are generally hopper-bottomed with GAC slurry pumps or eductors located directly below the hoppers so that the GAC drops vertically down into the GAC slurry pump suction or eductor. GAC slurry lines should be provided with a tap for introduction of supplemental transfer water. Stainless-steel storage tanks are recommended due to the corrosive and abrasive nature of GAC.

For facilities storing dry reactivated or make-up GAC, the GAC is usually stored in a hopper-bottomed tank similar to that described previously. Instead of dropping into a slurry pump or eductor inlet, the GAC can be loaded from the hopper into a transfer vehicle that delivers the GAC to the treatment units.

Make-up GAC Storage. Spent GAC can be reactivated multiple times; however, some GAC is damaged or lost during the process due to overreactivation or abrasion during transfer that creates GAC fines. GAC attrition rates associated with thermal reactivation were estimated previously at 5 to 15 percent, and GAC attrition resulting from handling and transportation can range from 2 to 6 percent (Sontheimer et al. 1988). To compensate for GAC losses incurred in transport and reactivation, make-up GAC must be added to the system.

It is preferable to use virgin GAC for make-up purposes. Where necessary, reactivated GAC can be used, as long as it is from the same potable water facility; this rule prevents liability issues associated with introducing foreign material from reactivated GAC from other source waters (AWWA B605). Virgin GAC is delivered into a hopper, which is equipped with an eductor to transport the GAC to the make-up storage tank. An eductor is ideal for transfer service because its aspirator action minimizes the escape of dust to the atmosphere. In addition, the eductor is also an excellent wetting device, which is important because GAC is difficult to wet. The eductor water supplies both the motive force for transport and also the medium for slurry transport.

Spent GAC Storage. Dewatering bins can be used to reduce the moisture content of the spent GAC to 40 percent by weight prior to reactivation. A typical dewatering bin is fabricated from either carbon steel protected with glass-flake epoxy or stainless steel. The bins should be fitted with both wedgewire screen drains and a top overflow arrangement. Bins have 60-degree hopper bottoms for free flow of GAC. The bottom outlet should be fitted with a knife-gate valve.

Reactivated GAC Storage. Some facilities choose to store reactivated GAC dry in sacks instead of in wet slurry tanks. Storing dry GAC in large sacks can make it easier to transport GAC to the treatment units if a slurry system is not in place or if the reactivation facility is far from the treatment units.

GAC Transfer

Methods of GAC transfer are described in detail in chapter 8. Note that only reactivated or virgin GAC can be moved using air; spent GAC is wet and needs to be moved as a slurry.

Dewatered GAC can be fed from the dewatering bins to the furnace by means of screw conveyors, which should be constructed entirely of stainless steel to resist corrosion. Screw conveyors should be equipped with variable speed drives so that the rate of GAC feed to the furnace can be accurately controlled. Bucket conveyors can also be used for dewatered GAC. Mechanical dewatering screws can remove additional water from the GAC before it enters the furnace. Vibrating screens installed between the dewatering screw and the furnace can also help remove water from the GAC.

Reactivated GAC can be transferred from the furnace to the storage tank by screw conveyors. Bucket conveyors can lift GAC to higher elevations and can also be used to move the reactivated GAC.

Auxiliary GAC Transfer Equipment. For facilities that use sand or gravel under the GAC, a desander can be placed upstream of the reactivation furnace to remove any sand or gravel that is mixed in with the GAC.

A sieve-classifier or vibratory screener located downstream of the furnace can be used to remove GAC fines and clumps from the reactivated GAC to maintain the design size distribution. The reject reactivated GAC can be crushed to create PAC that the facility can use when additional adsorption is needed, such as during T&O events, or it can be packaged for disposal.

Plant Service Water Requirements

Plant service water is needed for a number of functions in the spent GAC reactivation process:

- GAC transport
- Reactivated GAC quench
- Process steam/boiler makeup
- Venturi scrubber/spray tower scrubbant

Where possible, use unchlorinated water for service water for GAC reactivation to avoid the potential formation of chlorinated organic compounds in the furnace.

GAC Transport Water

Transport of GAC is accomplished by mixing the GAC with water to form a slurry. The slurry is then pumped to the desired location. Slurry water is required to convey GAC as follows:

- Spent GAC from a treatment unit to storage
- Spent GAC from storage to the furnaces
- Regenerated GAC from furnaces to reactivated storage
- Regenerated GAC from storage to treatment units
- Virgin (make-up) GAC from storage to treatment units

Approximately one gallon of water per pound of GAC is required for each slurry. The system should be sized to account for evaporation losses of the transport water.

GAC Quench Water

Newly reactivated GAC is cooled from the elevated temperature required for reactivation to ambient temperature by quenching it in water. Some of the quench tank water will be lost to evaporation as a result of the quenching process.

Process Steam/Boiler Makeup

Process steam is required for use in reactivation furnaces. The steam added to the furnace is discharged as part of the exhaust gas. Process steam requirements are approximately 0.5 to 2.0 lb of steam per pound of GAC reactivated, which corresponds to 0.06 to 0.26 gal of boiler feedwater per pound of GAC reactivated.

Venturi Scrubber/Tray Tower

Water is used in a Venturi scrubber or tray tower as a scrubbant for the removal of particulates, SO_2, and NO_x. As a result, the blowdown from these processes will have a depressed pH and increased levels of sulfates, nitrates, nitrites, and suspended solids. The temperature of the blowdown will be higher than at the scrubber inlet as a result of the transfer of heat from the exhaust gas.

Once-through systems (100 percent blowdown) are common for scrubber applications. The spent scrubbant is often recycled to the headworks of the treatment plant after removal of suspended solids. The scrubbant water is managed as a wastewater stream if its quality precludes recycling. Scrubbant water requirements are approximately 1.3 gal per pound of GAC reactivated.

Permitting

A typical GAC reactivation facility will have air and wastewater discharges. Federal, state or provincial, and local regulations may apply to these emissions and discharges. An overview of potentially applicable federal regulations is provided in this section.

Air Quality Permitting

Air emissions and pollution are regulated through the Clean Air Act, which was originally enacted in 1970 and amended in 1977 and 1990. The Clean Air Act requires preconstruction and operating permits for new or modified major air pollution sources. Some states have an option that combines the preconstruction and operating permits into a single permit. Regulations for air toxics and accidental release prevention might also apply to a new GAC reactivation facility.

Preconstruction Permits

Preconstruction permits for industrial projects that are major sources of air pollution are required under the USEPA's New Source Review (NSR). Preconstruction permits apply to new major sources and major modifications to existing sources of emissions. NSR requirements include (Clean Air Act 2008):

- Prevention of Significant Deterioration (PSD) permits for areas in attainment with the National Ambient Air Quality Standards (NAAQS).
- Nonattainment NSR permits for facilities in areas that do not meet the NAAQS.
- Minor source permits for facilities that do not require PSD or nonattainment NSR permits.

PSD permits are intended to preserve the air quality in areas that comply with the NAAQS and set maximum allowable increases in concentrations for a number of chemicals. PSD permits limit the emissions that a new or modified source can produce and require facilities to install the best available control technology (BACT) for each regulated pollutant (Clean Air Act 2008). Facilities are also required to monitor emissions and the effect on ambient air quality.

The Nonattainment NSR permit was developed as part of the 1977 Amendments to the Clean Air Act and applies to facilities that are located in areas that are not already in compliance with the NAAQS. Requirements for the Nonattainment NSR permit are more stringent than those for the PSD permit: Under the Nonattainment NSR permit, new projects must install technology that ensures the lowest achievable emission rate (LAER), provides offsets for any significant emissions resulting from the new or modified facility, and offer opportunity for community involvement (Clean Air Act 2008).

Operating Permits

Operating permits are required under the 1990 Clean Air Act Amendments. The permits include information on the types and quantities of emissions being released and the measures the owner must take to reduce emissions (Clean Air Act 2008). A plan for measuring and reporting emissions must be included in the permit. An annual fee is also associated with the operating permit.

Air Toxics/HAPs

If the GAC reactivation facility is classified as a major source of emissions of air toxics or HAPs, it will be subject to regulation by the 1990 Clean Air Act Amendments, which require the use of maximum achievable control technology (MACT). MACT is determined by the USEPA and represents the control technology that provides the maximum reduction in emissions while considering the cost of implementing the technology, as well as any "non–air quality health and environmental impacts and energy requirements" (Clean Air Act 2008). MACT may include reducing or eliminating emissions by process changes, substituting materials, enclosing processes, or capturing and treating emissions.

Accidental Release Prevention

Prevention of accidental release of air toxics and other extremely hazardous substances is regulated by the 1990 Clean Air Act Amendments. A list of regulated chemicals that are reasonably expected to cause acute adverse health effects in humans is provided by the USEPA. A threshold quantity is identified for each chemical, accounting for the chemical's "toxicity, reactivity, volatility, dispersibility, combustibility, or flammability" (Clean Air Act 2008). Any facility storing regulated chemicals in excess of the threshold quantity must complete a hazard assessment of the facility. The facility may also be required to submit a Risk Management Plan (RMP) for detection and prevention of the accidental release of the hazardous substances.

Solid Waste Handling Requirements for Spent GAC

Spent GAC reactivated on-site is recycled and, as such, is not considered a waste material (Liang et al. 1996). However, spent GAC that is reactivated at an off-site facility or removed from the site could be considered a waste material and could potentially be classified as a hazardous waste. Under current regulations, USEPA uses the following process to determine whether a waste is hazardous:

1. Determine whether the waste is "excluded by definition."
   a. Waste that is defined as not being a solid waste; and
   b. Solid waste specifically defined as non-hazardous.
2. Determine whether the waste is "listed."
3. Determine whether the waste exhibits hazardous "characteristics."

Spent GAC is neither excluded from hazardous waste regulations nor is it a listed hazardous waste (Liang et al. 1996). Therefore, an analysis of the spent GAC characteristics is required to determine compliance with regulatory limits. Some of major GAC manufacturers can perform an analysis of the WTP's GAC to determine the adsorbate profile, while others may require the WTP to characterize the spent GAC as hazardous or nonhazardous prior to accepting spent GAC for reactivation.

Process Water Control and Discharge

Process water is used at reactivation facilities for a number of purposes as noted earlier. Some of these waters can be reused, others are lost to the atmosphere through steam, and some must be handled as a waste stream. A study of GAC reactivation found that approximately one-third of the process water used for reactivation was nonrecoverable and must be handled as a waste stream. The principal component of the nonrecoverable waste stream is blowdown from the air pollution control device (i.e., Venturi or scrubber). Comparison with existing discharge permits or requirements for new permits for waste stream discharges need to be considered.

Scrubber Blowdown

Scrubber blowdown rates can be reduced by recirculating the scrubbant water. As a result, contaminant concentrations increase proportionately to the decrease in blowdown rate. For example, at 10 percent blowdown rate, the contaminant concentration increases by a factor of 10. The number of cycles of concentration will normally be limited by the saturation concentration of the contaminants and the deposition of scale. Assuming the spent scrubber water stream will be discharged to the sanitary sewer system, the waste stream will need to comply with local sewer system contaminant limits, temperature limits, and pH limits.

Spent GAC Transport Water

Normally, the spent GAC transport water can be recycled within the plant. However, if the spent GAC is classified as a hazardous waste, as may be the case for some remediation applications, the spent GAC transport water may also be classified as hazardous. The spent GAC can be removed from the transport water by settling or other techniques. Although significant desorption of contaminants is unlikely, the water should undergo toxicity testing to determine if it has been contaminated by contact with the spent GAC. If the water is classified as a hazardous waste, it will be subject to federal, state or provincial, and local regulatory requirements.

Operational Considerations

Reactivation facilities are equipped with meters, instrumentation, and interlocks to regulate parameters such as furnace temperatures, GAC feed rate, steam flow, airflow, and natural gas flow. High-temperature thermocouples can be used to regulate temperatures in the furnaces. It is recommended that dual thermocouples be installed wherever accurate temperature readings are critical because calibrating the thermocouples can be difficult.

Operators should be available to handle any regular or emergency maintenance items while the reactivation facility is in operation. Depending on the size and complexity of the facility, one or more operators might need to be dedicated solely to the operation of the reactivation facility.

GAC reactivation furnaces are most efficient when run continuously because starting up a furnace that has been out of service takes a significant amount of energy and time. Ideally, the reactivation facility would operate 24 hours per day, 7 days a week for several months at a time. It is important that the GAC feed rate to the facility be consistent to ensure continuous operation of the furnaces. Another benefit of operating the furnaces continuously is that it reduces the expansion and contraction fatigue effects that occur with starting and stopping the furnaces.

Having a redundant furnace is beneficial if one of the operating furnaces is taken out of service for regular or emergency maintenance. If one of the active furnaces fails suddenly or experiences a significant reduction in capacity, the standby furnace can be placed into operation.

The start-up of an inactive multiple-hearth furnace, however, cannot be accomplished instantaneously. The furnace should be heated slowly (about 100°F/hour) to prevent damage to the refractory lining of the furnace and rabble arms. Consequently, a minimum of 18 hr are required to bring a cold furnace to an operating temperature of 1,800°F. Facilities with multiple-hearth furnaces might want to provide additional storage for spent GAC to account for start-up time.

The furnaces should be occasionally removed from service for longer periods of time. Repair of refractory lining and rabble arms is required on an annual or semiannual basis, and downtimes for major refractory lining repair can extend for several weeks. It should not be necessary to provide additional GAC storage for routine maintenance because planned maintenance procedures can be accommodated through planning around water production schedules.

GAC Safety

Refer to chapter 8 for safety issues related to storage and handling of GAC.

References

Adams, J.Q., R.M. Clark, B.W. Lykins, Jr., J. DeMarco, and D. Kittredge. 1988. GAC treatment cost experience at two drinking water utilities. *Jour. Environ. Engrg.*, 114(4):944–961.

AWWA (American Water Works Association). B605: AWWA Standard for Reactivation of Granular Activated Carbon. Denver, Colo.: AWWA.

Cannon, F.S., V.L. Snoeyink, R.G. Lee, G. Dagois, and J.R. DeWolfe. 1993. Effect of calcium in field-spent GACs on pore development during regeneration. *Journal AWWA* 85(3):76–89.

Clean Air Act. 2008. Title 42 U.S. Code, Chapter 85.

DeMarco, J., R. Miller, and D.J. Hartman. 1988. Discovery and elimination of dioxins from a carbon reactivation process. *Journal AWWA*, 80(3):66–72.

Eberle, D. 2009. Interview by Brian J. Leto. *GAC reactivation energy.*

Inhoffer, W.R., and F.K. McGinnis, 1980. Regenerating granular activated carbon at Passaic Valley. In *Proc. Amer. Water Works Assoc. 1980 Annual Conference, Atlanta, Georgia*. Denver, Colo.: AWWA.

Karanfil, T., and J.E. Kilduff. 1999. Role of granular activated carbon surface chemistry on the adsorption of organic compounds. 1. Priority pollutants. *Environ. Sci. Technol.* 33(18):3217–3224.

Knappe, D.R.U., V.L. Snoeyink, G. Dagois, and J.R. DeWolfe. 1992. Effect of calcium on thermal regeneration of GAC. *Journal AWWA*, 84(8):73–80.

Lambert, S.D., G. San Miguel, and N.J.D. Graham. 2002. Deleterious effects of inorganic compounds during thermal regeneration of GAC: A review. *Journal AWWA* 94(12):109–119.

Liang, S., D.K. Welch, D.M. Owen, and M.D. Beuhler. 1996. Siting a large GAC regeneration facility. *Journal AWWA* 88(2):86–99.

Miller, R., and D.J. Hartman. 1982. Feasibility study of granular activated carbon adsorption and on-site regeneration. Cincinnati, Ohio: Municipal Environmental Research Laboratory, Office of Research and Development, United States Environmental Protection Agency.

Moore, B.C., F.S. Cannon, J.A. Westrick, D.H. Metz, C.A. Shrive, J. DeMarco, and D.J. Hartman. 2001. Changes in GAC pore structure during full-scale water treatment at Cincinnati: A comparison between virgin and thermally reactivated GAC. *Carbon*, 39:789–807.

Norit Americas, Inc. 2001. Granular activated carbon evaluation. NA00-4. 2001. Marshall, Texas.

San Bernardino County Air Pollution Control District (SBCAPCD). 1991a. Draft staff report: Regulation XCIII new source review. (Unpublished)

SBCAPCD. 1991b. Meeting with Malcolm Pirnie. (Unpublished)

Scott, R. 2009. Interview by Brian Leto. Superintendent of water treatment operations, City of Glendale, Ariz.

Shovlin, M.G., and S.S. Tanaka. 1990. Risk communication in Los Angeles: A case study. *Journal AWWA* 82(11):40–44.

Shrive, C.A., J.A. Westrick, D.H. Metz, J. DeMarco, W.F. Reeves, and D. McMullen. 1998. The use of GAC at the Cincinnati Water Works: A balancing act between water quality and furnace operations. *Proc. AWWA Water Quality Technology Conference, San Diego, Calif.*

Sontheimer, H., J.C. Crittenden, and R.S. Summers. 1988. *Activated carbon for water treatment,* 2nd ed. Karlsruhe, Germany: Druckerei G. Braun GmbH.

Speitel, G.E., and F.A. DiGiano. 1987. The bioregeneration of GAC used to treat micropollutants. *Journal AWWA*, 79(1):64–73.

Strack, B. 1976. Operation, problems and economy of activated carbon regeneration. In *Translation of Reports on Special Problems of Water Technology.* EPA-600/9-76-030. H. Sontheimer (Ed.). U.S. Environ. Protection Agency, ORD, Cincinnati, Ohio, 284–311.

United States Army Corps of Engineers (USACE). 2001. *Engineering and design adsorption design guide.* DG 1110-1-2.

USDOE (US Department of Energy). 2007. *Improving process heating system performance: A sourcebook for industry,* 2nd ed. Golden, Colo.: National Renewable Energy Laboratory.

USEPA (US Environmental Protection Agency). 1995. AP 42: Compilation of air pollutant emission factors, 5th ed. Office of Air Quality Planning and Standards.

USEPA. 2000. Wastewater technology fact sheet: Granular activated carbon adsorption and regeneration. EPA 832-F-00-017.

von Dreusche, C. 1978. Process aspects of regeneration in a multiple-hearth furnace. In *Carbon Adsorption Handbook*, P.N. Cheremisinoff and F. Ellerbusch (Eds.). Ann Arbor, Mich.: Ann Arbor Sci. Publ. Inc., 923–953.

Waer, M.A., V.L. Snoeyink, and K.L. Mallon. 1992. Carbon regeneration: Dependence on time and temperature. *Journal AWWA*, 84(3):82–91.

Wood, S.C. 1994. Select the right NOx control technology. *Chem. Engrg. Progress*, pp. 32–38.

Zanitsch, R. 1991. Personal communication. Calgon Carbon Corporation.

Chapter 12

PAC Feed System Design

This chapter addresses the use of PAC in water treatment systems. PAC was first used in water treatment in the mid-1950s and was used primarily to treat T&O problems (Hansen 1972). Today, PAC continues to be used seasonally for mild to moderate T&O treatment and increasingly for TOC removal. Some utilities keep PAC on hand for use in case of a spill event. The primary advantages of PAC are the low capital cost and ease of application. PAC can be employed at an existing facility quickly and with little modification to the existing treatment process. The disadvantages of using PAC are that more mass of activated carbon is generally used per volume of treated water compared to GAC, and PAC is a single-use material that increases plant residuals. In general, PAC is typically used as a reaction to a specific water quality event rather than a broad-spectrum proactive approach for enhancing water quality.

This chapter first considers issues with the design and construction details of PAC facilities, and then the operational issues with using PAC are addressed. More information on PAC process planning is presented in chapter 4.

Design and Construction

The design phase for the design of PAC treatment systems consists in determining the key design parameters and any modifications to the existing treatment process.

Review Key Design Parameters

When beginning a design for a PAC system, first review the recommendations from the water treatment plant's master plan. If a PAC treatment technology is selected, the first step of the design phase is a review of the key design parameters:

- Design flow rate
- Influent concentrations of target and background DOC
- Treatment objectives
- PAC type
- Design dose and duration
- Contact time and mixing

Design Flow Rate

The minimum and maximum design flow rates for the PAC process affect the dosing requirements, the size and number of storage vessels, and amount and size of feed equipment. The maximum design flow rate is typically equal to the plant design capacity, but it may be determined based on future design capacity. The minimum design flow rate is typically equal to the minimum plant flow rate.

Influent Concentrations of Target and Background DOC

Influent concentrations of the target contaminant and the background DOC determine the PAC dose that is needed. Dose-response curves are generally developed in the process planning stage for each of the target compounds prior to design.

Treatment Objectives

The target contaminants (e.g., T&O, synthetic organic compounds, or DOC) are identified prior to the design process. The target effluent concentration corresponds to the concentration desired in the treated water for the process to meet water quality treatment goals. The treatment objective ultimately determines the dose and, in turn, the size of facility required.

PAC Selection

PAC is typically classified by the source of its manufacture (e.g., bituminous, lignite, or coconut shell). Because reactivation is not practiced with PAC, many PACs are available with different surface treatments (e.g., acid washed or chemically activated) that may affect the cost of the PAC and should be weighed against the superior performance. The type of PAC and PAC treatment technology recommended by the master plan or other studies conducted prior to design help guide PAC selection. However, in general, evaluation of different PAC products is much simpler than for GAC. The PAC type affects the design dose and the design of the storage and feed system.

Design Dose and Duration

The PAC dose for a given facility depends on what contaminants are being targeted for removal and the level of removal required. Jar testing provides a reasonable estimate of the required PAC dose for a given source water and can be designed to implicitly include competitive adsorption and interference from other chemicals. Typical PAC doses range from 5 to 50 mg/L, with higher doses needed periodically to treat unusually high T&O or TOC concentrations.

Some facilities feed PAC continuously, but the majority only feed periodically to treat seasonal T&O issues or in emergency response to an organic chemical spill (AWWA/ASCE 2012). The PAC storage and feed system should be designed in accordance with the anticipated frequency of use. Facilities using PAC seasonally are typically responding to the increased MIB and geosmin concentrations that occur during the summer when warm water temperatures allow algae to flourish. Seasonal events like storms or fall runoff that increase the TOC loading to the plant might also trigger the need for PAC.

Water treatment plants begin or adjust their PAC feed rates based on one or more of the following criteria:

- Threshold odor test
- Customer complaints
- Flavor profile analysis (FPA)
- Water temperature
- Storm and spill events in the source water
- Organic compound concentration (TOC or specific target compound)
- T&O compound concentrations (e.g., MIB and geosmin)
- Dose-response curve

Contact Time and Mixing

Contact time is a key parameter in the design of PAC treatment because adsorption equilibrium is not typically reached, making mass-transfer kinetics rate limiting. Contact time and mixing conditions are generally fixed for PAC applications in existing treatment trains. Contact time depends directly on the point of application. For an example, consider the most common PAC application scenario with PAC addition at rapid mix and removal during sedimentation. In rapid mix, mixing conditions are very good, but contact time is short. Flocculation basins provide a decent balance between mixing and contact time, but interference from coagulants and flocs may hinder adsorption. Finally, sedimentation potentially provides a lot of contact time, but the lack of mixing causes the PAC to readily settle, which limits additional adsorption. Construction of a PAC contact basin can provide optimal contact time and mixing but sacrifices the advantage of low capital cost. PAC addition and removal points are generally specified in the master plan, and more information can be found in chapter 4.

Construction Details of PAC Systems

A PAC system should "deal with how best to provide for a chemical that is dirty, difficult to handle, and difficult to store for long periods, but is needed on a moment's notice" (AWWA/ASCE 2012). Thus, the design of a PAC system can be a challenge in itself. In this section, design options and considerations for PAC systems are addressed.

System Type

PAC can be stored either wet or dry. The choice affects design of the entire storage and feed system. Regardless of whether the system is designed for storage of wet or dry PAC, the feed rate is set by the required dose and the plant design flow, which also determine the physical size of the system. The design feed rate determines whether it is more economical or convenient to use bulk storage tanks or smaller pallets or bags of PAC.

PAC is available in 1,000-lb bags, pallets of 50-lb bags (Figure 12-1), or in bulk form (typically 40,000-lb deliveries). Bags are usually preferred for smaller dry systems, while bulk deliveries are typical for wet systems and large dry systems (AWWA/ASCE 2012).

Figure 12-1. Pallets of 50-lb PAC bags at a Wilmington, Del., WTP

Wet PAC Storage and Feed Systems

Wet PAC systems are ideal for facilities in which PAC usage is likely to be frequent and doses are high. Capabilities for storage and feed of bags of dry PAC can also include a wet system to allow facilities to respond to sudden, heavy dose requirements without requiring a bulk delivery (AWWA/ASCE 2012). Wet PAC is corrosive, and all surfaces in contact with the slurry should be constructed of suitable material, such as stainless steel. The following sections discuss equipment and materials for PAC storage and feed systems. A schematic of a wet PAC storage and feed system is shown in Figure 12-2.

Wet PAC Storage. For wet systems, PAC is stored in a tank as a slurry. The PAC is usually mixed with water to achieve a maximum concentration of 1 lb PAC/gal of water (10 to 15 percent by weight) (Nielsen and Whipple 1979). The slurry tank and mechanical mixer might make achieving a 5 percent solids (0.4 lb PAC/gal of water) more practical (Hassibi and Lemmon 2009). The slurry tank can be constructed of steel or concrete and should have a capacity of at least 20 percent more than the maximum delivery volume of PAC (AWWA/ASCE 2012) to account for the water added to make the slurry. The retention time should be at least 15 min to provide adequate mixing of the PAC and water (Hassibi and Lemmon 2009). A lining can be provided in the slurry tank to prevent corrosion, although it is not generally needed with concrete tanks. All metals in contact with the PAC should be stainless steel (AWWA/ASCE 2012). An example of a concrete slurry storage tank is shown in Figure 12-3.

Single-speed or two-speed mixers can be used in the slurry tanks to mix the PAC with water and keep it in suspension. Offsetting the mixer on one tank axis can create vortices that mix the PAC more thoroughly (Hassibi and Lemmon 2009). An air sparge system can also be used to prevent PAC from settling out of solution.

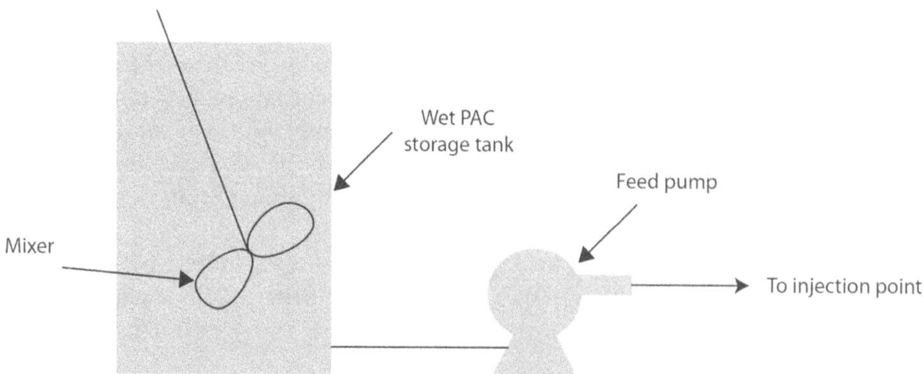

Figure 12-2. Wet PAC storage and feed system schematic

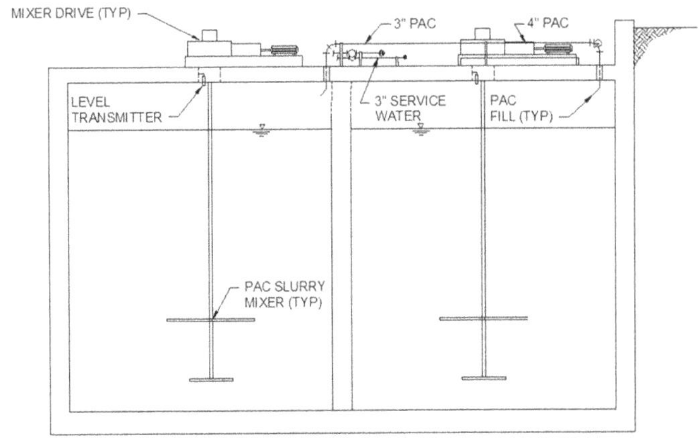

Figure 12-3. Concrete PAC slurry tanks

Slurry tank mixers are especially important for systems that are not operated continuously. If not mixed, the PAC can settle and solidify in the bottom of the slurry tank (AWWA/ASCE 2012). The PAC is difficult to remove once it has solidified. The amount of mixing required to prevent solidification appears to vary between water treatment plants and different PAC products. Some facilities mix continuously and still notice some solidification occurring, while others only need to mix once a month or for the day before the PAC is used. Make-up water may need to be provided to the slurry tank to account for water lost to evaporation or to heat energy from the mixer.

An alternative to mixing is to completely empty the slurry tanks when they are not in use. However, taking tanks out of service might prevent the facility from being able to respond quickly enough to address an event when it occurs (AWWA/ASCE 2012).

Wet PAC Feed Systems. PAC can be fed as a slurry using rotary volumetric feeders, eductors, centrifugal pumps, or positive displacement pumps such as peristaltic hose pumps (Figure 12-4; AWWA/ASCE 2012, Hassibi and Lemmon 2009). Proprietary slurry feeders are available, such as the Merrick OMEGA® volumetric feeder (Figure 12-5), which uses a metering mechanism composed of several dippers arranged on a rotating wheel. When pumps are used to feed PAC slurry, provide an automatic flushing sequence to prevent the PAC from solidifying in the pump and piping.

Figure 12-4. Peristaltic hose pump used to feed PAC

Figure 12-5. Volumetric feeder interior (courtesy Merrick Industries, Inc.)

Dry PAC Storage and Feed Systems

PAC systems can also be designed to store and handle dry PAC. Dry systems are most often used for facilities where PAC use is likely to be infrequent and the required feed rate is relatively low (i.e., less than a few hundred pounds per hour), although dry systems can also be used for higher feed rates (AWWA/ASCE 2012).

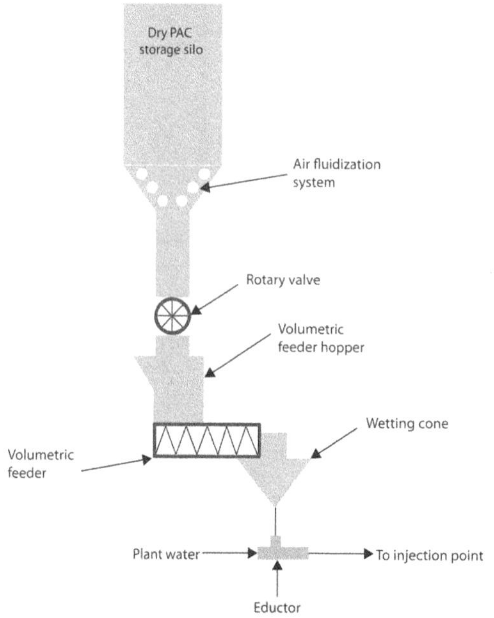

Figure 12-6. Dry PAC storage and feed system schematic

Figure 12-7. Dry PAC storage silos

The primary components of a dry PAC system are the storage tank or silo, wetting device, and feeder (Figure 12-6). These components are described in detail below.

Dry PAC Storage. PAC can be stored dry in an elevated tank or silo (Figure 12-7). The main benefit of using a dry PAC system is that the storage silos can be smaller for storing the same volume of PAC as a wet system. Dry PAC storage units can also be less expensive than slurry tanks and easier to locate closer to the injection points. Dry storage tanks also do not require mixing equipment. Construction of a silo for dry PAC might also be faster than constructing a slurry storage tank. Another potential benefit of a dry PAC system is that controlling the rates of PAC feed might be more accurate than for a wet system (Nielsen and Whipple 1979).

Bulk storage silos can be welded, carbon steel vertical cylinders with a conical bottom and the top sloped a minimum of 10 degrees for drainage. The cone of the silo should be designed with an angle of 60 degrees, at which the PAC flow is initiated easily and maintained with minimum air fluidization. The minimum cone angle is considered to be 50 degrees (Norit Americas 2003). An emergency pressure and vacuum relief valve should be installed on the silo roof to protect the silo from damage due to overpressurization or vacuum. Secondary pressure-relief devices such as a pressure relief access manhole (Figure 12-8) can also be incorporated into the silo design.

Small holes in the silo welds and space between the flanges can result in significant PAC leaks. Welds can be inspected during silo fabrication using dye-penetrant tests to identify leaks. Silo flanges can be sealed with a flowable sealant (Hassibi and Lemmon 2009).

Figure 12-8. Shand and Jurs model 94520 emergency pressure vent and manhole cover (courtesy L&J Technologies)

Dry PAC feed equipment is sensitive to the presence of moisture. Care should be taken to keep the dry-storage silo free of moisture, which can be achieved by providing an air dryer for the compressor that supplies air for moving and fluidizing the dry PAC. Minimizing the moisture that enters the silo also reduces the risk of corrosion in the silo. Because moisture is usually low in the silo, corrosion is less of a concern for dry-PAC storage silos than it is for slurry tanks. Carbon steel silos have a good record of service for storing dry PAC, and urethane coatings can further extend their expected life span. Stainless steel and aluminum can also be used (Norit Americas 2003).

Note that the weight of the dry PAC may increase by up to 10 to 15 percent during transfer from the delivery truck to the silo as the PAC adsorbs moisture from the ambient air, especially in areas of high relative humidity. The additional moisture weight should be taken into consideration when designing the storage structure (Norit Americas 2003).

The volume of the storage silo should be also carefully selected. As the PAC is transferred pneumatically, it increases in volume due to the air added to move the PAC (Norit Americas 2003). Norit Americas (2003) recommends sizing the silo for 1.5 times the volume normally received in a single bulk delivery, which results in a volume of about 3,000 ft^3. Sizing the silo with additional volume allows complete unloading of the PAC from the delivery truck without stopping to let the PAC settle. It also allows the facility to receive a full delivery without having a completely empty tank.

The level of the stored PAC can be measured and controlled with instrumentation such as a radar level detector or tuning-fork type-level switch. Weight load cells that measure the mass of PAC in the silo may provide a more consistent measurement of how much PAC is stored because the volume of PAC tends to compress over time (Fuerman 2003).

Figure 12-9. PAC storage and feed system for 1,000-pound bags

Dry-PAC storage silos should also include dust control devices, which are discussed in the Additional Design Considerations section.

Dry PAC Feed Systems. For small systems (10 to 400 lb/hr), several manufacturers can provide a packaged metering system that draws PAC from 1,000-lb bags, mixes it into a slurry, and conveys it to the injection point (Figure 12-9). These units include a bulk bag support system, volumetric feeder, wetting device such as an eductor or wetting cone, and controls (Calgon Carbon Corporation 1994, Norit Americas 2007b). Power, compressed air, and service water connections should be provided for the system. Plant operators remove and replace the bulk bags using a skid-mounted monorail. The skid can be installed on a concrete pad indoors or outdoors, although pipes should be protected from freezing for outdoor installations (Norit Americas 2007b). The PORTA-PAC system provided by Norit Americas has a footprint of 6 ft × 6 ft and a height of 16 ft. For a facility with a low PAC feed rate, a packaged metering system might be an economical and simple way of handling PAC.

Larger dry PAC facilities are designed as a system of components that measure the PAC and move it to the injection point. The components of a dry PAC feed system are described in this section.

Live Bottoms. Moving PAC out of the storage silo can be a challenge because PAC tends to either bridge inside the silo or flood out of it. Live bottoms are one way of controlling and maintaining the PAC flow out of the storage silo. Live bottoms are installed in the cone of the silo and consist of a mechanical or pneumatic means of moving and agitating the PAC, such as through vibrations or screw conveyors. To prevent both flooding and compaction, the live bottom's on-off cycle should match the feeder's feed rate (Hassibi and Lemmon 2009). The live bottom operates for a given amount of time, filling either the feeder's inlet hopper or an intermediate tank. The feeder draws from the hopper at a continuous rate. Operating the live bottom for too long causes PAC to back up in the intermediate hopper and into the silo, resulting in compaction. The timing of the live bottom's cycles should be set to supply enough PAC for the feeder to operate continuously without being overwhelmed.

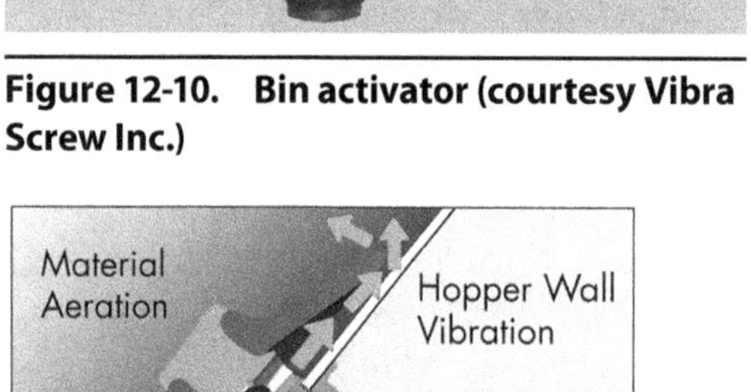

Figure 12-10. Bin activator (courtesy Vibra Screw Inc.)

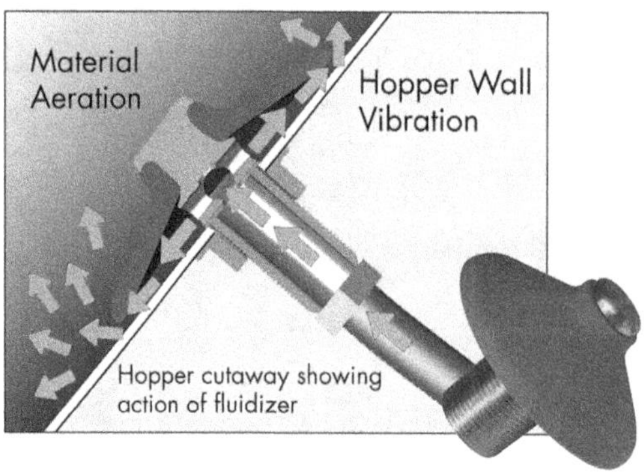

Figure 12-11. Silo fluidizer disk (Solimar Pneumatics, Minneapolis, MN)

Bin activators are one type of live bottom that are cone-shaped and include a gyrator that applies vibrations to the material in the silo (Figure 12-10).

Another type of live bottom supplies air to fluidize the media, either through intermittent high-pressure pulses or a continuous low-pressure air stream (Hassibi and Lemmon 2009). Fluidizers mounted on the silo wall (Figure 12-11) may also provide some vibration effects (Solimar Pneumatics 2008). The fluidizing air supply should be dry, and the air pressure should be less than 10 psi for low-pressure fluidization (Hassibi and Lemmon 2009). Fluidization is not required during the silo filling process, unless PAC is being fed from the silo at the time of filling (Norit Americas 2003).

Air fluidizers are typically pads or discs installed in the cone of the silo (Figure 12-11 through Figure 12-13). Pad-type

Figure 12-12. Air fluidizer

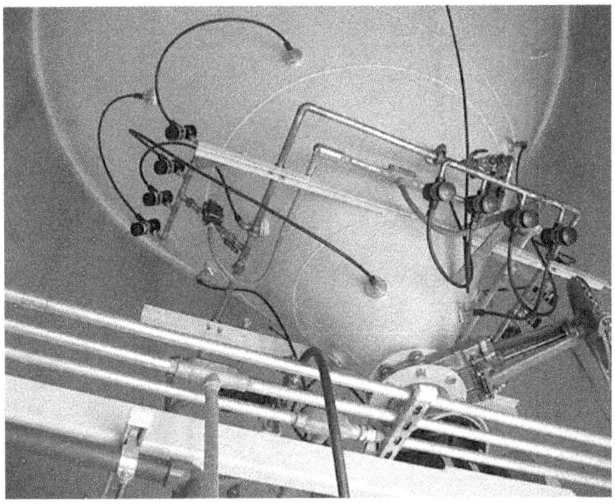

Figure 12-13. Live bottom with air fluidization connections

air fluidizers (Figure 12-12) require an air pressure of 5 psi or greater but have a relatively low cost per unit and simple maintenance. Over time, the PAC will plug holes in the pad, which results in a decreased fluidization capacity (Norit Americas 2003).

Dry Feeders. Volumetric feeders, such as rotary airlocks and screw feeders, are commonly used to feed PAC. Note that equipment in contact with PAC should be constructed of abrasion-resistant material, such as high-chrome and Ni-hard steel (Hassibi and Lemmon 2009).

Rotary airlocks are valves with several vanes mounted on a rotor inside a housing, as shown in Figure 12-14 and Figure 12-15. Rotary airlocks are mounted below the silo discharge cone. Some facilities use rotary airlocks to meter the PAC from the silo; they can be used for high-volume feeding at rates of 100 ft^3/hour or more (Hassibi and Lemmon 2009). However, rotary airlocks can be difficult to adjust for accurate metering (Norit Americas 2003).

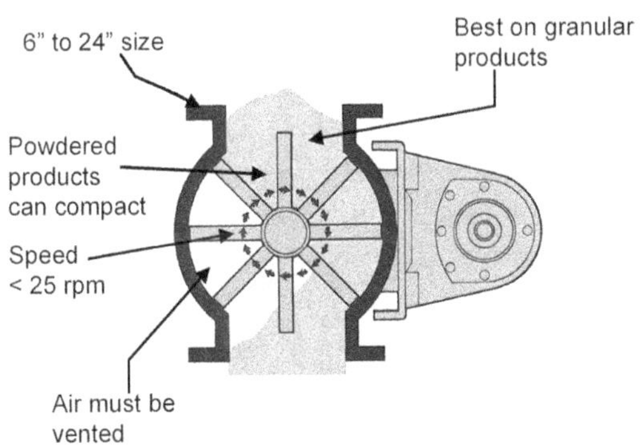

Figure 12-14. Rotary feeder schematic (Hassibi and Lemmon 2009)

Figure 12-15. Donaldson® Torit® cast-iron rotary valve (courtesy Donaldson Torit)

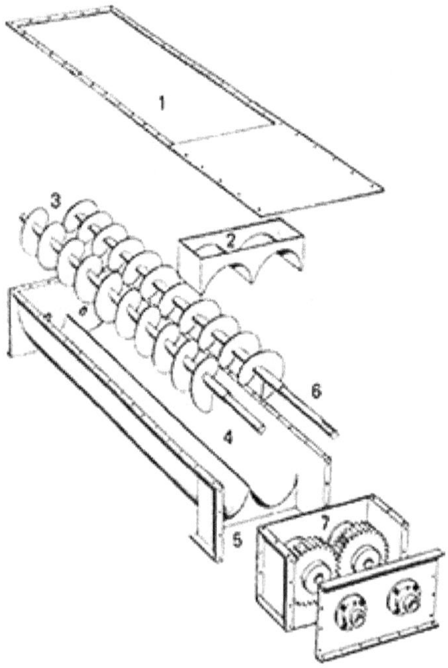

Figure 12-16. Screw feeder (KWS Manufacturing Co. 2008)

Figure 12-17. Volumetric screw feeder

Another option is to use the rotary airlock to transfer PAC from the silo to a metering device with a higher precision, such as a screw feeder (Norit Americas 2003). The rotary airlock serves to prevent flooding of the downstream feeder and is used at rates less than 100 ft^3/hr. A small hopper can be installed between the airlock discharge and the feeder inlet so that the rotary airlock can be cycled on and off while the feeder operates continuously. If a live bottom is used in the silo, it should be cycled on and off with the rotary airlock (Hassibi and Lemmon 2009).

A volumetric screw feeder consists of a screw enclosed in a tube (Figures 12-16 and 12-17). Similar to an Archimedes screw, the vanes of the screw feeder push material along the length of the screw toward the discharge. Screw feeders are typically used for feed rates less than 100 ft^3/hr. Screw feeders with closed flights ("solid-shaft screw") are preferable to open flights ("helix") if there is not a rotary airlock ahead of the screw feeder because uncontrolled PAC sometimes flows outside of the open flights. Using a hopper vibrator with the screw feeder is not recommended because it can compact the PAC and damage the equipment (Hassibi and Lemmon 2009).

Loss-in-weight feeders control the PAC feed rate based on weight instead of volume (Figure 12-18). Loss-in-weight feeders consist of a hopper, scale, and feed mechanism such as a screw conveyor.

Pre-Wetting. Although dry PAC can be fed directly to the application point, it is most often converted to slurry form before injection. The PAC is first pre-wetted by drawing it out of the silo and wetting it in a wet-down cone or an eductor (Nielsen and Whipple 1979) (Figure 12-19 and Figure 12-20). Pre-wetting the PAC prior to it entering the slurry tank is recommended because it minimizes dust generation and increases the contact time between water and PAC (Hassibi and Lemmon 2009).

Figure 12-18. Loss-in-weight feeder (courtesy Merrick Industries Inc.)

Figure 12-19. Wetting cone

Figure 12-20. Inside of wetting cone

In a wet-down cone, the inside surface of the cone is coated with water to prevent the PAC from sticking to it. Water jets spray into the cone, providing wetting and mixing of the PAC. From the wetting cone, the PAC can be sent either directly to a slurry mixing tank or an eductor for further mixing prior to the slurry tank (Hassibi and Lemmon 2009). The slurry might remain in the slurry tank for as little as a few minutes or as long as a few years, depending on the demand for PAC.

For feed rates of 100 ft^3/hr or more, an eductor can be installed after the wetting cone to improve the premixing (Hassibi and Lemmon 2009). The wetted PAC flows from the wetting cone, through the eductor, and into the slurry tank or to the injection point.

PAC Handling

PAC delivery, storage, and feed can be very dirty processes. Properly designed storage and transfer facilities can minimize the difficulties associated with handling PAC.

PAC Delivery. PAC can be delivered in 1,000-lb sacks, pallets of 50-lb bags, or bulk delivery (typically 40,000-lb) trucks. The quantity of PAC needed per shipment will determine the method of delivery. Bulk tanker trucks are the most common way to deliver PAC to large facilities. Some facilities feed PAC manually with bags, although it is a messy process.

PAC Storage. Where possible, PAC should be stored in a separate, climate-controlled building. Storing PAC outside on pallets is not recommended (AWWA/ASCE 2012). Dry PAC is typically stored in silos, while wet PAC is stored as a slurry in a concrete or steel tank with a mixer.

PAC Transfer. For wet-PAC systems, dry PAC is usually transferred from the truck using an eductor, which wets and mixes the PAC during transfer to the slurry tank. An on-site water supply is required for the eductor.

For dry-PAC systems, dry PAC can be transferred pneumatically from bulk delivery trucks to storage silos (Nielsen and Whipple 1979). It is important to note that transferring dry PAC using air can generate a significant amount of dust, which must be controlled (refer to section on dust control under "Additional Design Considerations"). Delivery trucks are typically equipped with compressors that supply pressurized air (5 psig) to the truck and blow the PAC to the storage silo or slurry tank. The pressurized air system is sized to move the PAC up to 60 ft vertically and up to a few hundred feet horizontally. PAC can be transferred pneumatically at a rate of about 10,000 lb/hr (Norit Americas 2003). Increasing the air pressure to transfer PAC at a faster rate can result in excessive wear on the fill line and fittings (Hassibi and Lemmon 2009).

Transferring PAC pneumatically generates static electricity, which can cause a spark and, potentially, combustion of the PAC. Ensure that the storage silo and fill piping are properly grounded to prevent static charge build-up and potential fire (Hassibi and Lemmon 2009).

Delivery trucks are also usually equipped with lengths of hose with quick-connect fittings for use in transferring the PAC from the truck to the storage silo. Check with potential PAC suppliers during design to identify the correct connections (size, type, and male/female) required at the silo. Also ensure that, between the PAC supplier and the facility owner, enough hose is available to transfer the PAC between the truck and the storage silos.

PAC Piping Systems. PAC slurry tends to clog piping, which can be a major maintenance concern. When PAC settles, it can form a hardened deposit that cannot be removed by flushing; instead, the pipe or fitting should either be cleaned using mechanical means or replaced (Hassibi and Lemmon 2009). Several steps can be taken during design to minimize clogging:

1. *Slope:* Slurry pipe should slope downward toward the application point (AWWA/ASCE 2012).
2. *Length:* Minimize piping length between the slurry tank and the injection point.
3. *Velocity:* Maintain a velocity of at least 5 ft/sec in slurry piping to prevent PAC from settling (AWWA/ASCE 2012).
4. *Eliminate locations for PAC accumulation:* Eliminate dead ends and replace tees with 3-way valves. Minimize bends and use long-radius elbows when bends are unavoidable (Figure 12-21). Hose connections can be used to change between multiple PAC feed points to reduce the number of dead ends and valves needed (Figure 12-22).
5. *Drain and flush:* Provide cleanouts or flushing connections so that slurry piping can be drained and flushed with water when taken out of service. The

piping can be equipped with a flushing line that automatically flushes the piping for at least 60 sec after PAC feed is stopped (AWWA/ASCE 2012).

Slurry piping should be corrosion and abrasion resistant; suitable materials include rubber, plastic, and stainless steel (AWWA/ASCE 2012). Using long-radius bends can also reduce wear on the interior surfaces of the piping resulting from abrasion (Norit Americas 2003). Elbows on slurry piping should have a radius of at least 4 ft (Hassibi and Lemmon 2009).

Figure 12-21. Long-radius bend on PAC piping

Site Design

The PAC facility layout should consider ease of access to the facility by delivery or maintenance vehicles and personnel, distance from the injection point, and aesthetics to neighbors. The PAC storage and feed equipment should be installed as close to the application point as possible to minimize pipe runs (AWWA/ASCE 2012). If the appearance of the water treatment plant is an issue to surrounding property owners, it might be advantageous to locate the PAC facility where it is screened from view (Nielsen and Whipple 1979). Installing a new PAC facility at an existing water treatment plant creates additional layout considerations, especially if space is limited.

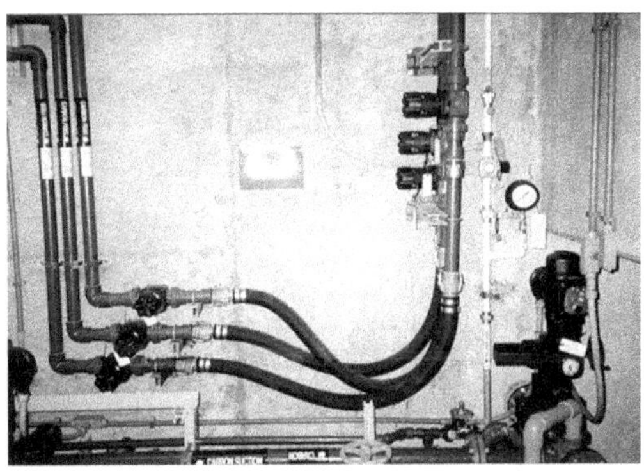

Figure 12-22. Flexible tubing used for PAC

Provide truck access and parking near the storage silos or tanks. The transfer process takes several hours, so the truck parking area should be out of the way of other vehicles accessing the site (Bryant 2009). Driveways accessing the PAC facility should be designed to accommodate delivery trucks. Norit Americas (2003) recommends providing driveways with a turnaround circle with a diameter of 100 ft for a bulk delivery truck that is 65 ft long.

PAC systems are fairly compact; silos and slurry tanks are taller than they are wide, which minimizes the required footprint. Wetting and feeding equipment associated with dry-PAC systems are typically located directly below the silo, which also helps to keep the footprint small.

Additional Design Considerations

Additional design considerations include dust control, residual solids handling, and fire protection.

Dust Control. Dust control during PAC delivery and operation can be a major design consideration. PAC dust pollution might be very objectionable to nearby property owners in addition to being a nuisance and health issue for plant operators. One way of controlling dust in the PAC feed system is to install a dust collector system that pulls a vacuum at any point where dust is expected to occur, such as valves, storage silos, and wetting cones (Nielsen and Whipple 1979). Note that a dust collection system may require an air permit.

Figure 12-23. Round silo dust collection system, exterior (C&W Manufacturing and Sales Co.)

Figure 12-24. Round silo dust collection system, interior (C&W Manufacturing and Sales Co.)

Storage silos should be equipped with bin vents that allow the air used during filling operations to exit the silo. The bin vents can be equipped with dust collection systems such as bag or cartridge filters (Figure 12-23 and Figure 12-24). Dust collectors include a fabric filter on which a dust cake forms as particles in the transfer air are captured and a shaker or pulse-jet mechanism to dislodge the dust from the filter back into the silo. The dust cake itself provides some of the filtration, but it should be removed periodically when the head loss across the filter reaches a certain setpoint (e.g., 10 in. of water column). The pulse-jet mechanism operates continuously or periodically to remove dust from the filter throughout the filling process, while the shaker system operates only after filling has been completed (Hassibi and Lemmon 2009).

The size of the dust collection system depends on the amount of air used for transferring the PAC and, if regulated, the required quality of the air discharged from the system. Dust collection systems may include cloth bags, paper cartridges, or plastic foam to remove particulates from

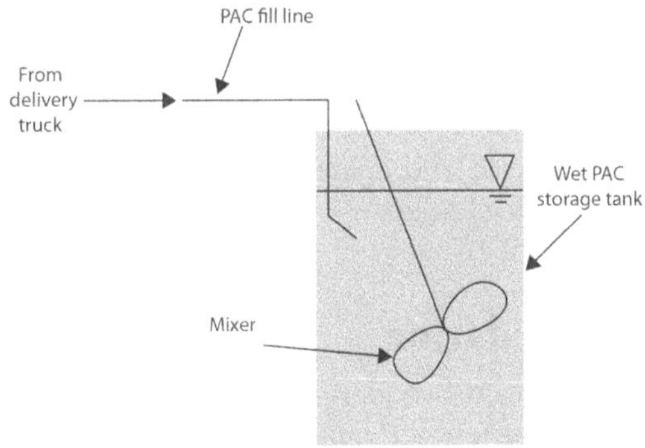

Figure 12-25. Schematic, PAC fill line below water surface

the air. The *air-to-cloth* ratio is one parameter that determines the size of the dust collection system and represents the airflow (in acfm) to surface area of the filter (in ft^2). A larger air-to-cloth ratio means that the unit can treat more air with the same amount of fabric than a unit with a lower air-to-cloth ratio. Typical air-to-cloth ratios for dust collection systems are: cartridge = 1:1, bag = 3:1, and foam = 4:1. Thus, a cartridge filter would require more surface area than a bag filter to treat the same flow, although both could have acceptable treatment efficiency (Norit Americas 2003).

For systems in which dry PAC is transferred from the delivery truck to a storage tank containing water or slurry, dust can be minimized by discharging the PAC below the water line in the storage tank, as shown in Figure 12-25 (AWWA/ASCE 2012).

Dry-PAC processes are dusty and dirty, and even when dust collectors are in service, dust tends to be pervasive. NEMA 4 or 4X is usually acceptable for electrical enclosures. Electric motors should be TEFC (totally enclosed fan cooled) or TENV (totally enclosed nonvented) (Norit Americas 2008).

Residual Solids Handling. Settled PAC is removed from the sedimentation basins and sent to the solids handling facility, where it is treated with or separately from other sludges. The PAC sludge can be thickened and dewatered using typical solids handing equipment such as centrifuges, gravity thickeners, belt presses, and sludge drying beds.

PAC removed in the filters is transferred to the filter backwash waste flow and recycled through the plant's solids handling process.

Fire Protection. Activated carbon dust is classified as combustible but not explosive by National Fire Protection Association (NFPA) Standard 820 (NFPA 2008), which states that fire protection is not required for activated carbon processes. However, it is good practice to take additional precautions to prevent a dangerous dust condition from occurring. To create a potential dust explosion hazard, dust must be present in air in sufficient concentrations (greater than 0.140 g/L) and an ignition source must also be present (Avantor Materials 2011). While these conditions are not commonly present at PAC storage systems, sources of sparks, open flames, and smoking should be prohibited near the dry-PAC storage silo, especially during the filling process. The design of the silo should prevent static charge build-up in the fill piping and include a dust collection system, as previously discussed.

Operation and Maintenance

Operational Considerations

When seasonally or periodically operated PAC systems are shut down, flush and clean out all piping and equipment, including pumps and hoppers. Otherwise, settled PAC can settle and harden and is difficult to remove.

Maintenance Requirements

Maintenance issues with PAC relate primarily to ensuring that the equipment and piping do not become clogged with settled PAC. PAC feed and storage systems should be inspected and operated regularly, even when PAC feed is not required, to ensure that the feed lines are clean and flushed and that PAC is still in suspension (AWWA/ASCE 2012).

Preoxidation

As with GAC, oxidants such as chlorine and ozone react with PAC and limit TOC adsorption rates (Najm et al. 1991). Water treatment plants using prechlorination prior to installing PAC feed systems should reevaluate preoxidation when deciding to use PAC. One water treatment plant found that PAC was "relatively ineffective" for TOC removal and TTHM reduction when prechlorination was used (Carns and Stinson 1978). Another consideration when using both PAC and chlorination is that the reaction of chlorine with PAC reduces the amount of chlorine in the water; additional chlorine might need to be added to maintain a residual (Najm et al. 1991).

Prechlorination might also affect the adsorption of T&O compounds. Gillogly et al. (1998) found that the presence of chlorine reduced adsorption of MIB onto PAC. Not only did chlorine oxidize adsorption sites prior to MIB occupying them; it also released adsorbed MIB back into solution. Delaying chlorine addition after PAC addition can help maximize the amount of MIB that is adsorbed, but the optimum adsorption is achieved by preventing contact between PAC and chlorine as much as possible (Gillogly et al. 1998).

PAC Safety

Activated carbon products have some safety issues users should acknowledge. Some information from material safety data sheets for PAC manufacturers illuminates the issues with respect to oxygen depletion, fire, and airborne dust.

Activated carbon (especially when wet) can deplete oxygen from air in enclosed spaces, and dangerously low levels of oxygen may result (Norit Americas 2007a). Before entering such an area, sampling and work procedures for low oxygen levels should be taken to ensure ample oxygen availability, observing all local, state or provincial, and federal regulations (Calgon Carbon Corporation 2008).

As with most organic solids, fire is possible at elevated temperatures or by contact with an ignition source. Activated carbon is difficult to ignite (the ignition temperature is greater than 220°C) and tends to burn slowly (smolder) without producing smoke or flame. Carbon monoxide and carbon dioxide gas may be generated if combusted. If PAC is allowed to smolder for a long period of time, the amount of carbon monoxide produced may approach the lower explosive limit (Norit Americas 2007a). Contact with strong oxidizers such as ozone or liquid oxygen may cause rapid combustion (Calgon Carbon Corporation 2008).

Airborne dust from PAC can be a respiratory hazard. Manufacturers recommend that PAC handling procedures be conducted to minimize production of dust and respirators be worn if necessary to protect workers.

References

Avantor Materials. 2011. Material safety data sheet: Charcoal activated. Center Valley, Pa.: Avantor Performance Materials, Inc.

AWWA/ASCE (American Society of Civil Engineers). 2012. *Water treatment plant design,* 5th ed. S.J. Randtke and M.B. Horsley, eds. New York: McGraw-Hill Professional.

Bryant, Dana N., interview by Brian Leto. 2009. PAC Systems.

Calgon Carbon Corporation. 1994. PAC metering system. Equipment bulletin. EB-1026-02/94. Pittsburgh, Pa.

Calgon Carbon Corporation. 2008. Material safety data sheet: WPH. Pittsburgh, Pa.

Carns, K.E., and K.B. Stinson. 1978. Controlling organics: The East Bay Municipal Utility District experience. *Journal AWWA,* 70(11):637–644.

Fuerman, B. 2003. Interview by Chris Hill. Dry PAC feed system.

Gillogly, T.E.T., V.L. Snoeyink, A. Holthouse, C.M. Wilson, and E.P. Royal. 1998. Effect of chlorine on PAC's ability to adsorb MIB. *Journal AWWA,* 90(2):107–114.

Hansen, R.E. 1972. Granular carbon filters for taste and odor removal. *Journal AWWA,* 64(3):176–181.

Hassibi, M., and R. Lemmon. 2009. Equipment solutions for storing and feeding a hard-to-handle material such as activated carbon. Monongahela, Pa.: Chemco Systems, L.P.

Najm, I.N., V.L. Snoeyink, B.W. Lykins Jr., and J.Q. Adams. 1991. Using powdered activated carbon: A critical review. *Journal AWWA,* 83(1):65–76.

NFPA (National Fire Protection Association). 2008. NFPA 820: Standard for Fire Protection in Wastewater Treatment and Collection Facilities. Quincy, Mass.: NFPA.

Nielsen, H.L. and D.E. Whipple. 1979. An improved powdered activated carbon feed facility. *Journal AWWA,* 71(6):324–327.

Norit Americas, Inc. 2003. Information sheet: Commonly asked questions about powdered activated carbon storage and handling. Marshall, Texas.

Norit Americas, Inc. 2007a. Material safety data sheet: Hydrodarco. Marshall, Texas.

Norit Americas, Inc. 2007b. PORTA-PAC for wet PAC addition. Information sheet. 00-906-IS. Marshall, Texas.

Norit Americas, Inc. 2008. Design considerations for environments containing virgin powdered activated carbon. Information sheet. 00-902-IS.

Solimar Pneumatics. 2008. Silo fluidizer. Product bulletin. Minneapolis, Minn.

Chapter 13

Activated Carbon System Implementation and Economic Considerations

Today, even more so than in the past, water utilities are continually seeking ways to make their operations more effective and efficient while maintaining their overall mission to protect public health and the environment. For most utilities, the debt service costs associated with infrastructure projects represent a significant part of their annual budgets and rates. In addition, the implementation of capital projects often presents many unique challenges.

Construction of a new treatment plant, or the work required to provide granular activated carbon treatment at an existing plant, can result in a significant project, especially the installation of GAC post-filter contactors with or without on-site GAC reactivation facilities. Reliable cost estimating is also important in a successful project.

In search of the most cost-effective pathways for fulfilling these needs, water utilities are using various approaches for procurement and project delivery. This chapter discusses the basic goals for the delivery of capital projects along with the issues often raised when alternative capital project delivery methods are considered. The chapter also presents methods of procuring activated carbon and estimating the project costs. While there is no simple, single method to estimate construction, project, and operation costs, the most realistic approach is to begin with a carefully developed feasibility analysis, accounting for the many variables a project is likely to encounter during development.

Capital Project Delivery Goals and Issues

Utilities typically have four basic goals for the delivery of capital projects:

- Ensuring the quality of the designed and constructed project.
- Controlling life-cycle costs (capital costs and long-term operating costs).
- Creating a project that will permit effective and efficient operation and maintenance.
- Completing the project within the required schedule and budget.

The following issues are often raised when capital project delivery methods are considered:

- Legal authority
- Degree of desired utility control
- Quality of completed project
- Capital and operating costs and efficiencies
- Schedule
- Risk management and risk allocation
- Internal and external influences
- Utility experience
- Market share, availability, and adaptability of potential contractors
- Labor–management relationships

The order in which these issues are listed is not intended to reflect the importance of one issue over another. Experience has shown that the relative importance of these issues varies from project to project and from utility to utility.

Pathways to Higher Performance (The Synergy of Integration)

Four types of expertise are essential to delivering a complex capital project such as a treatment facility: design, construction, operation, and maintenance (including repair and replacement). The degree to which individuals with these kinds of expertise can work collaboratively—particularly during the design phase—significantly influences project success.

Historically, three separate parties have accomplished the traditional design–bid–build method of project delivery:

- A design consultant in collaboration with a utility's staff
- A low-bid contractor
- The utility's operation and maintenance (O&M) forces

The project's design phase usually includes some steps to incorporate construction expertise, either from within the design consultant's firm or through subconsultants, as well as some O&M expertise. Some of these steps—scheduled constructability reviews, operability reviews, and value engineering discussions—provide a degree of synergy among the three parties.

Nevertheless, the highest degree of synergy can occur when the contractor who will actually construct the project and the O&M staff who will ultimately operate and maintain the facility can work together as a team throughout the project's design and construction. During this collaborative effort, utility leaders and managers should also be integrally involved in project decision-making.

Numerous decisions affect a project's constructability and future operability, as well as its capital and operating costs. With the right leadership, good decisions usually result regardless of the delivery method chosen. Some of the questions typically encountered during the design phase include:

- Why are we embarking on this project?
- Are we solving the right problem?
- Is there an easier or better way to accomplish our goals?
- What is the right balance among design conservatism, system redundancy, and affordability?

- How can we best manage various design, construction, and operation risks?
- What level of risk is acceptable in relation to associated cost savings?
- Can we do without all or part of the project's components and still accomplish its functional objectives?
- Should the facility be staffed 24 hours per day or in some other manner?
- What are the most appropriate treatment technologies to be used?
- How can we use information technology to achieve reliability and operational efficiencies, while addressing regulatory requirements for staffed operation?
- How can we minimize conflicts during construction?

While the traditional design–bid–build method of project delivery is extensively used and has served utilities well, it typically does not allow inclusion of expertise from the contractor(s) who will construct the facility. The lack of contractor input is due to the fact that under this method the constructor is selected only after the design has been completed and the project has been competitively bid.

Capital Project Delivery Methods

Among the various options being used by utilities for delivering capital projects, four categories of methods predominate:

- Traditional design–bid–build
- Enhanced design–bid–build
- Construction manager at risk and construction manager for a fee
- Variations of design–build

These methods are discussed in the following sections.

Traditional Design–Bid–Build (DBB)

In the United States, the traditional method of designing, constructing, and operating water treatment facilities has been the design–bid–build method of project delivery, followed by utility operation of the facility. Basically, the DBB arrangement involves numerous parties: the utility, the designer, the construction contractor, and the utility operator or a third-party operator under contract with the utility. Under the DBB method, the utility does the planning and permitting, with or without the assistance of one or more consultants.

Although the traditional project delivery method has some disadvantages, it has successfully produced reliable infrastructure for utilities for decades. It is well understood by governing bodies, utilities, consultants, communities, and the construction industry.

Today, some utilities are rethinking their own practices along with the basic roles and responsibilities implicit in the traditional capital project delivery method. Some of the factors contributing to increased interest in alternative delivery methods include:

- A changing point of view about the capabilities of public utilities versus the capabilities of private entities.
- The inherent attractiveness of integrating design, construction, and operation.
- Concern about the use of low bid as the determining factor in selecting contractors for construction.
- The emergence of private enterprises that are actively pursuing design–build and even financing of projects.

- The enactment of enabling state legislation that allows for alternative project delivery methods.

The use of DBB still remains by far the most extensive approach for delivering water and wastewater systems and facilities. Virtually all public utilities have policies, procedures, and standard contract documents developed for delivering projects using the DBB approach. The DBB approach gained widespread acceptance decades ago because it simplified selection of construction contractors by having low cost as the main or even the only selection factor.

The participants and their relationships under a DBB delivery method are illustrated in Figure 13-1.

Enhanced Design–Bid–Build Delivery Methods

Recognizing the value of synergistic relationships among the designer, contractor, and operator, many utilities are enhancing the traditional design–bid–build delivery method under such labels as *enhanced design–bid–build*, *optimal design–bid–build*, and *collaborative design and construction*. The premise underlying each of these variations is to allow greater integration of technical expertise during design and construction.

When the completed facility is to be operated by the utility's existing operations staff, key representatives such as the lead operator, the supervisors of mechanical and electrical maintenance, and the individuals in charge of the supervisory control and data acquisition and instrumentation and control systems become essential members of the project team.

Because including the actual builder on the project team is not possible in many cases, some utilities have used a virtual contractor. The concept of using a virtual contractor recognizes that involving general contractors during the design process would likely give these companies an unfair advantage if they decided to bid on the project. In addition, such contractors might not be willing to share ideas that could make them more competitive at bidding time. To avoid these, former or currently

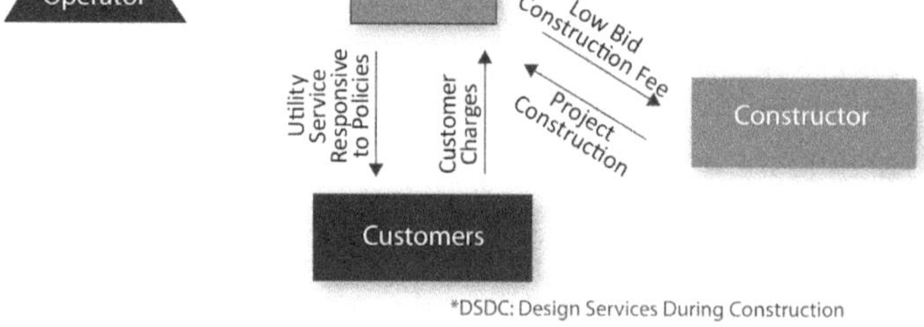

Figure 13-1. Participants and relationships in design–bid–build

practicing contractors who are not interested in bidding on the project can function as virtual contractors on the project team.

Virtual contractors provide expertise in foundation and excavation work, concrete work, HVAC system and mechanical work, scheduling, and cost estimating. Virtual contractors also review the major design elements of the project from a construction perspective and offer recommendations pertaining to constructability and safety that may be inherent in facility layouts being considered.

The enhanced design–bid–build method is not widely used, although many advanced or sophisticated utilities are using this approach for certain projects.

Construction Management at Risk (CM@Risk)

The CM@Risk arrangement can have a number of configurations depending on the desires of the owner and legal requirements. In one form of the CM@Risk arrangement, the owner selects a design firm through a qualifications-based selection process. The designer's responsibilities are similar to those under a traditional design–bid–build delivery method. During the early stages of design development (i.e., concurrent with design consultant selection or before 30 percent of the design has been completed), the owner also retains a construction manager (CM) or construction contractor through a qualifications-based selection process similar to that used to select the design firm.

The CM's scope of work encompasses collaborative participation in the design process, including reviewing the design documents and offering suggestions to increase value, reduce construction time, and improve the quality of the constructed project. A primary purpose is to bring "contractor sense" into the design phase.

Some other specific services provided by the CM/contractor during the design phase typically include performing constructability reviews of the design as well as offering recommendations to the design engineer and the utility owner as to how the design might be completed to provide the highest-quality facilities for the lowest construction costs. The CM also develops both a proposed project construction schedule and a proposed construction cost model that is used by the utility, the design engineer, and the CM to make decisions regarding the final design of the project that will produce the desired goal of highest quality for lowest cost and will stay within the guaranteed maximum price once it is negotiated.

Under the CM@Risk arrangement, the CM provides the owner a guaranteed maximum price (GMP) or lump-sum price, usually at the point when the design is 60 to 100 percent completed. The CM then solicits competitive bids from subcontractors and major equipment suppliers for various work packages. A CM might self-perform some of the work or manage subcontractors to complete the work. If a specific CM@Risk project is based on a GMP, the contract with the owner typically includes a cost-sharing provision in which the CM and the owner share any savings that accrue if the actual construction cost comes in below the GMP. If the actual construction cost exceeds the GMP and no changes have been made in the scope of work, the CM absorbs the cost overrun. The CM assumes construction liability, and the owner retains design and performance liability and responsibility for operating and maintaining the facility.

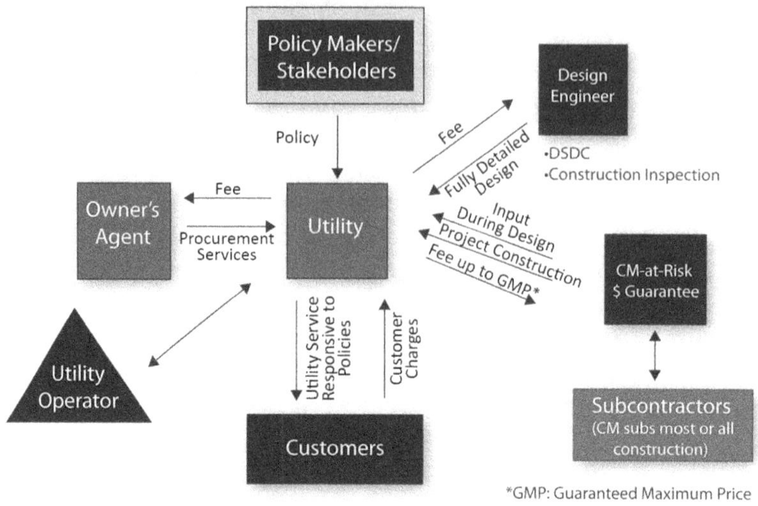

Figure 13-2. Participants and relationships in CM@Risk

The CM@Risk approach has gained significant acceptance and is being widely used where state legislation allows such delivery methods.

The key participants and their relationship under the CM@Risk delivery method are illustrated in Figure 13-2.

CM Services Offered by a Consultant for a Fee

Construction management services offered by a consultant for a fee are not the same as the CM@Risk delivery method. Instead, the owner pays the CM a fee to manage construction, and the owner assumes responsibility for any cost overruns that occur. Sometimes the CM's contract will have incentive clauses that are linked to cost and schedule controls.

CM/General Contractor Services for a Fee (CM/GC)

CM/GC delivery method is commonly referred to as *construction manager/general contractor*. CM/GC is also not a CM@Risk delivery alternative. In the CM/GC arrangement, three major participants enter into the contract with the utility:

- The CM contracted by the utility provides coordination services in lieu of a general contractor and contributes construction expertise by providing input and assistance during the design process.
- The design engineer, also under contract with the utility, is responsible for the design.
- The CM assists with selecting construction contractors. The utility contracts with the prime construction contractors, who are responsible for their respective areas of specialty during the project's construction.

The utility typically selects the construction manager at the same time as or shortly after selecting the design engineer. The prime construction contractors are selected after the design phase and are not involved in project design.

The CM/GC method is more prominent in industries other than the water industry.

Design–Build

Design–build has many faces or alternative configurations. The most frequently used types of design–build delivery methods are:

- Basic design–build
- Design–build–maintain
- Design–build–operate
- Design–build–own–operate–finance

A critical consideration in any design–build effort is whether the owner will ultimately be operating a facility that another entity designed and constructed. The owner's operation and maintenance experience and expertise should be incorporated through the procurement documents used to select the designer–builder and throughout the design process.

Basic Design–Build. In the basic design–build (DB) delivery method, a single entity is contracted to provide both design and construction services. If the project owner desires a greater measure of control over the process, this method is sometimes called *owner-guided design–build*. The owner develops performance requirements to be used in securing a designer–builder. Independent technical, legal, and financial consultants may serve as the owner's agents in managing or assisting the procurement process, establishing performance criteria, and monitoring performance. The designer–builder assumes liability for design and construction, contracts directly with subcontractors, and is responsible for the project's delivery and performance. The primary objective of the DB delivery method is to transfer liability for the project's design and construction to the single-entity designer–builder.

The participants and their relationships under the DB delivery method are illustrated in Figure 13-3.

Basic design–build is the most popular of all the various design–build alternatives and may be the most popular of all alternative project delivery methods.

Design–Build–Maintain. The design–build–maintain (DBM) method is similar to the design–build method except that longer-term maintenance of the facility is combined with design and construction into a single contract. The DBM method is used when the utility wants to maintain a public workforce but also wants to benefit

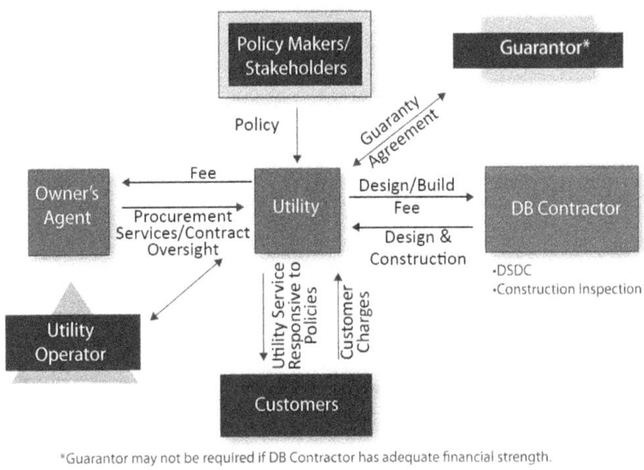

Figure 13-3. Participants and relationships in design–build

from using the private sector for maintenance. An independent consultant can serve as the utility's agent to help with establishing performance and maintenance requirements and with contract monitoring. The primary purpose of an extended maintenance period (typically five to ten years) is to ensure the quality of the equipment and its installation. The DBM arrangement moves the project delivery method closer to the design–build–operate model except that it allows the utility to operate the facility.

Under the DBM method, the utility has the ability to transfer a significant number of responsibilities and risks associated with a constructed project to a single point of responsibility. The single party is responsible for integrating facility design, construction, and maintenance for a significant period of time in accordance with performance criteria established by the utility.

The use of DBM has been very limited in the water industry largely because of difficulties in allocating responsibility between public operators and private-sector maintenance personnel.

Design–Build–Operate. The design–build–operate (DBO) method is similar to the design–build method except that long-term operation and maintenance of the facility (typically 10 to 20 years) is combined with design and construction into a single service contract. Independent technical, legal, and financial consultants may serve as the owner's agents in managing the procurement process, establishing performance criteria, and monitoring performance.

The single DBO entity is responsible for integrating design, construction, and operation and maintenance expertise into the development of the facility in accordance with performance criteria established by the owner. The single DBO entity guarantees the performance of the full-service contract and has the incentive to balance capital cost efficiencies with long-term operation and maintenance costs and to optimize life-cycle costs.

The participants under the DBO delivery method are illustrated in Figure 13-4.

Design–Build–Own–Operate–Finance. The design–build–own–operate–finance (DBOOF) method is similar to the DBO method except that the facility is owned and financed substantially by the DBOOF entity over the long-term period of operation. Under the DBOOF delivery method, the project is financed by the

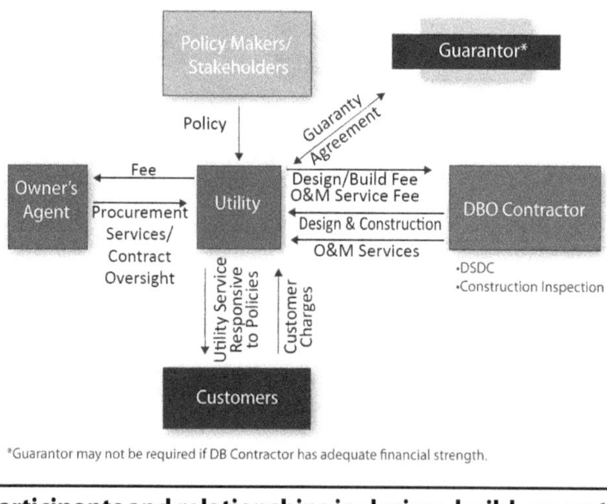

Figure 13-4. Participants and relationships in design–build–operate

DBOOF entity. Independent technical, legal, and financial consultants may serve as the owner's agents in managing the procurement process, establishing performance criteria, and monitoring performance.

Under the DBOOF delivery method, the owner procures a long-term service contract with a DBOOF entity for construction of a privately owned facility. In accordance with the service contract, the owner pays the DBOOF entity an annual service fee, which includes an allowance for the service provider's debt service as well as its operation and maintenance costs.

The DBOOF entity, as the recipient of the tax benefits associated with the project, typically depreciates the project and contributes equity, which can be used to offset the amount of debt needed to finance the project. The DBOOF will own the project when the service contract expires or otherwise terminates, and the owner can either renew the contract or purchase or rent the facility at fair market value if the owner wishes to continue receiving service from the facility. During the life of the contract, the owner typically retains responsibility for rate-setting, billing, collections, and administrative services.

The DBOOF entity assumes the financing risk during project construction, but the owner typically pays a premium because of the higher cost of capital associated with private-sector financing and the transfer of the financing risks to the contractor.

DBOOF has not been widely used in the United States because public financing using public tax-exempt bonds is significantly more cost-effective than private-sector taxable bonds. Elsewhere in the world, where tax-exempt financing is not available, DBOOF is commonly used.

Pathways for Selecting an Appropriate Delivery Method

Utilities have used various means to evaluate and select the capital project delivery method that best fits its particular needs and objectives. In some instances, the selection is merely a decision made by the head of the utility or the governing body to which the utility reports. In many instances, several or all of the alternative methods are evaluated. Important lessons learned from the experience of other utilities in evaluating alternative capital project delivery methods include:

1. Before evaluating alternative delivery methods, clearly define the following:
   a) Goals and objectives for the project, including facility performance, available budget, and deadline.
   b) The legal framework for the procurement.
2. Thoroughly understand and evaluate all of the alternative capital project delivery methods that are applicable to your utility's project, and select an option that is consistent with the utility's risk tolerance and desire for control.
3. Involve representatives of all stakeholder groups in the process used to support decision-making, and ensure that they accept and support a decision to proceed with an alternative project delivery method selected.
4. Carefully evaluate both short- and long-term (i.e., life-cycle) financial implications in the decision-making process.
5. Determine up front who will make the final decision regarding the use of an alternative delivery method and what information will be required to support that decision.

Table 13-1. Key features of delivery methods

	DBB	CM@Risk	DB	DBM	DBO	DBOOF
Who selects the design engineer?	U	U	C	C	C	C
Who designs it?	U/AE	U/AE with CM input	C	C	C	C
Who operates it?	U	U	U	U	C	C
Who owns and finances it?	U	U	U	U	U	C
When is the construction price fixed?	100% design	Varies	Typically when DB contract is signed		Typically when DBO contract is signed	
When is the Operations & Maintenance price set?	NA	NA	O – NA M – Typically at contract signing		Typically at DBO contract signing	

U = utility, C = contractor, AE = architect/engineer, NA = not applicable, CM = construction manager

6. Communicate with representatives of various stakeholder groups who will be affected by the decision to use an alternative delivery method, and include them in developing and, when appropriate, implementing the evaluation process.
7. Exercise care to avoid real or perceived conflicts of interest in choosing independent advisors or consultants who might be involved in evaluating alternative delivery methods or in any private-party procurement process related to your utility's project.

The key features of delivery methods described are shown in Table 13-1.

The major advantages and disadvantages that utilities have experienced under the various delivery methods discussed in this chapter are illustrated in Table 13-2.

Experience with Capital Project Delivery Methods

While the traditional design–bid–build method of capital project delivery remains the choice of most water utilities, there is considerable experience with alternative project delivery methods for both water and wastewater utilities.

A list of water and/or wastewater utilities using alternative capital project contracting approaches is available from the Design–Build Institute of America, utility websites, Water Design–Build Council, the Water Partnership Council, and other selected sources. This list of sources is not meant to be comprehensive, but instead it should serve as an aid to utilities seeking references for projects similar to ones they are considering.

Case studies are included of the experience of two water utilities using design–build–operate for large water facilities:

- San Diego County Water Authority (Case Study No. 4)
- City of Phoenix (Case Study No. 8)

Table 13-3 contains a list of the water treatment plants identified in chapter 4 of this book and the delivery method used for their GAC treatment facilities.

Table 13-2. Comparison of delivery methods

	Advantages	Disadvantages
DBB	• QBS for designer; cost-based selection of constructor • Well-known method • Owner-controlled • Well-defined project (low-risk premium)	• No design efficacy guarantee • No single point of responsibility • Cost uncertain until bids received • Change orders • Low bidder • No O&M guarantees
CM@Risk	• QBS for designer • Select CM on qualifications and price • CM@Risk bids out work • GC input during design • GMP established	• No design efficacy guarantee • No single point of responsibility • Scope and quality should be well defined when GMP established • Extensive coordination required • Additional cost of CM • No O&M guarantees • Additional procurement effort
DB	• Select on qualifications, price, and other criteria • Single point of DB responsibility • Design efficacy risk transfer for testing/ warranty period • Potential for schedule reduction (overlap D&B) • GC input during design • Guaranteed price early in project development	• Scope and quality should be well defined before design complete • Reduced control • Critical to shift design risk and have warranty period; market may resist • No O&M guarantees • Additional procurement effort
DBO	• Select on qualifications, price, and other criteria • Focus on life-cycle cost • Complete risk transfer • Single-source responsibility • Long-term performance guarantee • Potential for schedule reduction (overlap D&B) • GC and operator input during design • Guaranteed price early in project development • Long-term costs known and controlled	• Scope and quality should be well defined before design complete (including O&M) • Least control • Long-term commitment to DBO contractor • Current operator/ labor unions may oppose • Additional procurement effort

Pathways to Creating Lasting Solutions

A well-developed service contract for all but the traditional design–bid–build method of project delivery is critical to establishing a project's performance requirement. The traditional design–bid–build contracts are usually prescriptive-based rather than performance-based. The service contract is typically developed by the water utility based on the procurement steps, including issuing solicitation requests for qualifications or

Table 13-3. Alternative delivery methods for chapter 4 GAC Systems

System	GAC Treatment	Delivery Method
Cincinnati, Ohio	Post-filtration	DBB
Phoenix, Ariz.		
Lake Pleasant WTP	Post-filtration	DBO
Deer Valley WTP	Filter adsorbers	CM@Risk
Union Hills WTP	GAC biofilters	CM@Risk
24th Street WTP	GAC biofilters	CM@Risk
Val Vista WTP	Post-filtration	CM@Risk
Scottsdale, Ariz.		
CAP WTP	Post-filtration	DBB
Chaparral WTP	Post-filtration	DBB
Glendale, Ariz.		
Cholla WTP	Filter adsorbers	DBB
Oasis WTP	Filter adsorbers	DBB

requests for proposals, evaluation of the proposals of respondents, and negotiations with the selected proposer.

Specific details for developing solicitation documents, performance standards, and service contracts are outlined in *The Changing Water Utility* (Westerhoff et al. 1998), which provides specific features and contents of procurement documents and service contracts.

Increased experience with private-sector services in utility operations and in capital project delivery has resulted in complex contractual arrangements. In the traditional design–bid–build method of capital project delivery, utilities maximize the results of contracted services by carefully monitoring compliance with design and contract requirements.

Many of the initial DB, DBO, and operations service contracts typically did not include sufficient levels of detail in performance specifications and were not carefully monitored as a result of the belief that because more services, including professional services, were being allocated to the private sector, there would be little need for detail in contracts or for independent third-party monitoring. Utilities are finding that more specificity in service contracts and contract monitoring is essential. Still, utilities should exercise care not to overspecify requirements and thus inhibit private-sector creativity.

Activated Carbon Procurement Considerations

In many ways, the technical specification for activated carbon is very similar to specifications for other granular media (e.g., anthracite and sand) with respect to content and arrangement. However, the unique features of activated carbon should be considered when developing technical specifications for GAC or PAC.

Granular media technical specifications detail the specific physical requirements that are most important for determining filtration performance (i.e., particle removal), including size, shape, density, and hardness. Beyond these physical properties are performance-related considerations unique to activated carbon. For example, GAC will likely be used for both filtration and adsorption in a filter application; however, not all GACs will perform similarly with respect to adsorption. Therefore, in order to select a media for adsorption performance, additional components to the specifications will be necessary beyond physical properties. Ideally, the procurement process accounts for the fact that activated carbons may (and often do) perform differently; a properly developed procurement process identifies the best-performing media on a cost basis. Moreover, activated carbon has specific physical properties that should also be detailed beyond those included for other granular media.

Physical Properties

Beyond the traditional physical media properties of size, density, shape, and hardness, activated carbon has additional physical characteristics that are usually included in the technical specification.

Iodine number is a traditional requirement in GAC specifications, and it is a relative measure of the micropore volume in GAC. Given that TOC adsorption mostly occurs within micropores, it is generally believed that micropore volume strongly influences TOC adsorption performance. The idea behind including a minimum iodine number in a GAC specification is that it will exclude products with low micropore volume that are not likely to perform well. However, the current practice is more toward using a performance-based specification for GAC, and consequently iodine number is no longer typically used in the specification process.

Technical specifications in the United States typically specify that bituminous products used must be reagglomerated. The reagglomerated characteristic refers to the manufacturing process that bituminous coal-based activated carbons undergo, and it is specified to eliminate products that were manufactured through direct activation. Offshore products often are produced through direct activation, which begins with an inexpensive raw material and proceeds directly to crushing, sizing, baking, and activating. To save production costs, the pulverizing, binding, and reagglomeration steps are eliminated. While direct activation results in a lower-price-per-pound media, it compromises long-term product performance in most applications. Products manufactured in the United States typically undergo the complete activation process, including the reagglomerating steps. It is important to note that reagglomeration only applies to bituminous coal-based activated carbons and not to lignite-based, wood-based, or coconut shell-based materials.

When specifying more than one base material activated carbon type (such as bituminous coal-based and lignite-based), it is important to include separate tables of physical properties for each activated carbon type. In the case of organics removal, more than one type of activated carbon may perform adequately.

Special Considerations in Developing Technical Specifications for PAC

The use of PAC has been detailed in an earlier chapter (chapter 12), and key issues concerning the use of PAC for removing T&O compounds were addressed. The variability among commercially available activated carbons when it comes to removing T&O was discussed; which is an important consideration as a utility selects a PAC.

In order to address that issue, special considerations should be included in the PAC procurement process.

PAC Performance Testing

Performance testing may be used in the procurement process to help a utility improve the effectiveness and efficiency of its PAC application. The most appropriate method for evaluating PAC performance is through the use of jar tests. The protocol for the jar test should match full-scale conditions as closely as possible, especially with respect to contact time, mixing velocities, presence and dose of other treatment chemicals, and conditions in settling basins that can lead to desorption. Particular attention should also be paid to using similar source quality water with respect to NOM and odorant concentrations. As T&O events can often be seasonal, a utility may need to spike geosmin or MIB as part of the testing process. A case study is presented below to highlight the use of performance testing in the PAC procurement process.

PAC Example. The Manatee County Public Works Department Water Treatment Plant (MCPWD-WTP), in collaboration with the Water Research Foundation (formerly Awwa Research Foundation), developed a procedure to use site-specific bench-scale tests to predict PAC performance for the selection of PAC type (Graham et al. 2000). The bench-scale protocol may be performed in most drinking water facilities because it is, in general, a modified jar test. Bench-scale testing was used rather than full-scale studies or pilot-scale studies for a variety of reasons, most importantly because bench-scale studies can be performed in relatively short times with low costs, and variables can easily be controlled in a laboratory setting. By contrast, full-scale studies are impractical, and pilot-scale studies can be difficult to correlate to full-scale adsorption performance.

As part of the procurement process, site-specific bench-scale tests using the candidate PACs are conducted by the utility. The bench-scale test results are then screened to determine the PAC that provides the most adsorption, and these results are incorporated into the selection process. Results may be screened using either of two methods: a minimal method and a more in-depth procedure.

The minimal method uses the site-specific bench-scale protocol results to determine the PAC that adsorbs the most odorant (e.g., geosmin, MIB) at a specific PAC dose and then applies bid factors to single out the most cost-effective type of PAC for odorant removal. In this example, three bid factors were determined for four different types of PAC: (1) concentration bid factor, (2) percent odorant removed bid factor, and (3) flavor profile analysis bid factor. The concentration bid factor is based on concentration of odorant removed, where the PAC removing the most odorant concentration is assigned a factor of one. The factor is determined for the remaining PACs by taking the highest concentration of odorant removed divided by the concentration of odorant removed for each of the remaining PACs. The second bid factor, percent factor, is obtained by first determining the percentage of odorant removed for each PAC used and then taking the percentage of odorant removed by the PAC removing the most divided by the percentage of odorant removed for each of the remaining PACs. The concentration bid factor and the percent/odorant removed bid factor will yield the same results. The final bid factor is the flavor profile analysis factor, which is determined by taking the PAC with the lowest flavor profile analysis intensity value divided by the flavor profile analysis intensity values for the remaining PACs.

In each bid factor calculation, the PAC that removes the most odorant will always be given a factor of one, and the remaining PACs will be above one, depending on their relative odorant removal capacity. Each of these bid factors can then be multiplied by the PAC costs from activated carbon suppliers to determine the most cost-effective PAC based on performance.

The second method to screen the site-specific bench-scale protocol is a more in-depth procedure that considers the nonlinear relationship between odorant removal and PAC dose. There are four major steps to this process: (1) developing dose-response curves for all PAC screened, (2) establishing performance goals, (3) developing dose-equivalent factors for each PAC screened, and (4) applying the dose-equivalent factors to PAC costs. MCPWD-WTP used this process in their PAC selection, comparing three PAC types for geosmin removal. Using bench-scale tests, the dose-response curves for geosmin removal were generated. These dose-response curves consisted of at least four PAC doses per product evaluated, with the PAC doses plotted on the x-axis and the concentration of odorant (geosmin in this example) remaining on the y-axis. The second step was to establish PAC performance goals; for the MCPWD-WTP, the performance goal was set at 7 ng/L of geosmin, as this is the complaint threshold for this odorant for their system. In general, the performance goal should be utility-specific based on the concentration of odorant above which customer complaints become unacceptable.

Dose-equivalent factors are then calculated by determining the dose required to reach the performance goal for each type of PAC tested. The lowest dose of PAC required is assigned a factor of one, and the other PAC types tested are assigned dose-equivalent factors based on the ratio between their dose needed and the lowest dose of PAC. Similar to the minimal method, the dose-equivalent factors are then multiplied by the PAC costs from activated carbon suppliers to determine the cost-effectiveness of each different PAC and the actual treatment cost for the utility.

This protocol can be used for conventional surface water plants using coagulation, flocculation, sedimentation, and filtration. The protocol has been verified by several full-scale studies and can effectively be used to select the most cost-effective PAC for geosmin and MIB removal.

Special Considerations in Developing Technical Specifications for GAC

Similar to the PAC specifications, developing a technical specification for GAC includes a number of special considerations beyond those of traditional filter media. Firstly, not all activated carbon companies are able to provide both installation and disposal services for GAC, potentially requiring an additional agreement with a separate contractor. While this would likely not be a concern for traditional granular media such as sand or anthracite, GAC may need to be replaced more frequently in order to maintain adsorption capacity. The general specifications must include the complete scope of work for both installation and disposal services for the GAC.

Operational circumstances may also dictate which vendors are available for selection; it is important to consider a vendor that will be able to deliver the activated carbon in the time frame required. For example, a system with multiple adsorbers that depends on frequent GAC replacement to achieve water quality treatment goals will require a reliable supply of GAC. In recent years, higher demand for activated carbon (as a result of the Stage 2 D/DBP Rule and a heightened awareness of the potential unintended consequences of converting to chloramines) has increased the lead time for delivery. Systems requiring frequent GAC replacement should also consider on-site GAC storage so that a missed delivery does not compromise treatment performance.

GAC Performance Testing

Performance tests may be used to approximate the adsorption performance of an activated carbon, allowing a utility to best determine the most appropriate GAC for their goals. There are a number of performance tests available, from the desktop to the full-scale. Prequalifying activated carbon through performance testing and incorporating those results into the procurement process is a relatively new approach for GAC. A case study is presented below for a water utility in the United States that used performance test results within its procurement process.

At the desktop level, mathematical models can simulate the behavior of an adsorber using physical and chemical properties of the contaminants combined with performance estimations. One such alternative, AdDesignS,™ interfaces several fixed-bed adsorption models, model parameter estimation methods, and isotherm and adsorbent databases. AdDesignS yields the best results when the adsorption characteristics of individual products and/or contaminant properties are used. As with any model, the results can be unreliable if incorrect inputs are used, and full-scale performance may not match the modeling results. Batch experiments, pilot-scale experiments, and model simulations using AdDesignS were performed to determine the geosmin removal performance and bed life of a GAC filter adsorber at the St. Paul Regional Water Services (SPRWS) in St. Paul, Minn. Results revealed that the actual bed life (>714 days) was much longer than the simulated values because bed life was extended by biological degradation of geosmin. Activated carbon vendors also provide adsorption isotherms based on batch studies. Isotherm testing is performed to determine the relationship between adsorption capacity of a GAC and concentration of an adsorbate under specific background water conditions. Isotherms should be used with caution; they do not necessarily reflect flow-through performance because kinetic effects are not properly represented.

At the bench scale, mini-column tests, or rapid small-scale column tests (RSSCTs) can quickly (typically within weeks) provide accurate predictions of full-scale performance. These tests, however, are not easy to conduct or interpret and require a qualified laboratory. The most useful adsorption performance data would be obtained from the pilot- or full-scale operation of GAC. However, these data take much longer to produce than data from RSSCTs, and thus they are not normally gathered as part of the bidding process.

GAC Project Example. A water utility in New York was planning to use GAC for the specific goal of reducing TOC levels in the finished water during the third quarter for Stage 2 D/DBP Rule compliance, with GAC replacement occurring annually. Adsorption performance, therefore, became a very crucial component of the GAC specification. The technical specification therefore included prequalification testing prior to the bidding process such that only products that meet or exceed a minimum performance standard could be included in the bidding.

In the approach used for this example, the first step was to determine the minimum performance standards for the GAC that still meet the utility's water quality goals. Pilot-scale testing had been completed the previous year to evaluate GAC for

TOC removal. These pilot-scale results were used to develop minimum performance standards.

The performance standards were then incorporated into a prequalification criteria notice submitted to activated carbon vendors mandating that each interested bidder submit a product to be tested against the minimum standards. The prequalification notice outlined both the performance standards as well as the process to be followed. Using this approach, the utility specified that the basis for awarding the project comprised two distinct selection criteria: (1) prequalification of the GAC material by providing a GAC that can achieve a minimum performance standard, as determined via performance/qualification testing (see below), and (2) providing the lowest total bid price.

Interested bidders were required to provide 1 lb of the specified GAC product to a preselected laboratory for performance testing. As outlined earlier, various testing methodologies are available to approximate adsorption performance. For this particular case, RSSCT testing was conducted to ascertain the TOC adsorption characteristics of the proposed GAC. The testing protocol was defined by the utility and specified an empty-bed contact time.

Acceptance criteria were specified in the prequalification notice. Because the utility was not able to perform the prequalification tests during the same time of year as the pilot study (the results of which were used to establish the prequalification requirement), the performance requirement was based on normalized breakthrough profiles in terms of C/C_0. Using pilot-scale and RSSCT results for the water, it was determined that the normalized breakthrough profiles corresponding to influent TOC concentrations of 1.8–2.5 mg/L were virtually identical. Therefore, it appeared that slight changes in the influent TOC concentration would not have a significant effect on how quickly the effluent TOC profile reaches 35 percent breakthrough (the basis of the performance requirement). The bed life will vary more with influent TOC concentration when the influent TOC concentration is highly variable (i.e., if it varies by more than 2–3 mg/L).

The minimal standards set for this example, and included in the prequalification notice, were as follows: An acceptable activated carbon shall generate an effluent TOC profile that does not exceed 35 percent breakthrough ($C/C_0 \leq 0.35$) after processing 5,000 bed volumes. The influent TOC concentration will likely be between 1.8 and 2.5 mg/L. These performance goals are based on the results of pilot-scale testing conducted at the plant. TOC levels were monitored in the effluent until the profile reached at least 35 percent breakthrough. Results from the performance testing were evaluated to determine eligibility for bid submittal. GAC vendors with eligible GACs were notified of the results and then invited to submit a bid. The supplier providing the lowest total bid price for a product that met the performance standard was then awarded the bid.

Additional Performance Test Considerations

The previous case study described how RSSCTs could be used to prequalify GACs for the bid process. An alternative approach would be to directly compare the RSSCT results for various products in a manner similar to that described above for PACs. However, this can prove challenging for the following reasons. First, inherent variability of the RSSCT test can make it difficult to accurately predict the number of bed volumes that correspond to specific effluent TOC targets. That is, duplicate

RSSCTs may not produce identical profiles, though a central lab specializing in RSSCTs can obtain reproducible results. Furthermore, significant gaps can exist between data points in a TOC profile generated by an RSSCT, requiring interpolation to determine the number of bed volumes corresponding to a specific TOC concentration. A large utility in the Southwest used the approach of directly comparing GAC performance via RSSCTs. Because of the challenges noted above, it was difficult to differentiate the performance of the candidate products, and ultimately the process was abandoned in favor of a standard bidding approach (using standard, nonperformance specifications).

Additional Installation Considerations

Beyond the physical and performance properties of activated carbon, the technical specification should also detail additional installation considerations such as disinfection, media compatibility, and delivery.

- *Disinfection.* Unlike anthracite or sand, GAC cannot be disinfected via chlorination after installation. GAC rapidly neutralizes chlorine (negating any disinfection) and loses adsorption capacity in the process. Some GAC vendors have developed disinfection procedures using sodium hydroxide or hydrogen peroxide; however these are not commonly employed. Rather, the empty filter cell is disinfected prior to GAC installation. Disinfection of the filter cell can be performed in accordance with AWWA C653. Additionally, as the GAC will not be disinfected, it must be handled carefully during the installation process.
- *Media Compatibility.* In filters that contain both GAC and sand, the required sand size/type may be different than in an anthracite/sand filter; due to the low bulk density of GAC, a smaller and/or lighter sand will be necessary to allow for uniform fluidization of the GAC and sand layers during backwashing. Moreover, the characteristics of the GAC and sand should be coordinated to avoid excessive mixing of the layers.
- *Delivery.* When evaluating delivery methods (tanker trucks vs. super sacs), consider any reasons why one approach is preferable to the other. For example, drainage limitations may render super sacs impractical, and limited truck access may make tanker trucks impractical.

Considerations for Developing Cost Estimates

Cost estimating at various milestones from concept development through completion of final design is an important element in the implementation of a PAC or GAC project. Typically, cost estimating consists of three components: construction cost, project costs, and operating and maintenance costs. The major divisions of project cost estimates include design, financial, legal, inspection during construction and start-up, appropriate contingencies, land acquisition, public information, and other nonconstruction cost components.

The utility, along with elected and appointed officials, needs realistic and reliable cost estimates when making a decision on whether to proceed with a project and then as implementation of the project proceeds. It is important to recognize that many activated carbon facilities will need to be integrated into existing treatment plants. The estimates should allow for the inherent difficulties of working in and around existing structures and the cost of maintaining required treatment operations in service during construction.

Experience indicates that additional items of work are often added into the project as it proceeds. This is reasonable because it is often appropriate to expand, repair, or improve other facilities at the treatment plant concurrently with other projects. Either adequate contingencies must be included in project estimates to include such work expansion or they should be accounted for as the project proceeds.

No reliable cost curves adequately reflect all of the variables involved with the implementation of an activated carbon project. Example cost curves are presented in McGivney and Kawamura (2008); however, numerous site-specific factors can significantly affect cost estimates. These can include foundation conditions, construction workforce availability and wage rates, competition in bidding environment, and changes in costs between original estimates and actual time of construction. Often these are accounted for as project contingencies, but they can be substantial and should not be overlooked in the cost estimating process. Estimating construction costs for a PAC system is more straightforward than estimating costs of a GAC system; however, site-specific considerations may still need to be addressed and included in the estimate.

It is also difficult to approach an estimate of operation and maintenance costs using a cost curve. The best practice is to develop a staffing plan and O&M schedule. In a PAC system, the plan should include the dose and cost of the powdered activated carbon; and in a GAC system, the plan should include the reactivation frequency, reactivation cost, and make-up carbon cost. If the GAC facility is not equipped with an on-site reactivation facility, the operational cost should include the cost of replacement carbon. In addition, labor associated with replacing media (or a service contract for media replacement) should also be included in the O&M cost estimate for GAC.

References

Graham, M., I. Najm, M. Simpson, B. MacLeod, S. Summers, and L. Cummings. 2000. Optimization of powdered activated carbon application for geosmin and MIB removal. AWWA [now Water] Research Foundation; Denver, Colo.

McGivney, W., and Kawamura, S. 2008. *Cost estimating manual for water treatment facilities*. John Wiley and Sons, Hoboken, N. J.

Westerhoff, G.P., D. Gale, P.D. Reiter, S.A. Haskins, and J.B. Gilbert. 1998. *The changing water utility: Creative approaches to effectiveness and efficiency*. J.B. Mannion, ed. Denver, Colo.: AWWA.

Appendix A

Example Environmental Life-Cycle Assessment

The concept of using a life-cycle assessment (LCA) as a systematic approach to quantify potential environmental burdens of a product or process was presented in the book introduction. An LCA follows the International Organization of Standardization (ISO) 14040 standard using standard software tools.

This appendix contains an example to illustrate the use of LCA to evaluate the environmental impacts on typical processes to reduce disinfection by-products for complying with the USEPA Stage 2 D/DBP Rule. Three treatment technology scenarios were selected for the example: (1) GAC filter adsorbers, (2) GAC post-filter contactors, and (3) enhanced coagulation followed by disinfection using chloramines (EC/chloramines).

System Boundaries

In this example, the system boundaries are defined for each scenario evaluated and include raw material extracted from the environment (GAC, chlorine, etc.), water treatment plant (WTP) production, and end of life (emissions, waste disposal, etc.). Figure A-1 is a simplified version of the major components of the GAC filter adsorber, GAC post-filter contactors, and EC/chloramines scenarios. Processes common to all scenarios are not included in the boundaries as they would not demonstrate environmental differences between the scenarios. For example, power used for mixing and flocculation will be similar across scenarios and therefore is not included.

Key Assumptions Used in Developing the LCA

1. *The LCA only evaluates differences in processes.* For example, other chemicals (e.g. pre-chlorine, polymer, fluoride, etc.) required for the production of the potable water that are similar to each scenario are not included.
2. *The LCA assumes all plant equipment necessary to support the treatment processes exists and has the same life-span.* This is a good assumption for the GAC filter adsorber and EC/chloramines scenarios because, in most cases; a conventional WTP needs minimal modifications to implement the processes. Most conventional WTPs do not have post-filter contactors and would need to significantly modify the structure to accommodate GAC post-filter contactors. However, because the life of the filter basins relatively long (assumed at 150 years), building material and construction of these basins will have minimal environmental impact compared to operational parameters.
3. *The analysis includes sludge handling to the exit of the plant.* Sludge disposal is not included in the LCA boundaries.
4. *The resulting water quality of each scenario is considered to be equal (to meet Stage 2 D/DBP Rule).* The LCA does not evaluate impacts of human consumption

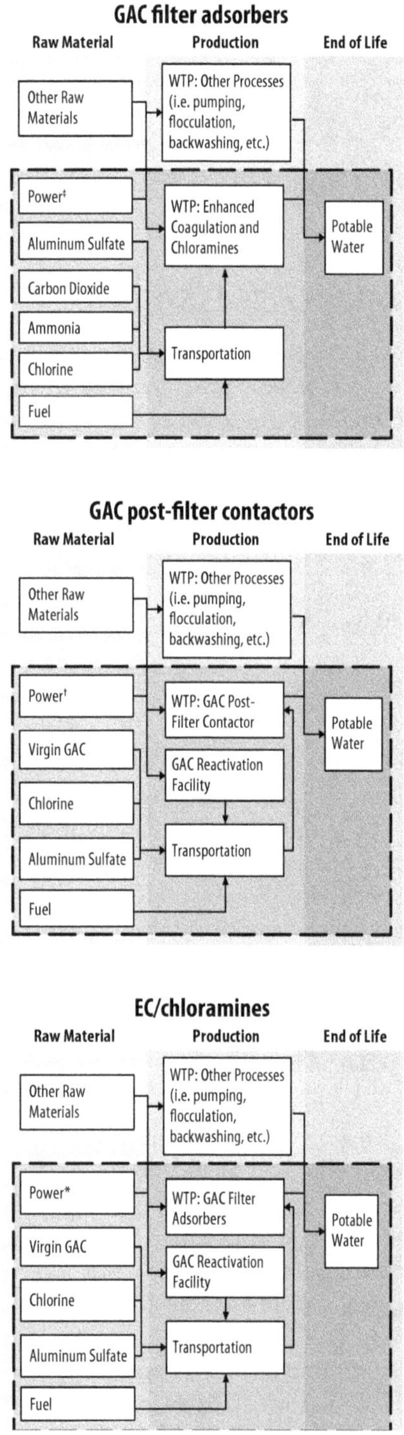

Figure A-1. Major scenario components

of the product water. Impact category parameters (e.g., human toxicity potential) include some regulated contaminants (e.g., chloroform and VOCs); however, they do not contain known emerging DBPs of chloramines, which may differ from DBPs of GAC-treated waters and may have human health effects.

5. *GAC is considered to be reactivated with the exception of 5 percent, which is burned off during reactivation and supplemented with virgin GAC.* The LCA does not account for a plant that decides to use virgin GAC for each change-out. The total organic carbon (TOC) sent to scrubbers is not included in the LCA boundaries.

Scenario Descriptions

This example assumes that the end product of all processes is potable water with comparable water quality sufficient to meet the Stage 2 Disinfectants and Disinfection By-Products (D/DBP) Rule. The functional unit for this example is one day of production at a 20-mgd flow rate. The following are descriptions of each process evaluated:

GAC Filter Adsorbers

GAC is a highly porous media used to adsorb organic and inorganic constituents from water. When the GAC is no longer able to adsorb materials, it is "spent" and needs to be reactivated. GAC adsorbs certain DBP precursors, thus rendering them unavailable to react with chlorine or other disinfectants. If a treatment plant wishes to use this technology, GAC filter adsorbers would replace conventional media filters. This scenario includes aluminum sulfate addition for coagulation, sedimentation, filtration with GAC filters, and chlorine addition for disinfection, as seen in Figure A-2. This scenario also assumes that there is minimal retrofitting construction required.

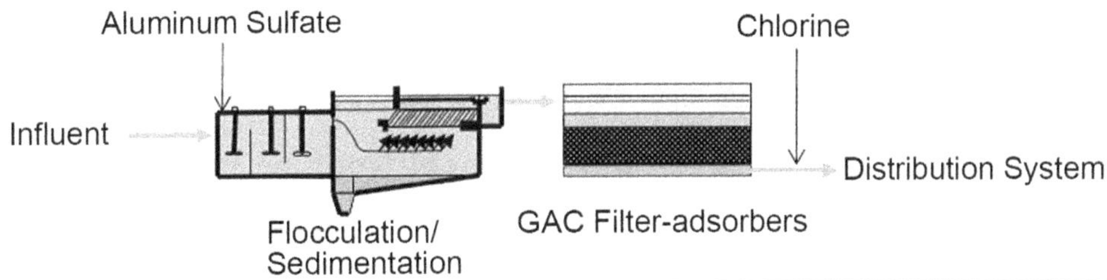

Figure A-2. GAC filter adsorber schematic

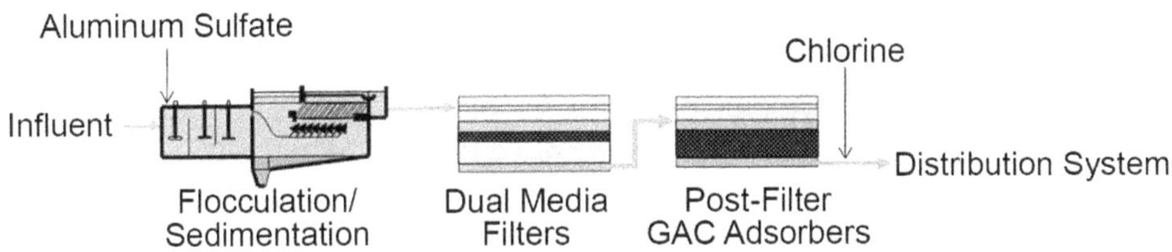

Figure A-3. Post-filter contactor schematic

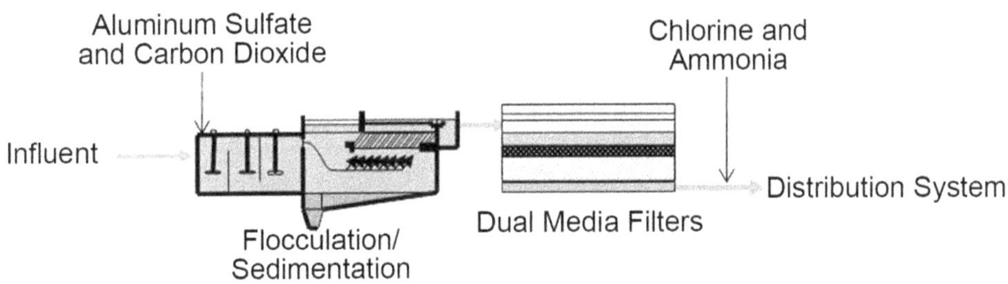

Figure A-4. EC/chloramines schematic

GAC Post-Filter Contactors

GAC post-filter contactors are basins or contactors placed after conventional media filters. Typically they are used to reduce DBP precursors or other contaminants, such as taste and odor compounds. This scenario assumes that the post-filter contactors already exist. This scenario includes aluminum sulfate addition, sedimentation, GAC post-filter contactors, and chlorine addition for secondary disinfection, as seen in Figure A-3.

Enhanced Coagulation and Disinfection Using Chloramines

Enhanced coagulation is the process of obtaining improved removal of DBP precursors by conventional treatment. The enhanced coagulation process analyzed in this scenario includes the following two-step process: (1) increasing aluminum sulfate dose and (2) changing pH using carbon dioxide. Chloramines are formed in water undergoing chlorination with ammonia. Chloramines have a lower tendency to create chlorinated DBPs regulated under Stage 2 D/DBP Rule. This scenario includes aluminum sulfate and carbon dioxide addition, sedimentation, and chlorine and ammonia addition, as seen in Figure A-4.

Life-Cycle Inventory Analysis

Operational data were primarily obtained from utilities in the Phoenix, Ariz., metropolitan area, the authors' experiences with these utilities, datasets available in the GaBi 4 software package, and literature sources. Transportation distances were obtained from authors' information and vendors. Table A-1 is a summary of inventory data used for all LCA scenarios.

Impact Assessments of Example

While many environmental impact categories can be used for an LCA, eight typical categories were selected for the purposes of this example. A description of each category is briefly listed below:

1. *Global Warming Potential* (GWP) is a global measure of greenhouse gases that contribute to global warming over a specified time period, usually 100 years.
2. *Ozone Depletion Potential* (ODP) is a global measure of ozone degradation in the stratosphere.
3. *Acidification Potential* (AP) is a regional measure of the decrease in pH of rainwater and fog due to the washing out of acidifying air pollutants, such as sulfur dioxide (SO_2) and nitrogen oxides (NOx).
4. *Eutrophication Potential* (EP) is a local measure of excessive nutrients entering water and land from substances such as phosphorus and nitrogen from agricultural runoff, air pollutants, and wastewater effluents.
5. *Photochemical Ozone Creation Potential* (POCP) is a measure of ozone formation (summer smog) in the troposphere by sunlight instigating photochemical reaction of NO_x with hydrocarbons and VOCs.
6. *Human Toxicity Potential* (HTP) is a local measure of potential continuous human harm caused by a unit of chemical released into the environment.
7. *Aquatic Ecotoxicity Potential* (AETP) and
8. *Terrestrial Ecotoxicity Potential* (TETP). Ecotoxicity potential (ETP) is a measure of continuous toxicological impact on water and soils.

Water use, sludge production, and energy were also quantified, as these are significant issues that affect the local environment.

Example Results

Results for one day of product water are presented in Table A-2. All data were compared using generally accepted LCA units of measure.

There is no one scenario that ranked higher or lower in all impact categories analyzed. Each scenario performed within 25 percent (reasonable estimate for percent accuracy of data) of each other for all categories, with the exception of acidification potential (AP), eutrophication potential (EP), and human toxicity potential (HTP).

- AP in GAC post-filter contactors was 33 percent and 24 percent higher than EC/ chloramines and GAC filter adsorbers, respectively. This is primarily due to the increased energy required to backwash post-filter contactors.
- EP in GAC filter adsorbers was 26 percent higher than EC/chloramines. This is primarily due to additional thermal energy required to reactivate carbon.

Table A-1. LCA Inputs, functional unit = 1 day at 20 mgd

	GAC Filter Adsorbers	GAC Post-Filter Contactors	EC/ Chloramines	Distance (km) to WTP (40-ton truck)
Granular activated carbon[1]				1,685
Virgin GAC (mg/L)	2.2	1.8	0	
Reactivated GAC (mg/L)	2.6	34.0	0	
Chlorine[2]				32
Dose (mg/L)	3	3	4.6	
Ammonia[3]				571
Dose (mg/L)	0	0	1.7	
Aluminum sulfate[4]				599
Dose (mg/L)	24	24	35	
Carbon dioxide[5]				1,889
Dose (mg/L)	0	0	30	
Additional energy inputs[6]				
GAC reactivation, transfer, and backwash (for contactors) (kWh/day)	7,277	12,633	0	
Sludge handling (kWh/day)	106	98	146	
Off-gas production from reactivation facility (kg/day)[7]				
Hydrocarbons	0.12	0.10	0	
Carbon monoxide	0.14	0.11	0	
PM10	0.05	0.04	0	
NO_x	1.12	0.89	0	
Sulfur dioxide	0.23	0.18	0	

1. Manufacturing locations and reactivation locations were assumed to be in the potential reactivation facility in Blue Lake, Calif. (Calgon Corporation is currently considering building this facility). Carbon is assumed to be reactivated until burned off during reactivation process (5 percent). LCA data source: PE International.
2. Chlorine is widely manufactured across the United States. Manufacturing location for chlorine is in the Phoenix metropolitan area and provided by Hills Brothers Chemical Company. Product is ultimately dissolved in the water. LCA data source: PE International (chlorine mix).
3. Manufacturing location for ammonia is in Industry, Calif., and provided by Hills Brothers Chemical Company. Product is ultimately dissolved in the water. LCA data source: PE International (ammonia mix).
4. Manufacturing location for aluminum sulfate is in Los Angeles and provided by General Chemical. Product is ultimately dissolved in the water and is contained within the sludge. LCA data source: PE International (aluminum sulfate mix).
5. Manufacturing location was provided by a MPI LCA evaluation for Peoria, Ariz. Manufacturing distance was based on the distance from Houston, Texas, to the WTP. Product is ultimately dissolved in the water. LCA data source: PE International (carbon dioxide mix).
6. Additional energy inputs for GAC evaluations include sludge handling, reactivation, GAC transfer pumping, and additional backwashing (for GAC contactors). Additional energy inputs for enhanced coagulation and chloramines include sludge handling. LCA data source: PE International (US Power Grid mix).
7. Off-gas production was estimated based on MWD Granular Activated Carbon Regeneration Study, 1992. This is an output that had to be calculated based off MPI data, not included in GaBi software. LCA data source: Granular Activated Carbon Regeneration Study. Prepared by Malcolm Pirnie for the Metropolitan Water District of Southern California. Carlsbad, CA. 1992. Unpublished study.

Table A-2. Overall LCA Results for 1 day of product water (20 mil gal)

Impact Category	GAC Filter Adsorber	GAC Post-Filter Contactors	EC/ Chloramines	Units
AP	90.05	117.95	79.32	kg SO_2-equiv.
EP	6.52	7.76	8.82	kg phosphate-equiv.
AETP, freshwater	42.37	49.53	49.47	kg DCB-equiv.
GWP, 100 years	26,191	31,173	29,699	kg CO_2-equiv.
HTP	836	1,087	776	kg DCB-equiv.
ODP, steady state	0.003	0.003	0.003	kg R11-equiv.
POCP	6.16	7.75	6.19	kg ethene-equiv.
TETP	23.97	26.70	31.90	kg DCB-equiv.

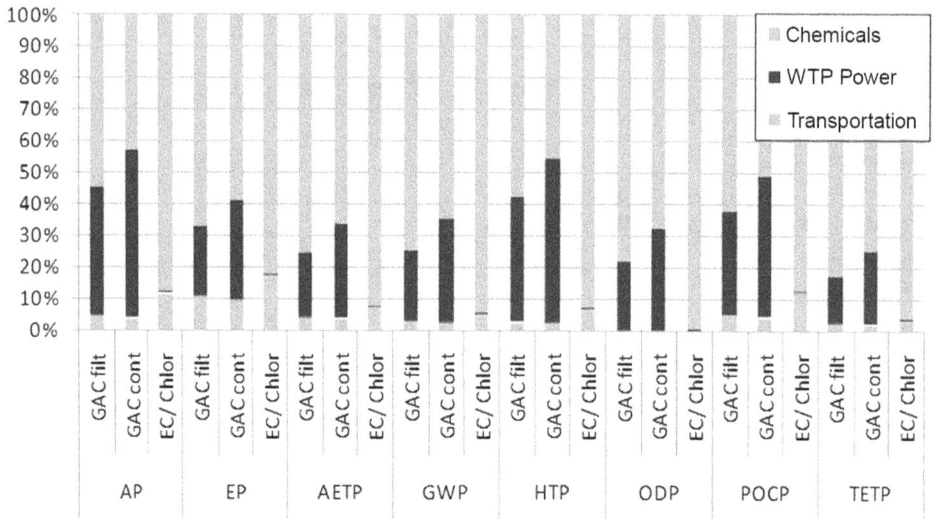

Figure A-5. Analysis of contributions by activity

- HTP in GAC post-filter contactors was 29 percent higher than EC/chloramines. This is also primarily due to the increased energy required to backwash post-filter contactors.

Chemicals have the largest effect on all of the environmental impact categories for each scenario as shown in Figure A-5.

Specifically, chlorine had the greatest effect on all impact categories, followed by power consumption. A sensitivity analysis was performed to measure the effect of total plant electricity (assumed at 26 MWh/day) on the impact categories. It was found that all the data were within 25 percent of each other and the largest contributor for all the environmental impact categories was energy.

There is a strong relationship between total organic carbon (TOC) and chemical usage (i.e. GAC, alum, chlorine, etc.). The source water in this project was assumed to have high TOC (3–5 mg/L). For waters with lower TOC values, chemical doses would be lower and therefore chemical usage would have a lower environmental impact.

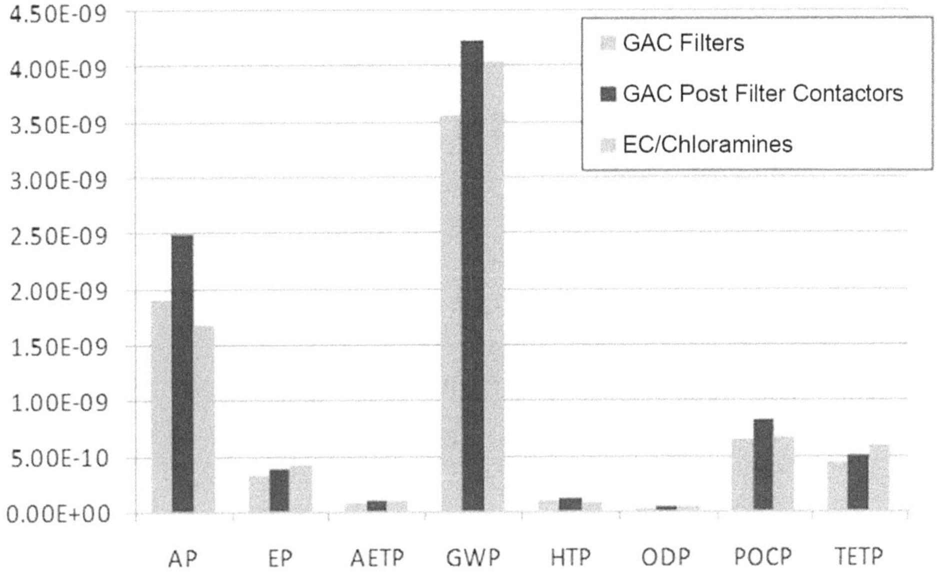

Figure A-6. LCA results normalized to North American emissions 2001

Table A-3. Additional sustainability measures for 1 day of product water (20 mil gal)

Sustainability Measure	GAC Filter Adsorber	GAC Post-Filter Contactors	EC/Chloramines	Units
Energy consumption	436,487	502,406	491,450	MJ
Water use	60,409	80,232	70,161	gal
Sludge production	4,900	4,900	6,700	lb

To give a frame of reference, Figure A-6 shows environmental impacts normalized to North American emissions in 2001 as provided in the GaBi software. As shown in Figure A-6, the three alternatives considered in this analysis were very similar with respect to environmental impacts within the categories available in the model; however, it should be noted that the categories available in the model are not fully capable of comprehensively capturing the environmental impacts. For example, the use of GAC either as post-filter contactor or as filter adsorber is expected to produce less toxic DBPs compared with a chloraminated option, but the HTP category is not capable of addressing this aspect.

The impact categories above are typically analyzed in a LCA. However, local issues, such as water use, may greatly affect the decision-making process and should also be considered. Table A-3 is a summary of additional sustainability measures for the three scenarios. Data for the additional sustainability measures were obtained by GaBi outputs, plant data, and literature. Energy consumption and water use for all scenarios were within 25 percent of each other. The EC/chloramines scenario produced 27 percent more sludge than the other scenarios. This is expected because the additional alum used in the EC/chloramines scenario is known to coagulate additional particulates in the water.

Appendix B

Case Studies

References to case studies appear throughout this book to supplement the text or to illustrate a point. In many cases, considerable information about these projects is available on the utilities' websites, which are listed and current as of publication date.

Case Study No.	System	Location
1.	Post Filtration GAC with On-site GAC Reactivation	Greater Cincinnati Water Works, Cincinnati, Ohio
2.	Water Quality Master Plan and GAC Implementation	Phoenix, Ariz.
3.	Water Quality Road Map	Birmingham Water Works Board, Birmingham, Ala.
4.	Design-Build-Operate WTP	San Diego County Water Authority, San Diego, Calif.
5.	PAC Treatment	Columbus Division of Power and Water, Columbus, Ohio
6.	Continually Mixed PAC Contactor	Consolidated Mutual Water Company, Lakewood, Colo.
7.	Post-Filtration GAC Treatment	Scottsdale, Ariz.
8.	Design-Build-Operate WTP	Lake Pleasant WTP, Phoenix, Ariz.
9.	GAC Filter Adsorbers	Cholla WTP, Glendale, Ariz.
10.	Ozone-Enhanced GAC Biofiltration	Greenway WTP, Peoria, Ariz.
11.	Groundwater GAC Demonstration Facility	Fair Lawn, N.J.
12.	Groundwater GAC Treatment	Dover Township, N.J.
13.	Groundwater GAC Treatment	Flowing Wells Irrigation District, Tucson, Ariz.
14.	Wellhead GAC Treatment	Metro Water, South Shannon Wellhead, Tucson, Ariz.
15.	Post-Filtration GAC Treatment	Northern Kentucky Water District, Erlanger, Ky.
16.	GAC Filter Adsorber Demonstration Project	Mohawk Valley Water Authority, Utica, N.Y.
17.	Post-Filtration GAC Treatment	Delphos, Ohio

Case Study No. 1

Post-Filtration GAC Treatment with On-Site GAC Reactivation

Greater Cincinnati Water Works
Cincinnati, Ohio

Background

Greater Cincinnati Water Works (GCWW) placed its 175-mgd post-filtration GAC treatment system with on-site GAC reactivation in operation in 1992. The system is designed to remove a wide spectrum of natural and synthetic organic contaminants from the Ohio River raw water source and to facilitate compliance with disinfectants and disinfection by-product (D/DBP) regulations. It consists of 12 large gravity downflow GAC contactors, each containing approximately 720,000 lb of GAC, that provide an empty-bed contact time of 15 min at design flow. The GAC reactivation system consists of two multi-hearth furnaces with afterburners and wet scrubbers, each designed to regenerate 50,000 lb/day of GAC.

Benefits

Now, after 20 years of continuous operation, the GAC system is providing key benefits for the GCWW. The system

- Provides peace of mind for the system's leaders and operation by providing a significant barrier for removal of a broad spectrum of organic contaminants that might be present with raw water supply
- Provides a high level of public trust from elected officials, the local media, and customers based on the state-of-the-art technology used to provide high quality water
- Helps GCWW meet disinfection by-products regulations
- Reduces the use of chlorine for disinfection, resulting in significant savings
- Avoids the use of chloramines to meet DBP regulations
- Assists compliance with the Lead and Copper Rule
- Improves control of tastes and odors
- Improves turbidity removal

Costs

The cost of the 175-mgd GAC facility with on-site regeneration was $63.9 million, in 1992 dollars. The current cost of treating water with GAC is less than 20 cents per 1,000 gal, or about $5.00 per quarter for both debt service and operation and maintenance, on an average total water bill of $53.83 for a single-family household with a 5/8-in. meter.

Utility Website

www.cincinnati-oh.gov/noncms/blended/water

Case Study No. 2

Water Quality Master Plan and GAC Implementation

Phoenix, Ariz.

Background

The city of Phoenix owns and operates four water treatment plants with a combined capacity of 620 mgd. A fifth water treatment plant that supplies up to 80 mgd of water to the Phoenix system is owned by the city and operated privately. A pathway has been established for the utilization of GAC at five of the WTPs with a combined capacity of 570 mgd.

Water Quality Master Plan

A Water Quality Master Plan was prepared for the city in September 1989. This was a time when the US Environmental Protection Agency (USEPA) was in the process of implementing the 1986 Amendments to the Safe Drinking Water Act of 1974. Because health effects research was continuing and the direction of future water quality regulation was uncertain, developing a single course of action for the long-term improvement of the quality for the city's drinking water system was impractical. Rather, a Water Quality Strategy for the surface water supplies was developed that could be implemented to provide increasingly higher-quality water as determined desirable by the city and/or required to meet more stringent state and federal water quality regulations. Microbiological quality and disinfection by-products were the most critical water quality parameters affecting the development of the strategy.

An update to the Water Quality Master Plan was completed in July 2009, establishing a water quality pathway to compliance with the Stage 2 D/DBP Rule.

GAC Facilities

The master plan focuses on the use of activated carbon either in the post-filter contactors, in a filter adsorber, or in a biologically active mode for compliance with USEPA Stage 2 D/DBP Rule. In addition, the city is in the process of identifying, prioritizing, and implementing optimization projects to reduce water age and lower DBP concentrations in the distribution system. Following is the present status of GAC use:

- Lake Pleasant WTP (80 mgd): post-filter GAC contactors with on-site GAC reactivation are in operation
- Deer Valley WTP (100 mgd): GAC filter adsorbers are in operation
- Val Vista WTP (220 mgd): post-filter GAC contactors are in operation
- 24th Street WTP (140 mgd): biologically enhanced activated carbon filters are in operation
- Union Hills WTP (160 mgd): biologically enhanced activated carbon filters are in operation

Powdered activated carbon is also available at four WTPs (Deer Valley, Val Vista, 24th Street, and Union Hills) and will be used on an as-needed basis when additional TOC removal is needed to meet TTHM goals.

Utility Website

phoenix.gov/waterservices

Case Study No. 3

Water Quality Road Map

Birmingham Water Works Board

Birmingham, Ala.

Background

The Birmingham Water Works Board (BWWB) uses conventional water treatment at its four surface water treatment plants having a combined treatment capacity of 188 mgd. The system has 50 finished-water storage tanks (74.5 mil gal capacity) and approximately 3,800 mi of water main. Free chlorine is used for primary and secondary disinfection at all four plants. Historical water quality data indicate that current DBP control practices are not sufficient to meet BWWB's internal water quality performance goals or regulatory requirements for Stage 2 D/DBP Rule.

BWWB has developed a unique road map to move efficiently and effectively from existing operation to rehabilitation of facilities that will enhance removal of DBP precursors and improve water quality. The road map is described in chapter 4.

GAC Facilities

A major GAC biofilter demonstration plant was constructed and operated to aid in the decision-making process. The plant, consisting of four 4-ft^2 filter boxes, concluded its year-long operation in the spring of 2009. One of the main goals of this plant was to demonstrate the effectiveness of GAC biofiltration without the use of ozone to enhance biological growth. Partners in the demonstration plant project were the Water Research Foundation and the Universities of Alabama and Colorado. Descriptions of the demonstration plant are included in chapter 4. Results from the Demonstration Plant trials were published by the Water Research Foundation in 2010 (*Cost-Effective Regulatory Compliance with GAC Biofilters*; Chowdhury et al.).

Utility Website

www.birminghamwaterworks.com/

CASE STUDY NO. 4

Design-Build-Operate WTP

SAN DIEGO COUNTY WATER AUTHORITY
SAN DIEGO, CALIF.

Background

The San Diego County Water Authority (Water Authority) implemented a new membrane filtration water treatment plant through the design–build–operate (DBO) approach. This is the first treatment project the Water Authority has undertaken and is now the largest DBO water treatment facility in California, with a design capacity of 100 mgd. The Water Authority used a consultant as the owner's agent to manage the comprehensive DBO procurement and provide as-needed assistance and oversight in the design, construction, and operation of the facility. The Water Authority also used a board of senior consultants to review solicitation documents and provide input into the procurement process.

The procurement consisted of a two-step process whereby a request for qualifications (RFQ) was issued, followed by issuance of a request for proposals (RFP) to a short list of qualified respondents to the RFQ. Proposals were evaluated based on qualifications (70 percent) and price (30 percent), and the preferred proposer was selected for execution of the DBO contract, which incorporated the technical approach and conceptual design submitted as part of the proposal. The DBO contract documents included legal, financial, and technical requirements that governed the specific tasks of the DBO contractor while designing, constructing, and operating the 100-mgd water treatment facility.

DBO Project

Giving due consideration to source water quality, treated water quality goals, site constraints, and environmental impacts, two alternative treatment trains were developed to assist with site planning and permitting prior to soliciting DBO proposals: conventional treatment with intermediate ozonation and ultrafiltration with granular activated carbon adsorption. These two alternative trains were determined to be able to produce water that would consistently meet the water quality goals with respect to disinfection, DBPs, and T&O control. Conceptual designs for the two alternative trains consisted of plant hydraulic profiles, design criteria, site layouts, and cost opinions.

The final design for the treatment facility was prepared by the DBO contractor and included fine screening, submerged membrane filtration, ozonation, and biologically enhanced activated carbon contactors.

Utility Website

www.sdcwa.org/

Case Study No. 5

PAC Treatment

Columbus Division of Power and Water
Columbus, Ohio

Background

The Hap Cremean Water Plant in Columbus, Ohio, uses a conventional two-stage lime softening process with recarbonation and sand filtration to treat raw water from the Hoover Reservoir. The plant has a maximum treatment capacity of 100 mgd. The watershed has significant agricultural activity resulting in agricultural runoff into the reservoir, which includes atrazine, a possible human carcinogen, ranging from < 0.2–12 μg/L on an intermittent basis. The MCL is 3 μg/L, based on running annual average of four previous quarterly grab samples.

PAC Facilities

Facilities for feeding PAC at a maximum dose of 50 mg/L were placed in operation in 1998. Jar tests performed by the city demonstrated that a dose of 30 mg/L of PAC can achieve up to 80 percent removal of atrazine. Additional benefits of using PAC include reduction of taste and odor compounds, removal of SOCs and VOCs that might result from a potential chemical spill into Big Walnut Creek, and reduction of DBP precursors.

The PAC facility includes six 40,000-gal tanks for storing carbon slurry. Six metering pumps feed carbon slurry to three separate feed points: the raw water intake, the first-stage rapid mix (prior to the sedimentation basins), and the lime channel upstream of the softening basins. The metering pumps are sized to feed a maximum of 50 mg/L at the maximum treatment rate of 100 mgd.

Utility Website

publicutilities.columbus.gov/

Case Study No. 6

Continually Mixed PAC Contactor

Consolidated Mutual Water Company
Lakewood, Colo.

Background

Consolidated Mutual Water Company (CMWC), a not-for-profit utility, serves approximately 50,000 customers in the city of Lakewood, Colorado. In 2005 CMWC adopted a plan for a new 15-mgd, expandable to 18-mgd, water treatment plant to address water quality issues including seasonal manganese, reactive total organic carbon, seasonal turbidity, cold water temperatures, algae blooms, and taste and odor for consumers. The facility plans included pre-oxidation (potassium permanganate, chlorine dioxide), chemical addition (PAC, ferric, polymer), PAC contactor, rapid mixing, flocculation, sedimentation, low-pressure membrane filtration, post-filter disinfection, and chemical feed systems, an 800,000-gal reservoir, pump station, and connection to the water distribution system. A detailed evaluation of technologies, capital and operation and maintenance costs, and plant constructability issues, as well as a pilot testing program, resulted in the selection of membrane microfiltration.

The project was divided into three phases. The first two phases were placed in service in September 2008. The third and final phase, the pretreatment area, was under design while the Phases 1 and 2 were under construction. Because the pretreatment facilities are located where the existing water treatment plant was located, Phases 1 and 2 had to be placed in service and delivering water to allow the demolition of the existing water treatment plant. Phase 3 was placed in operation by August 2009. The estimated total project cost is $25.5 million.

PAC Facilities

Based on laboratory studies at the University of Colorado, a 170,000-gal continually mixed contactor is included in the third phase for use when T&O and other high-organic-episodes occur in the raw water supply. The contactor provides a contact time of 15 min at 15-mgd flow rate. The contactor is bypassed when not required.

The project delivery method was construction manager at risk (CM@Risk) in which the owner (CMWC) selected the design consultant and the construction contractor based on qualifications and experience. Along with the CMWC management and operating staff, all parties worked closely together during all phases of the project to ensure near- and long-term effectiveness and efficiency.

Utility Website

www.cmwc.net/

Case Study No. 7

Post-Filtration GAC Treatment

Scottsdale, Ariz.

Background

The city of Scottsdale operates two surface water treatment plants with a combined capacity of 105 mgd where post-filtration GAC facilities have been installed to reduce disinfection by product precursors and provide the best available treatment (BAT) and highest-quality water. The Central Arizona Project (CAP) WTP has a capacity of 75 mgd, and the Chaparral WTP has a capacity of 30 mgd.

GAC Facilities

The GAC facility at Scottsdale's CAP WTP includes 11 post-filter contactors, each containing 11.3 ft of GAC bed depth, providing an EBCT of 20 min at design flow. The GAC facility at the Chaparral WTP includes 10 post-filter contactors, each containing 8.1 ft of GAC bed depth.

Stainless-steel underdrain systems are used at both facilities. To prevent air entrainment within the media, a seal well provides a fail-safe way to maintain the water level within the contactors. Similar to granular media filters designed for turbidity removal, the post-filter GAC contactors are designed with contactor-to-waste facilities. The contactor-to-waste and backwash water flows can also be equalized and either sent to the residual handling facilities or recycled to the head of the plant.

In addition, the facility designs include a network of fixed GAC transport eductors, piping, valves, and loading stations for hydraulic transfer of media. At the CAP WTP, the site layout accommodates staging of multiple tanker trucks for GAC delivery and spent GAC removal. However, site access at Chaparral WTP is limited, requiring more coordination for staging GAC delivery and removal. Truck drainage and washdown water are fully contained and sent to the equalization facilities.

The Chaparral WTP was completed in 2006, and the GAC facility at the CAP WTP was completed in September 2008. Together, these facilities put Scottsdale ahead of the curve in responding to the requirements of the Stage 2 D/DBP Rule and substantially eliminate the potential for taste and odor issues.

Utility Website

www.scottsdaleaz.gov/water/services

Case Study No. 8

Design–Build–Operate WTP

Lake Pleasant Water Treatment Plant
Phoenix, Ariz.

Background

The city of Phoenix saved $28 million by replacing conventional design and construction methods with a design-build-operate approach to create its new Lake Pleasant WTP. A team of consultants helped the city in selecting a firm to construct the 80-mgd plant (expandable to 320 mgd). The $336 million project provides enough water to supply 400,000 households in the growing northwest Phoenix area.

DBO Project

The project began in early 1999, when the city decided to investigate alternative delivery methods for its largest-ever capital project. The city used a multidisciplinary team of legal, financial, construction, and engineering consultants to evaluate conventional design-build and other alternative delivery, ownership, and financing systems. The team recommended employing the DBO method – contracting with a private entity to design, build, and operate the treatment plant for a 15-year period. This approach promised potential savings of up to 16 percent over traditional delivery methods, and in 2000 the Arizona legislature enacted legislation allowing the city to proceed with its innovative strategy.

To realistically identify and evaluate benefits from the DBO process, the city developed a benchmark and estimated costs for designing and constructing the facility and running it as a city-operated water treatment plant using a conceptual design for the benchmark facility. A two-step procurement process was employed to arrive at a service agreement with a DBO entity. After issuing an RFQ, six responses were received and evaluated by a selection committee based upon identified and published criteria, in accordance with Arizona statutes. Three teams were short-listed and sent an RFP, to which they responded with technical proposals, including preliminary design to meet water quality goals. After clarification meetings, final technical proposal and separate price proposals were submitted.

Based on a combination of price proposal scores and technical scores, a selection committee recommended awarding the project to the firm with the overall highest score and a net present value cost of $336,700,000—7.6 percent below the benchmark—over the life of the contract.

The plant has been constructed and is in operation. The plant includes ballasted flocculation (i.e., ACTIFLO®), ozonation, biologically active filtration, granular activated carbon contactors (for additional TOC removal), and UV disinfection.

Utility Website

phoenix.gov/waterservices/

Case Study No. 9

GAC Filter Adsorbers

Cholla Water Treatment Plant
Glendale, Ariz.

Background

The city of Glendale operates the 30-mgd Cholla WTP to provide drinking water to the central and southern portion of its service area. Because of the long water age and elevated water temperatures, improvements at the plant were needed to comply with the Stage 2 D/DBP Rule. Glendale opted for a filter retrofit and expansion to convert existing dual media granular filters to GAC monomedia filter adsorbers. This achieves removal of TOC in addition to particulate removal within a single process, resulting in lower capital investment than a separate post-filter process. Implementation of GAC at the Cholla WTP and at the newly constructed Oasis WTP provides the city the means to stay in compliance with the D/DBP Rule and supply its citizens with high-quality water.

GAC Facilities

Cholla's four granular media filters were converted to GAC monomedia filter adsorbers. Backwash troughs were raised to increase maximum bed depth from 39 in. to 68 in., the underdrain system was replaced with a stainless-steel underdrain that attached directly to the filter floor slab, and the carbon-steel surface wash system was replaced with a stainless-steel system. Two additional filter adsorbers were constructed to reduce surface loading rates on the filters and to provide an empty-bed contact time of 10 min at design flow. Each filter adsorber contains about 120,000 lb of GAC.

The six GAC filter adsorbers have two cells in each unit. Each cell (35 ft × 11 ft) has 68 in. of GAC media. For a typical operation at 28 mgd, the surface loading rate with all six units in operation is 4.2 gpm/ft^2, and the corresponding EBCT is 10.1 min. For the same flow with one unit out of operation, the surface loading rate increases to 5.1 gpm/ft^2. The GAC used in the filter adsorbers is bituminous coal derived GAC with an 8 × 20 mesh size. The GAC filter adsorbers are able to remove both organic compounds and particles, and thus reduce DBP concentrations and mitigate taste and odor problems.

In recent years, the plant has found it necessary to change out the media in nine filters each year because of higher TOC concentrations with SRP canal water than in the past.

The Oasis Water Campus also has GAC filter adsorbers. Each of the five units contain about 82,000 lb of GAC and treat an average annual daily flow of approximately 5 mgd. The GAC in each filter is replaced about twice per year.

Utility Website

www.glendaleaz.com/environmentalresources/waterquality.cfm

Case Study No. 10

Ozone-Enhanced GAC Biofiltration

Greenway Water Treatment Plant
Peoria, Ariz.

Background

The city of Peoria's Greenway WTP needed a way to reduce DBP precursors for compliance with Stage 2 D/DBP Rule. The 16-mgd plant uses preozonation followed by conventional flocculation/sedimentation, biologically enhanced activated carbon filtration, and chlorination. The low dose of ozone applied at the presedimentation basin is believed to be aiding the removal of taste and odor (T&O) compounds and enhancing biofiltration. Because of the presence of moderate concentrations of bromide in the source water, ozone doses are kept to a minimum to control the formation of bromate below the city's target. Raw water supply is from the Arizona Canal and contains total organic carbon concentrations between 2 and 7.6 mg/L.

GAC Facilities

The treatment plant has six declining-rate GAC filters (24 ft × 24 ft) with 5 ft of GAC over 1 ft of sand. The loading rate at design flow with one filter out of service is 3.9 gpm/ft^2, resulting in an EBCT of 10 min. Backwash of the filters is initiated when the runtime exceeds 80 hr, turbidity rises above 0.15 ntu, or the flow decreases below 500 gpm. Because the backwash flow is controlled by a gate valve, the flow varies, but reaches a maximum rate of 14 gpm/ft^2.

Historical data on exhausted GAC filters indicate that the plant is able to remove 10 to 20 percent TOC through biofiltration. The city is currently replacing GAC in two filters per year and evaluating the feasibility of utilizing a bromate control strategy to allow for higher ozone doses and consequently increase biofiltration.

Utility Website

www.peoriaaz.gov/newsecondary.aspx?id=1241

Case Study No. 11

Groundwater GAC Demonstration Facility

Fair Lawn, N.J.

Background

The borough of Fair Lawn and New Jersey Department of Environmental Protection (NJDEP) have teamed together to take a proactive step in furthering the protection of human health and drinking water in New Jersey by developing the first treatment facility in the state specifically designed to demonstrate the ability of GAC to remove a host of unregulated contaminants from groundwater supply sources. The goals of this project are to demonstrate the ability of GAC to provide this type of broad-based treatment and to serve as a benchmark for future work in this area.

This full-scale demonstration project will aid NJDEP in promoting the use of GAC elsewhere in the state as a cost-effective and efficient means of removing contaminants from potable water sources that could potentially pose a long-term threat to human health. NJDEP had two main goals for this project: (1) Use this facility for data gathering on performance of GAC for unregulated organic contaminant removal, and (2) Use as an example to demonstrate the viability, community acceptability, functionality, cost, and operability of these types of facilities.

GAC Facilities

The proposed GAC facility, which was designed in 2008–2009, has two GAC treatment trains. Each train has two contactors that can be operated in series or parallel to provide 20 min of EBCT at 900 gpm. The skid-mounted GAC contactors are sized and configured to provide operational flexibility by allowing the contactors to work in series or parallel and with different flow rates and thus EBCTs. The GAC treatment system is installed downstream of an existing air stripping tower and upstream of the finished water storage tanks.

Utility Website

www.fairlawn.org/dept/water.htm

Case Study No. 12
Groundwater GAC Treatment
Dover Township, N.J.

Background

Dover Township uses GAC to remove an unregulated semi-volatile organic compound in a contaminant plume from a Superfund site from its groundwater. Based on the results from treatability studies of membrane and activated carbon adsorption technologies, carbon adsorption was selected as the most effective and implementable treatment technology, and a 2-mgd GAC adsorption facility was designed. The design included pump replacement, preengineered GAC contactors, a temporary building, and approximately 1,500 ft of discharge piping. The system successfully provides the township water that meets all federal and state drinking water standards and received a 1998 ACEC New Jersey Engineering Excellence Award.

GAC Facilities

The facilities were placed into service in 1997. There are two GAC treatment trains, both with two GAC contactors that can be operated in series or parallel. Each contactor is a 10-ft-diameter steel pressure vessel containing 20,000 lb of GAC. The design flow at an EBCT of 15 min for each treatment train is 700 gpm, or 1,400 gpm for both trains. Air-stripped groundwater enters each vessel through the top and flows downward through the GAC bed. The treated water from each train is discharged into a common 8-in.-diameter discharge pipe.

A GAC contactor can be backwashed with potable water whenever the differential pressure across the contactor exceeds 8 psi or the flow through the contactor decreases. The backwash process must be manually initiated. Waste backwash water is discharged onto the ground outside the treatment facility.

Superfund Site Information

www.epa.gov/region2/superfund/npl/ciba/index.html

Case Study No. 13

Groundwater GAC Treatment

Flowing Wells Irrigation District
Tucson, Ariz.

Background

The Flowing Wells Irrigation District supplies water to a community of 16,000 people in Tucson. The source water comes from 10 wells; two of the wells (Wells 70 and 75) have elevated arsenic concentrations and are under the influence of a VOC contamination plume. Although the historical water quality data indicate only trace levels of VOCs in Wells 70 and 75, neighboring monitoring wells contain elevated levels of TCE, PCE, and MTBE. Radon, a radioactive gas, has been detected at both wells. DBPs, particularly TTHMs, have been detected in the Well 75 Reservoir. Periodic elevated levels of TTHMs observed at the site to date can be attributed to low turnover and long detention times in the finished-water reservoir at Well 75.

GAC Facilities

Design of the treatment system was completed in 2005, and the system became operational in December 2006. Liquid phase GAC for VOC treatment was integrated with granular iron media adsorption for arsenic treatment. Two 12-ft diameter liquid phase GAC vessels operate in series to treat 900 gpm of flow from Wells 70 and 75 prior to arsenic adsorption. GAC treatment is also mitigating radon and DBP formation.

Utility Website

www.fwid.org/

Case Study No. 14

Wellhead GAC Treatment

Metro Water South Shannon Wellhead
Tucson, Ariz.

Background

The South Shannon Well is located in Tucson. A 650-gpm Lowry Aeration Systems multistage diffused-bubble aeration (MSDBA) system was installed at the site in 1997 to treat trace amounts of VOCs, primarily PCE and TCE, in water produced from the South Shannon Well. As well water concentrations increased from 1997 to 2003, the MSDBA system had to be replaced.

GAC Facilities

In 2004, a design was completed to replace the MSDBA unit with two liquid-phase, 12-ft- diameter GAC contactors operating in series at a flow rate of 750 gpm. The system was placed in service in July 2006, and the results from recent sampling at the South Shannon Well indicate that the treated-water VOC goal was met. The MSDBA was retained to serve as a backup treatment unit and possible additional treatment if other less volatile species are detected upgradient (vinyl chloride and cis 1,2-DCE and 1,1-DCA) that adversely affect the efficiency of GAC adsorption. If the existing MSDBA system is returned to routine operation, a vapor-phase GAC adsorption system will be required to treat the MSDBA exhaust gas to remove VOCs.

Utility Website

www.tucsonaz.gov/water

Case Study No. 15

Post-Filtration GAC Treatment

Northern Kentucky Water District
Erlanger, Ky.

Background

Northern Kentucky Water District (NKWD) has three water treatment facilities located in Kenton and Campbell counties.

- Fort Thomas WTP, 44 mgd
- Taylor Mill WTP, 10 mgd
- Memorial Parkway WTP, 10 mgd

The Fort Thomas and Memorial Parkway plants receive raw water from the Ohio River. The Taylor Mill plant receives raw water from the Licking River, a tributary of the Ohio River.

NKWD explored various strategies for removal of organics at all three plants to control DBPs and comply with the Stage 2 D/DBP Rule for remote portions of the distribution system with longer travel and detention times. The alternative compliance strategies investigated include riverbank filtration, powdered activated carbon, sand ballasted flocculation (i.e., ACTIFLO®), enhanced coagulation, membranes, GAC, and magnetic ion exchange (i.e., MIEX®) processes.

GAC Facilities

NKWD evaluated post-filter GAC treatment for all three water treatment plants. The benefits NKWD saw of GAC treatment included reducing DBPs, control of tastes and odors, and the ability to provide an effective barrier for organic chemicals found in the Ohio and Licking rivers.

The post-filtration GAC alternative was selected after rapid small-scale column testing (RSSCT) showed it was effective at removing TOC from both water sources and thereby cost effectively meeting the DBP rules. In addition, GAC would provide a barrier for potential chemical spills and removal of potential endocrine disrupters, pharmaceutically active compounds, and personal care products in the raw water supply. Furthermore, this treatment technology is considered a best available technology by USEPA.

NKWD completed design of post-filtration GAC facilities for all three plants to lower concentration of TTHM and HAA5 that are formed when the organic precursor materials is contacted with free chlorine. GAC treatment also provides an effective means of mitigating seasonal taste and odor occurrences by removing 2-methylisoborneal (MIB) and geosmin.

The GAC facilities for the two plants, which receive water from the Ohio River, were constructed and are currently in operation.

Utility Website

www.nkywater.org/

Case Study No. 16

GAC Filter Adsorber Demonstration Project

Mohawk Valley Water Authority
Utica, N.Y.

Background

The Mohawk Valley Water Authority (MVWA) has completed a year-long, full-scale demonstration of GAC filter adsorbers in one of its four filters and plans to convert all of its filters to GAC for Stage 2 D/DBP Rule compliance if results are favorable and the project is deemed affordable.

The 32-mgd Hinckley WTP serves approximately 130,000 customers in the city of Utica and several other communities in the eastern portion of Oneida County, N.Y. The treatment plant is located in the town of Trenton, approximately 15 mi north of Utica, and treats raw water from the Hinckley Reservoir. The plant, which was placed in service in 1992, has four parallel treatment process trains and can provide the design capacity of 32 mgd with one process train out of service for maintenance or cleaning. The treatment process includes chemical pretreatment followed by upflow clarification and high rate filtration through mixed media filters. Final disinfection is provided by gas chlorination.

Compliance with the Stage 2 D/DBP Rule cannot be achieved via current treatment and distribution system optimization while using free chlorine as the system disinfectant. As such, a number of treatment strategies for reducing DBP levels were evaluated, including the use of chloramines and improving precursor removal prior to disinfection. GAC was identified as the most promising DBP control option for MVWA while retaining the use of chlorine. GAC post-filter contactors were deemed too expensive, but GAC filter adsorbers were identified as a potentially cost-effective compliance option.

Historical DBP data suggest that substantially reducing TTHM levels during the third quarter should allow for Stage 2 D/DBP Rule compliance. Filtered TOC levels are usually above average during the third quarter, resulting in the highest TTHM levels experienced throughout the year. Based on these trends, the proposed strategy is to annually install fresh GAC in the filters in June. A multi-year, multi-phase set of evaluations was conducted to assess the feasibility of this approach.

Bench-Scale Testing (Phase 1)

RSSCTs were used as a first step in this evaluation to predict the TOC removal performance of the proposed filter adsorbers. Results from the RSSCTs indicated that the Hinckley WTP's existing granular media filters may have sufficient depth to serve as GAC filter adsorbers that would be effective for removing TOC to levels necessary for compliance. The RSSCTs provided "proof of concept"–level information regarding TOC removal as a function of contact time and carbon type.

Pilot-Scale Testing (Phase 2)

Pilot testing was conducted at the Hinckley WTP using its four plexiglass filter columns (4-in.-diameter) to confirm the bench-scale results. Both dual-media and mono-media filter configurations were evaluated. The four pilot filters processed settled water directly from the full-scale plant, with no chemical adjustments or chlorine addition, so as to best match full-scale conditions. The results from the pilot testing indicated that, on a pilotscale, the use of GAC is a feasible alternative to the implementation of chloramines in an attempt to seasonally reduce DBP concentrations in the distribution system.

GAC Adsorber Demonstration Project (Phase 3)

This was a plant-scale, year-long demonstration of GAC filter adsorbers to determine the feasibility of this DBP control technology. The filter media in one of the WTP's four full-scale filters was removed and replaced with 63 in. of bituminous virgin GAC (8 × 30 mesh). The demonstration filter was placed online July 15, 2008, in order to capture the historically high TOC levels during the third quarter. Demonstration filter operations matched the other full-scale filters with unchlorinated clarified water and chlorinated backwash. A loading rate of 2.7 gpm/ft^2, resulting in an EBCT of 12.7 min at a plant flow of 20 mgd (average summer rate), was used for the study.

Results from the demonstration testing confirmed the viability of the proposed strategy, with TOC levels below 2.0 mg/L and TTHM and HAA5 levels below the MCLs and internal goals throughout the third quarter. Moreover, the GAC filter consistently achieved turbidity levels less than 0.3 ntu; both turbidity levels and headloss development matched well with the control filter. These results will be important as MVWA seeks to gain approval from NY Department of Health to use mono-media GAC filter absorbers.

GAC Procurement

For the demonstration project GAC procurement, MVWA used a unique process that included prequalification testing. Results from the pilot-scale work were used to determine minimum performance standards based on the water quality goals. Using this approach, the utility specified that the basis for awarding the project was two distinct selection criteria: (1) prequalification of the GAC material by providing a GAC that can achieve a minimum performance standard, as determined via performance/qualification testing, and (2) providing the lowest total bid price. This procurement process is further described in chapter 13.

Utility Website

www.mvwa.us

Case Study No. 17

Post-Filtration GAC Treatment

Delphos, Ohio

Background

The Delphos Water Treatment Plant is a 3.75-mgd lime softening plant with post-filtration GAC. The plant receives its raw water supply from a 450-mil-gal upground reservoir. The reservoir is maintained with water from the Little Auglaize River. The GAC system was installed to control taste and odor issues that result seasonally from algae in the raw water supply and to reduce DBP precursors in compliance with the Stage 2 D/DBP Rule. The plant was placed in service in 2007.

GAC Facilities

The GAC facilities consist of four fabricated steel contactors that can be operated in parallel or series. Each contactor has a capacity of 20,000 lb of GAC, providing an EBCT of 7 min at a 700-gpm flow rate. Water from the GAC contactors flows through UV reactors prior to fluoridation, corrosion control treatment, and post-chlorination. Three clearwells, with a total volume of 1.0 mil gal, provide in-plant storage for the finished-water supply prior to pumping to the distribution system.

Utility Website

www.cityofdelphos.com/government/departments/water

Index

Note: *f.* indicates table; *t.* indicates table.

www.ingramcontent.com/pod-product-compliance
Lightning Source LLC
LaVergne TN
LVHW060315240926
848814LV00005B/32